Environmental Science

Environmental Science

Physical Principles and Applications

Egbert Boeker
Free University of Amsterdam

Rienk van Grondelle
Free University of Amsterdam

JOHN WILEY & SONS, LTD
Chichester · New York · Weinheim · Brisbane · Singapore · Toronto

Baffins Lane, Chichester,
West Sussex PO19 1UD, England

National 01243 779777
International (+44) 1243 779777

e-mail (for orders and customer service enquiries): cs-books@wiley.co.uk
Visit our Home Page on http://www.wiley.co.uk
or http://www.wiley.com

Other Wiley Editorial Offices

John Wiley & Sons, Inc., 605 Third Avenue,
New York, NY 10158-0012, USA

Wiley-VCH Verlag GmbH, Pappelallee 3,
D-69469 Weinheim, Germany

John Wiley & Sons Australia Ltd, 33 Park Road, Milton,
Queensland 4064, Australia

John Wiley & Sons (Asia) Pte Ltd, 2 Clementi Loop #02–01,
Jin Xing Distripark, Singapore 0512

John Wiley & Sons (Canada) Ltd, 22 Worcester Road,
Rexdale, Ontario M9W 1L1, Canada

Library of Congress Cataloging-in-Publication Data

Boeker, Egbert.
Introductory environmental physics / Egbert Boeker and Rienk van Grondelle.
p. cm.
Includes bibliographical references and index.
ISBN 0–471–49576–X ISBN 0–471–49577–8 (pbk.)
1. Environmental sciences. 2. Physics. 3.Atmospheric physics. I. Grondelle, Rienk van. II. Title

GE105 .B65 2001
628—dc21

2001024907

British Library Cataloguing in Publication Data

A catalogue record for this book is available from the British Library

ISBN 0 471 49576 X (ppc) ISBN 0 471 49577 8 (paperback)

Typeset in 11/14 pt Times by Kolam Information Services Pvt. Ltd, Pondicherry, India
Printed and bound in Great Britain by Antony Rowe Ltd, Chippenham, Wilts
This book is printed on acid-free paper responsibly manufactured from sustainable forestry, in which at least two trees are planted for each one used for paper production.

Contents

Preface

This textbook is written for the science student who is interested in environmental problems and wants to get acquainted with a physics approach to them. He or she is at this stage not interested in the rigid and sometimes lengthy mathematical derivations that accompany the usual physics texts or does not have the time to study them. Therefore the mathematics is kept simple.

It remains, however, a physics text. This means that equations are used frequently as they summarize knowledge in a concise and unambiguous way. They usually are not derived but made plausible. For the derivations we refer to '*Environmental Physics*' [1] by the present authors or we will give references when appropriate. All references are listed at the end of the book.

Also, simplified and sometimes simplistic models are used as a first approach to a problem. This implies that, as a free spin-off, the reader gets acquainted with the physics way of thinking. It is understood, however, that motivation to read the book will be interest in the environment and not, initially, physics as a discipline.

Still, the book is a physics text and comprises many subjects of conventional physics courses. Consequently, the student may already have come across several of the topics discussed here. In that case, the present text may be used as a quick reminder and as a way of introduction to the notation used by the authors.

Environmental Physics as the authors see it, deals with problems arising from the interaction between man and his natural environment. It is therefore oriented towards understanding, analysing and mitigating problems, all from a physics perspective.

In Chapter 1 we start with a general introduction where most aspects of the book are dealt with briefly. It should give the student a general overview of the subject matter and of the way in which we made a selection out of the many topics that could have been discussed. In the other chapters we deal with climate (Chapter 2) and the human influence on it (Chapter 3). We discuss thermodynamics and conventional ways of harnessing energy for human ends (Chapter 4) and the new energies of the end of the 20th century: nuclear power and renewable energies (Chapter 5). Halfway through the text we draw up a provisional balance in Chapter 6. We continue with an elementary treatment of the transport of pollutants through the natural environment (Chapter 7) and give an account of the nuisance of noise (Chapter 8). Measurements are an essential part of the analysis of the quality of the environment and studying it over time. They are dealt with in two chapters: Chapter 9 on spectroscopy and Chapter 10 on geophysical methods.

One cannot decouple the interaction between man and environment from society as a whole. Therefore, we spend the final chapter on the societal aspect (Chapter 11). To illustrate the scientific or social context of the earlier chapters, we frequently insert a textbox with a brief edited quotation from a book or a recent article in general journals like Science, the New Scientist, Physics Today or the Scientific American.

So, in order to judge the 'truth' or relevance of the textbox, one should consult the source, of which the complete address or website is given in Appendix C. In this way, these 'appetisers' should urge the student to consult the library or the internet.

In some cases, less accessible sources for the appetisers were used as we found them illustrative. The appetisers run the risk of becoming outdated sooner than the main body of the text, but we believe that they are interesting, although we do not agree with all the statements quoted.

Besides the appetisers there are the usual illustrations in figures and tables. Most figures are black and white, but some colour plates are included when that adds to the information.

As the mathematics is simple, the text can be used in a first-year course in the physical sciences. It may also be used somewhat later in a non-physics study as a way to compare the results of that discipline with a physics approach. We assume that the student has at least a good background in secondary education in the natural sciences and mathematics and knows the symbols used there. To refresh the mind we have put some symbols and their numerical values in Appendix A. In Appendix B we give a quick introduction in some elementary vector algebra. The reader may consult that appendix when the need arises. An A level in mathematics according to the British tradition, or a comparable level in other educational settings, would be helpful for full appreciation of the text.

The references are given at the end of the book. They contain proofs or arguments on points discussed in the text. Some of them are useful background material for virtually an entire chapter or even more. In that case they have an asterisk (*) with some comments added. A few books are relevant for several chapters and may serve for background reading [2], [3].

The exercises at the end of the chapters are, generally, quite easy and should stimulate the student to have a closer reading of the chapter. Their purpose is not to train the student in problem solving. Some examples, which may be a little more complicated, are worked out throughout the text to emphasize the message. A manual for teachers, with worked exercises, is available from the publisher.

In science, theory has to be backed by experiments and computer simulations. That is the reason why some simple experiments are described briefly in Appendix D. They are worked out in detail on our website www.nat.vu.nl/envphys exp. On the same site some computer simulations of simple models, related to the theory of this book are given.

Egbert Boeker

Rienk van Grondelle

Acknowledgements

The authors are indebted to Dr Wim Sinke for producing the energy data used in Chapter 4; to Dr Konrad Bajer for his careful reading of a draft of Chapter 7; to Dr Arve Kylling and Dr Fred Sigerness for providing spectral data, used in Chapter 9; to Dr John Grin for commenting on Chapter 11 from a social science point of view; to Dr Rietje van Dam for her natural science comments on Chapter 11 and to Dr Nico Vermeulen for his briefing on dose-response relations.

The efforts of our student Jeroen van Zon in drawing and redrawing the figures and, finally, the encouragement and suggestions of the editorial team of John Wiley and the anonymous reviewers, made great improvements to the text.

Acknowledgements

[illegible]

Plate 1. Geyser in Yellowstone park. It is out of the question to use this geothermal heat to produce electricity (see also Fig. 1.6)

Plate 2. The many shades of green in a tropical rain forest on the Philippines (see also Fig. 9.1)

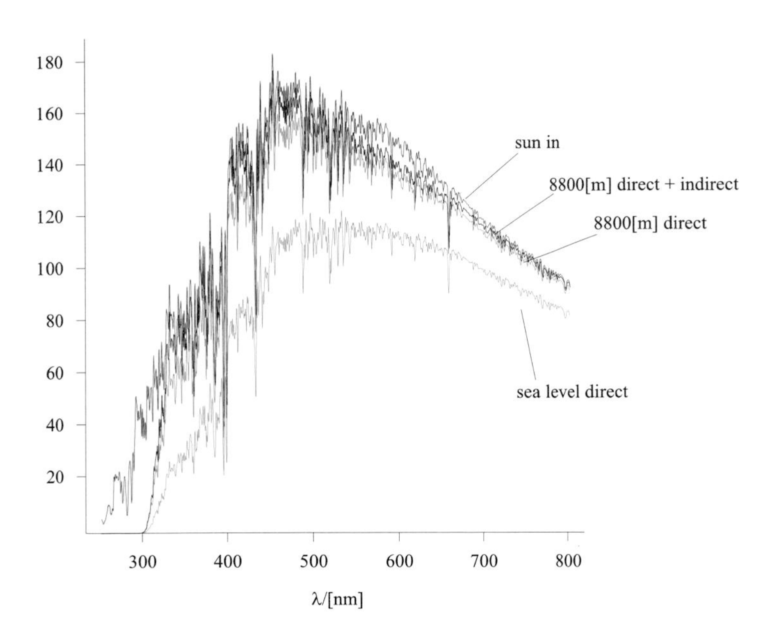

Plate 3. Voyage of solar radiation through the atmosphere. The top curve represents the incoming solar spectrum on top of the atmosphere. The curve below displays the summed direct and indirect radiation at 8800 [m], the Mount Everest level. The curve below that gives the direct (not scattered) radiation at 8800 [m] and the bottom curve the direct radiation at sea level (see also Fig. 9.6)

Plate 4. Fluorescence of a Chlorophyll *a* solution. From the left a blue 488 [nm] beam (coloured blueish white in the picture) is entering a cuvette. This contains a chlorophyll *a* bound to protein (the green material). The absorbed photons excite chlorophyll molecules that rapidly decay to their lowest excited state from which they emit red fluorescence with a maximum of about 670 [nm]. A small electric bulb in the background (not shown) gives the solution its green colour. (Photograph by Dr Bas Gobets and one of the authors (RvG) at their department) (see also Fig. 9.8).

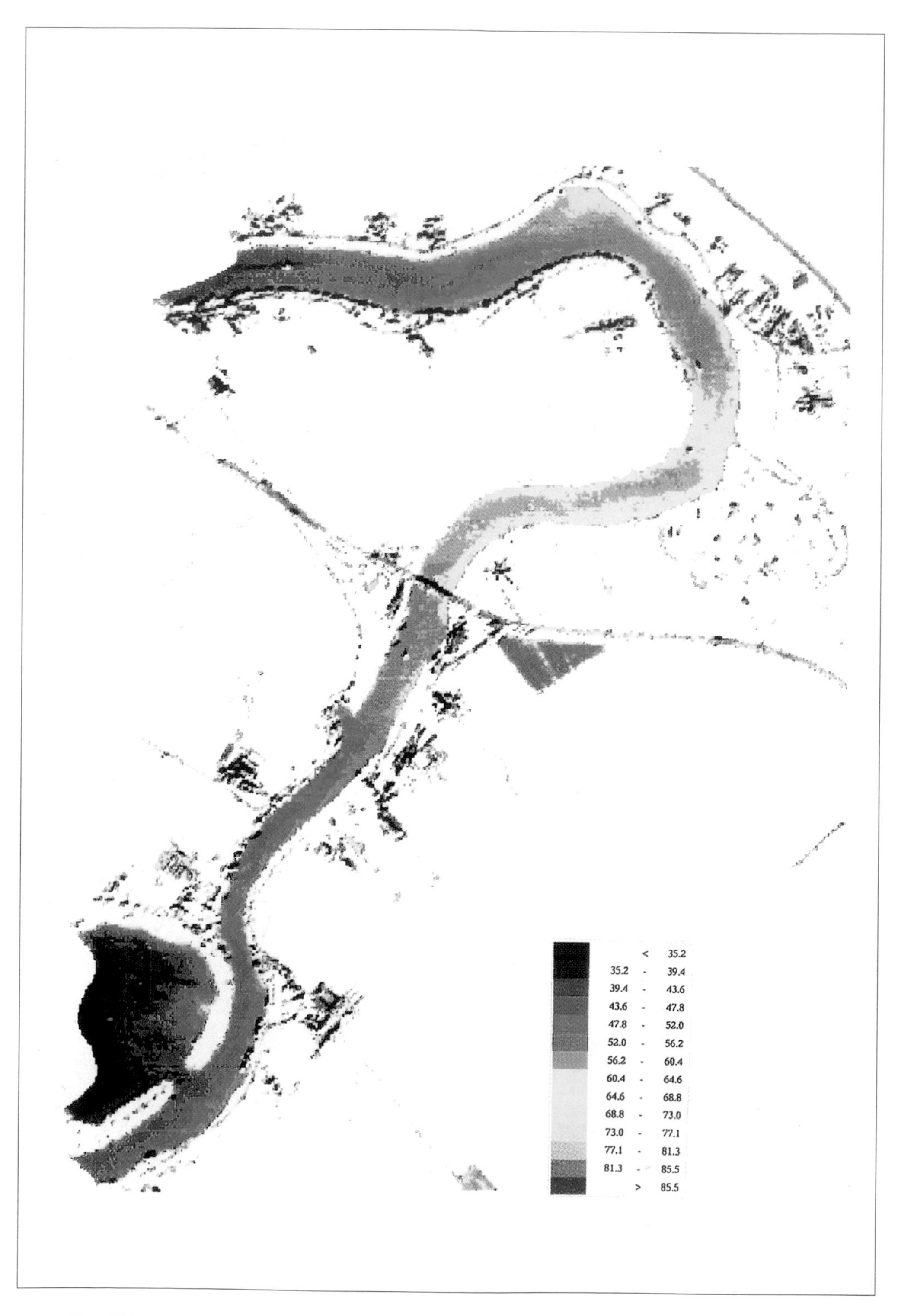

Plate 5. Chlorophyll *a* flows North into the river from the canal crossing the graph [106] (see also Fig. 9.15)

1 Introduction: A Physics Approach to Environmental Problems

Environmental problems arise from the interaction between man and nature and they have been with us for a long time. The unique and new aspect of modern times is the scale of the problems. These originate from two developments: the rapid expansion of the world population and the increase in consumption of energy and materials per person. This is made worse by an accumulation of damage from the past. Keeping the natural environment for future generations will demand efforts from many disciplines. In this textbook, the starting point is physics and in this first chapter we introduce the subject matter as simply as possible and use the opportunity to define a few concepts that we will need later on.

> Bern geo-chemists: 15000 years of atmospheric lead deposition at a bog in Switzerland's Jura mountains provide a record of lead melting and smelting extending back to before the Roman empire.
>
> *Science* **284**, 7 May 1999, 900–901

We number the references through the book in the order in which they appear. For ease of reference, we list them at the back under the chapter number in which they first appear. References of wider importance are indicated by an asterisk (*).

In Section 1.1 we start with a few examples of environmental issues and illustrate that different disciplines look at the same problems from a different perspective. Next, in Section 1.2, we consider economic and social aspects and then go on to physics in Section 1.3, which deals with sunlight and the solar spectrum. In Section 1.4 we discuss the greenhouse effect while in Section 1.5 we sketch the ways in which pollutants are transported. Section 1.6 deals with the circulation of carbon through the atmosphere and the land and sea masses; here we have chosen carbon, as that element is the main contributor to the greenhouse effect.

1.1 *WHAT ARE ENVIRONMENTAL PROBLEMS? – POLLUTION*

The science of ecology [4] may be helpful in defining what we mean by environment. In that discipline, the environment comprises all components of our surroundings. It contains the air which we breath, the habitat in which we dwell and the food and drinks we consume. Our food may be the milk of a cow whose energy originates from grass and crops, growing on the meadows. Ultimately the driving force of the process is the solar energy, converted to chemical energy by photosynthesis.

The point of ecology is that all are interconnected. An example is shown in Figure 1.1, where the waste that man produces may pollute the meadows and return to mankind by the food chain, unless precautions are taken. Also, different organisms (here man and cow) compete for the same resources (crops) and one organism (man) may act as predator for another (the cow). Because of the interconnections, the quality of our environment should be judged by the quality of the whole of the natural environment. Care for the environment may not stop at the national boundaries. Air circulation and river flows do not stop there anyway, since animals and people migrate and food is imported and exported on a large scale.

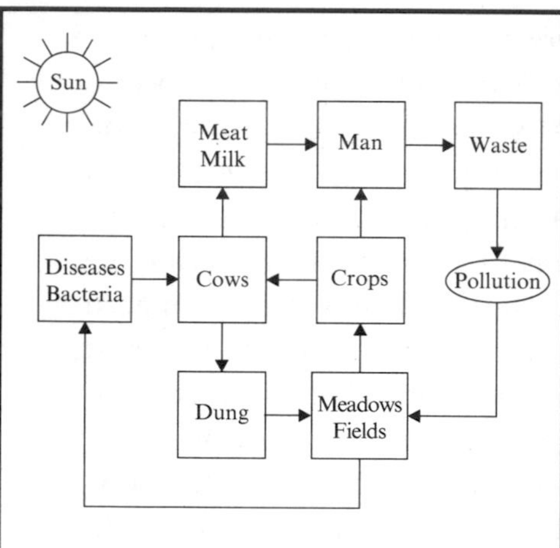

Figure 1.1 Example of ecology. Man competes with cows for food and produces waste, which may pollute his own agricultural land

1.1.1 Pollution

> Even Europe's remotest mountain lakes are chronically polluted. When researchers looked at sediments that had been laid down after 1830 in 10 remote lakes around Europe, they found significant concentrations of polycyclic aromatic hydrocarbons (PAHs). These chemicals are formed by burning wood or fossil fuels and some of them are carcinogenic.
>
> *New Scientist*, 22 April 2000, 20

We define pollution as the '*addition by man of any substance or energy to parts of the ecosystem that ultimately harm or damage man*'. To this may be added 'the removal of any benign parts of the ecosystem'. We have included 'by man' in our definition to distinguish it from natural damage to the ecosystem, e.g. by meteorites or by volcanoes.

The 'substance' in the definition may be in the form of particles or chemicals in any form or physical phase (solid, gas or liquid). It also may refer to microbiological organisms such as those occurring in human faeces. Often the important chemicals or organisms occur in very small concentrations and one needs special techniques to measure them accurately. Some of them are described in Chapters 9 and 10.

We also included 'energy' in the definition since the addition of surplus heat to rivers or lakes may change the local ecosystem. Finally, one may distinguish between *primary pollutants* which are brought into the ecosystem directly and *secondary pollutants* where the primaries have been changed by the chemical surroundings or by sunlight. We will include all these types with the term 'pollutants'. Several definitions are given in [5] page 94–5 for surface water pollution and in [6] page 17 for air.

Air pollution

Dry air consists of nitrogen (78%), oxygen (21%), argon (0.9%) and the rest is made up of small concentrations of many other gases of which CO_2 with 0.0358 percent is the most abundant. Air always contains some particles as well, from winds passing over the soil or plants and is never dry but always contains some water vapour (0.5%), most of it from the oceans. These percentages refer to volume fractions and they can be converted to weight fractions (see Figure 1.3)

Figure 1.2 The origin of ozone in the troposphere (lower atmosphere) as seen by an environmental scientist: the multitude of cars are to blame

> In 1866 a Finnish biologist made a list of the lichens growing on the trunks of trees in the Luxembourg gardens in Paris. He noted that they were poorly developed or sterile and deduced that lichen provided a measure of the purity of the air and constituted a kind of very sensitive instrument for measuring this quality.
>
> From [8] p. 19

The question of what level of additional substance or energy becomes harmful to the ecosystem, is difficult to answer. Some gases contribute to climate change and global warming; the mechanisms and the harm they do will be discussed in Section 1.4 and in Chapter 3. Other gases definitely are harmful in high concentrations. One usually assumes that there is a threshold concentration, below which the organism is able to repair the damage done to its cells.

In this chapter we will frequently use ozone as an example of a pollutant, which has chemical formula O_3. Its physics and chemistry will be discussed in Chapter 9, but from its chemical formula, it will be clear that ozone is an oxidant. It appears that high concentrations will damage plants and humans. Its particular effect on crops has been

Concentration in weights:
$[\mu g\ L^{-1}] = [mg\ m^{-3}] = 10^3[\mu g\ m^{-3}]$

Concentrations in volume fraction
$[ppmv] = [ppm] = 10^{-6}$
$[ppbv] = [ppb] = 10^{-9}$

$$1[ppb] = \frac{1 \text{ molecule gas}}{10^9 \text{ molecules air}} = \frac{N_A \text{ molecules gas}}{10^9 N_A \text{ molecules air}}$$

$$= \frac{M \text{ grammes gas}}{10^9(24.45)[L]\text{air}} = \frac{M}{24.45}[\mu g\ m^{-3}]$$

Figure 1.3 Concentrations of pollutants. Conversion from volume fraction to weight fraction. The result depends on temperature (here 21[°C]) and pressure (here 1 [atm])

measured as a function of exposure to a certain concentration during a period of time. The units of concentration are explained in Figure 1.3. We adopt a notation in which we show the unit in which we work between square brackets such as in [ppb]. Under controlled laboratory conditions a mean exposure of 50 [ppb] during summer would reduce the yield of spinach by 10 percent; at mean exposures of 100 [ppb] the yield of corn and wheat would be reduced by 20 percent ([7], p. 48, 49; [8], p. 88–96). The 'natural' background concentration is estimated at between 10 and 20 [ppb].

Surface water pollution

Surface waters like lakes and streams are the habitat of fish and their water quality is obviously of interest to the part of the ecosystem which feeds on fish, both humans and animals. Of the nine forms of pollution that one may distinguish ([5], p. 95) the most important one is the addition of biodegradable organic material. That means that bacteria will use the oxygen in the water to oxidize organic compounds like carbohydrates according to the reaction

$$C_6H_{12}O_6 + 6\,O_2 \rightarrow 6\,CO_2 + 6\,H_2O + 4.66 \times 10^{-18}\,[J] \qquad (1.1)$$

This reaction depletes the oxygen, which the fish need to breathe.

A similar form of pollution is the *eutrophication* of the waters. The word originates from the Greek '*eutrophos*' which means 'well-fed'. It refers to the fact that human society produces detergents, fertilizers and human or animal wastes. They add nutrients, especially P and N, to the waters, resulting in noxious algae blooms and toxic substances. The water is then no longer useful for fishing, drinking or swimming.

Petroleum products and chemicals often are stored in tanks above the ground. Ruptures or leaks can release chemicals, which then have the opportunity to seep into the ground. A serious case of groundwater contamination occurred in Indiana, USA, when one 200 litre tank of perchloroethylene was damaged by vandals and the contents leaked into the ground.

[9], p. 23

Last, but not least, the acidity of many surface waters is dramatically increased due to combustion of sulphur-rich coal. The resulting SO_2 in the air is transported over large distances and rained out as H_2SO_3 and H_2SO_4. It must be said that rainwater is always slightly acidic (with a pH somewhat below pH = 7). But the minerals in the soil are able to neutralize the 'natural' rain. The sulphur acids however cause much lower pH's, and the majority of fish are not able to survive in waters with pH lower than 5.0 or 4.5 ([7], p. 135). The change to low-sulphur coal in power stations, enforced by government regulations, fortunately reduced the acid rain problem in many countries.

Soil and groundwater pollution

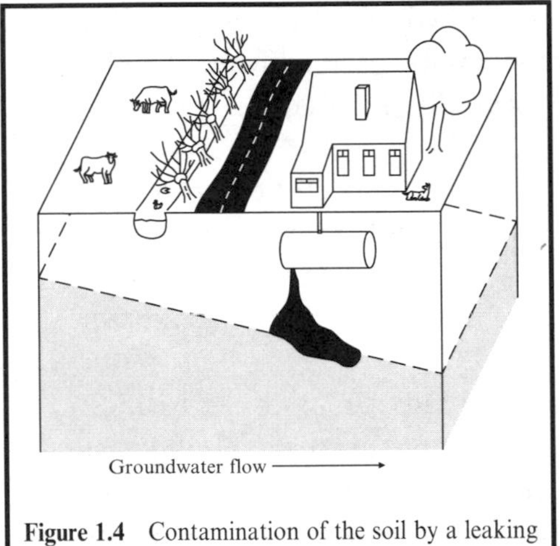

Figure 1.4 Contamination of the soil by a leaking fuel tank

Harmful liquids can leak into the soil in many ways. Figure 1.4 illustrates one of them: an underground fuel tank is corroded and slowly releases some of its contents down to the groundwater level. In the case shown, the released liquid apparently is lighter than water so it disperses on top of the groundwater, polluting the soil. In other cases it may be heavier and dissolve into the water.

The groundwater may be polluted in many ways [9]. For example, waste sites will get water from above in the form of rain; that water absorbs organic and inorganic materials from the dump and brings them down into the groundwater. A plastic cover on top of the dump may help for a while, but eventually the cover will break down. After some time, the position of the sites is forgotten anyway and a new generation of people will perforate the dump, looking for water or resources.

Groundwater, polluted or not, will slowly move through the soil. The soil particles may remove some of the solved pollutants from the water (sorption) or, when the water is clean but the soil is not, add contaminants to the water which then will be transported to another place. The groundwater by itself may not be pure, but contain small colloid particles, which may pick up contaminants in one position and deposit them at another. This mechanism of contaminant transport is slow but dangerous, as it occurs down in the earth away from direct human observation.

1.1.2 Standards

Standards for environmental quality have been developed by international organizations like the World Health Organization (WHO) and are designed as an advice to national governments. These may set their own standards for the quality of drinking water, outdoor air, surface water etc. Within the USA, states may develop stricter legislation than required on the USA level and, within the European Union, the member states may adopt stricter rules than the EU Directives.

The standards are usually very detailed and cannot be summarized here. Let us, however, give some data for ozone. For humans the US standard is set at 120 [ppb] as an hourly average which may not be exceeded for more than one day during a year. It is assumed that, below this concentration, eye irritations and breathing difficulties will not occur. For ozone or other air oxidants an alert would be called if a concentration of 100 [ppb] were expected to last for at least 12 hours. Warnings would be given when the concentration reaches 400 [ppb] and an emergency stage called for 600 [ppb], in each case when the weather forecaster expects the high concentration to last for at least 12 hours ([10], p. 253).

In all these cases one will try to reduce the high concentrations of pollutants by tackling their origin. For ozone that is mainly from the exhaust gas of cars and trucks in combination with atmospheric circumstances such as the absence of wind; in many countries one would temporarily reduce the number of vehicles on the roads.

In order to maintain environmental quality standards, one has to look at the origin of the pollutants. Besides the unintentional leakage of all kinds of containers, which we discussed above, the major sources are emissions from industry, agriculture, transportation, the services and households. So,

Figure 1.5 Industrial complex (Corus Skyline, IJmuiden, NL). Reducing emissions is costly. Reproduced with kind permission from Corus

governments are setting emission standards, usually very detailed for all kinds of equipment. The standards are not obtained by calculating backwards from the quality standards, as a scientist might assume. Instead, they are related to what may be achieved by the best possible technology against a reasonable cost.

When it is clear from experience that such an approach will not be sufficient to prevent recurrent pollution and health hazards to the public, a government may set standards for which the technology does not yet exist. This happened in California, where, around 1990, the state government set very strict emission standards for the private car, to be achieved by about 10 years later. The car industry of course protested strongly, in part because of the high development costs involved. Later they were cooperating and this hopefully may set a trend. So, the economic aspect of environmental protection is an important one.

1.2 *THE ECONOMIC AND SOCIAL CONTEXT*

> The market value of the services nature provides for us is estimated as $33 trillion a year. For example, the New York City watershed consists of 4000 [km^2] soil and forest to the north of the city. It effectively cleans the rainwater and provides drinking water for its huge population. Villages and farms are polluting it increasingly. To clean at source would cost $ 2 billion; a filtration plant cleaning polluted waters would cost $ 8 billion.
>
> *New Scientist*, 6 Febr 1999, 42–45

For the designers of environmental policies it is easiest when a monetary value can be set on the damage caused by pollution. Ozone is a good example, where this is possible to some extent. The decrease of yield of commercial crops as a function of concentration has been measured ([7], [8]). The monetary value of the damage can be calculated from the market prices of the crops and balanced against the cost of ozone abatement. For human health, such a calculation is more complicated. Usually, the persons afflicted first by air pollution are the young, the old and the sick. It is difficult to put a value on their additional suffering. So, in putting down quality criteria for the environment, political judgements are made. Therefore, it is useful to mention a few aspects of the environment, which should be taken into account in decision making.

1.2.1 *Environment as a resource*

The natural environment must be regarded as a finite resource. When all trees are cut down, one eventually will run out of timber. When the soil is polluted

The bark of *P. africana* contains an ingredient that reverses what Africans call 'old man's disease' – the swelling of the prostate gland. It is the basis of an ancient African remedy that in the 1960s was patented by a Frenchman as pygeum. The trees are felled and the powdered bark is sold to ageing men in Europe and North America. Men are about to lose this remedy, as the trees take 15 years to re-grow.

New Scientist, 29 April 2000, 6–7

in such a way that the crops are becoming inedible, one loses the source of food production. When the surface waters remain polluted, one will run out of fish – and lose another source of proteins. So there are plenty of reasons to take care of the environment.

1.2.2 Environment for enjoyment

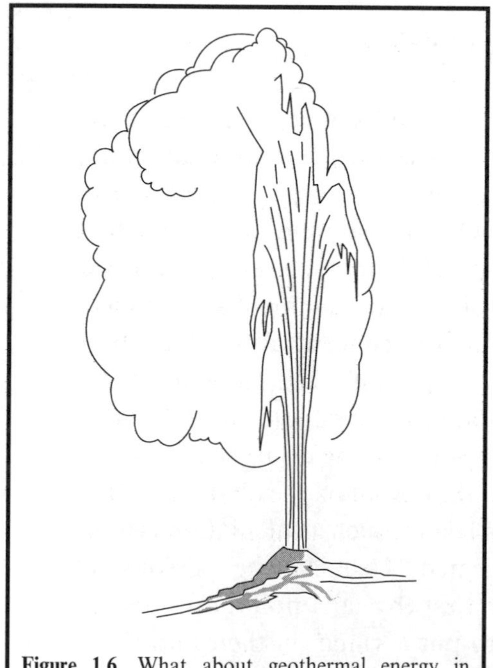

Figure 1.6 What about geothermal energy in regions of great natural beauty (Plate 1)?

A second important aspect is recreation. As we shall see later, there is a great desire to turn to renewable energy sources in order to reduce the burning of fossil fuels. Two of the options are an increase in hydropower installations and further use of geothermal energy of which the well known geysers are the most visible example (Figure 1.6 and Plate 1). In both cases one has to use land which has previously been used for recreation. Valleys with characteristic little villages and sometimes ancient monuments have to be filled with water to provide potential energy for the dams. Moreover, buildings, installations and power lines have to be set up around the dams.

Undoubtedly, one may perform the required changes in the landscape in an interesting way and construct alternative recreational possibilities. Still, other options for recreation are lost and some special examples of natural beauty then only can be seen on old videos.

1.2.3 Political choices

Man has always changed his natural environment; he has cut trees to make a fire for cooking and to build his towns. He has constructed dykes along rivers and seas to protect himself and his property against floods and to regulate the irrigation of his lands. In fact, since the beginning of agriculture some 7000 years ago in the Nile-Euphrates region; man has interfered with the natural environment to adapt it to his needs. So, we may not simply say that the environment should remain as it is now, or that we should return to the situation of, for example, 1950. The question of how to deal with the environment is a matter of political choice.

The value of the environment has to do with the way of life one finds acceptable. How much 'unspoilt' nature should be kept. What about the horizons changed by power lines? How does one value the constant hum of motorways which can be heard even when walking through the remnants of forests in densely populated countries. What air quality does one provide for people with respiration problems?

In these matters, political choices will be made. In the ideal case they are the outcome of a political process in which, not only professional politicians, but also members of the public participate as they may feel their interests are being challenged. They may form a citizens group to protect a piece of forest or even a single tree in a big city.

Many public interest groups work on local problems. Some groups operate on a national level, such as the American Sierra Club and similar conservation societies. Others concentrate on global issues such as the Greenpeace movement.

On a national or international level one has to realise that the environment is threatened in two ways: the waste and emissions of all our human activities together and dwindling resources. Concerning resources, classical economists usually believe that the market mechanism will 'automatically' turn to alternatives once the original resources become more expensive. Figure 1.7 (from [11]) shows the price of crude oil against cumulative production, where the total resource amounts to a hypothetical 1725 units. On the double logarithmic scale, one notices that the price only starts to increase rapidly when half of the resources are exhausted. It is generally assumed that, at that point, companies will start looking for cheaper alternatives to make new profits.

The question is whether, for a commodity like energy, society can afford to wait until conven-

> Our national energy plan is based on ten fundamental principles (of which we mention):
>
> 3. We must protect the environment
>
> 6. Reduce demand through conservation
>
> 7. Prices should generally reflect the true replacement cost of energy. We are only cheating ourselves if we make energy artificially cheap and use more than we really can afford.
>
> The Third Transition, US President Jimmy Carter, April 1977.
>
> *International Communication Agency*, USA

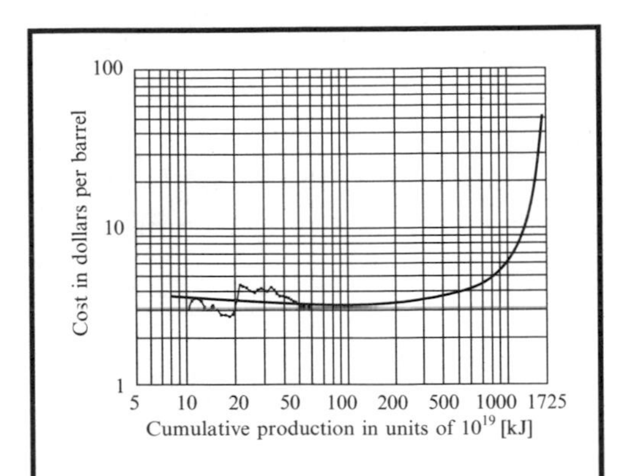

Figure 1.7 Cost of US crude oil at the well as a function of cumulative production with depletion of the resource at 1725 units. With permission of John Wiley from [11], p. 46

tional resources are running out. The development of solar electricity or other renewable energy sources is so expensive that hardly any company can make the necessary investments. Therefore, the practice is that governments are subsidizing the search for reasonably priced renewable energy sources, are taxing (slightly) the use of scarce or polluting energy sources and are putting limits on the emissions of equipment, cars, farms, industries etc.

Traditionally, companies are complaining about the cost of adhering to environmental regulations and taxation. They fear that their competitors will not follow them, so that they will lose profits or market share. This, of course, is a reason to enforce the regulations in large economic units like the USA or the European Union. Better still would be to have truly international agreements on the level of the United Nations. This sometimes works.

Enforcement of regulations is required. It is encouraging however, that shareholders are becoming interested in the environmental record of the companies in which they hold shares. The companies need environmental performance data for their internal management in any case. These are getting standardized and now are often published [12]. Apparently the shareholders are becoming convinced that for the long-term health of their investments, a sound environmental company policy is required.

1.3 SUNLIGHT AND THE SOLAR SPECTRUM

Sunlight is captured by photosynthesis by the inverse reaction of (1.1), which is

$$6\,CO_2 + 6\,H_2O + 4.66 \times 10^{-18}\,[J] \rightarrow C_6H_{12}O_6 + 6\,O_2 \tag{1.2}$$

On the right of Equation (1.2) one finds the carbohydrates in which the energy is stored. In our bodies the carbohydrates from our food are oxidized by means of the oxygen we breathe in. The reverse of (1.2), which is (1.1) provides the energy which is used to maintain our bodies and to perform external work. It therefore makes sense to have a closer look at the sunlight entering the earth.

1.3.1 The solar spectrum

By means of a prism one may divide the sunlight into a broad spectrum, which displays the intensity of the light I_λ as a function of wavelength λ. More precisely put $I_\lambda d\lambda$ is the energy entering per second per square meter [W m^{-2}] with a wavelength between λ and $\lambda + d\lambda$. Remember that we put units between square brackets where that is appropriate. Figure 1.8 shows the solar radiation I_λ [W m^{-2} μm^{-1}] as a function of wavelength λ [μm] outside the atmosphere and at sea level. Note that 1[μm] = 10^{-6}[m], 1[nm] = 10^{-9}[m] and that $I_\lambda d\lambda$ has the unit [W m$^{-2}\mu$m^{-1}] $\times$ [μm] = [Wm^{-2}].

Figure 1.8 contains a lot of information. We therefore will read the graph in detail and first look at the outer full curve, which represents the solar irradiation outside the atmosphere. The solar light that arrives at the top of the atmosphere has a very broad spectrum ranging from the ultra violet, ($\lambda < 0.4$ [μm]) abbreviated as UV, to the deep infrared ($\lambda > 0.8$ [μm]), abbreviated as IR. In between we have the visible region. The three regions are indicated on top of the graph. We notice that an appreciable part of the incoming sunlight has wavelengths in the visible region.

The radiation entering the earth's atmosphere has passed through the empty space between sun and earth unchanged. Therefore, it will show the solar spectrum as it leaves the outside of the sun. It is reasonable to assume that the structure in that spectrum is caused by the composition of the sun's outer atmosphere. In that case the main spectrum leaving the sun would be something like the smooth dashed curve, to be discussed later. The structure in the spectrum would then be due to the composition of the sun's outer atmosphere. In particular, free atoms may be around of which it is known

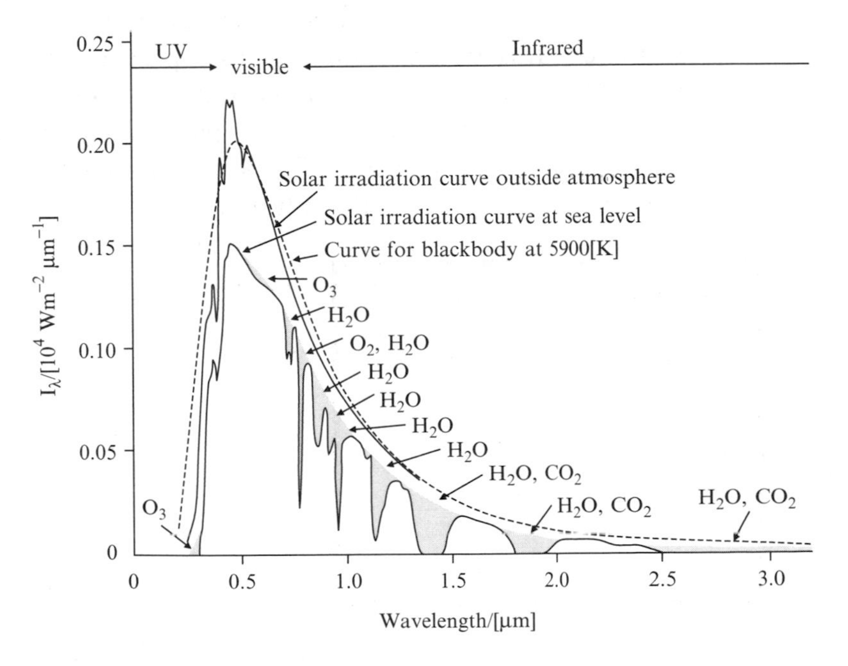

Figure 1.8 Spectral distribution of incident solar radiation outside the atmosphere and at sea level. On top of the graph the visible region is indicated with the Ultra Violet (UV) on the left and the Infra Red (IR) on the right. Major absorption bands of some of the important atmospheric gases are indicated. (Reproduced from [13], Fig. 16.1, p. 16.2). The emission curve of a black body at 5900 [K] is shown for comparison, with its peak adjusted to fit the actual curve at the top of the atmosphere

that they emit and absorb radiation at specific wavelengths. (In Chapter 9 we will discuss this feature of atoms and molecules more thoroughly.) This indeed is the explanation of the structure of the spectrum, as we will illustrate.

With a plastic triangle, one may verify that the spectrum exhibits a dip at 0.4 [μm] = 400 [nm]. From the study of atomic spectra it is known that Ca atoms at high temperature may emit or absorb at wavelengths of 0.393 [μm] and 0.397 [μm]. This is close enough to the 0.4 [μm] we have measured roughly, to account for the explanation that the dip in the spectrum means that the Ca atoms in this case are absorbing the radiation. A similar dip, a little to the right, at 0.42 [μm] also can be attributed to the presence of Ca atoms in the outer solar atmosphere. When one has measured the spectrum in much more detail than shown here, the spectrum exhibits the spectra of a series of atoms like a 'fingerprint'. It appears that many more than Ca atoms contribute to the observed structure.

Underneath the curve we have discussed so far, we notice another drawn curve, the solar irradiation curve at sea level. The fact that it is lower must be due to absorption by the atmosphere and the scattering of sunlight by the atmosphere back into space. Indeed, it has been measured that only about 47 percent of the incoming sunlight reaches sea level. At sea level the solar spectrum exhibits much more structure that on top of the atmosphere. The reason is absorption by molecules, indicated in Figure 1.8 by means of arrows. Besides this general lowering, the spectrum goes to zero at, for example, 1.4 [μm]. This is due to H_2O and CO_2 molecules, which absorb over a broad region of wavelengths and particularly in the regions which are indicated. Note that H_2O and CO_2 molecules do not absorb in the visible region, but in many regions in the Infrared.

Finally, we draw attention to the arrows indicated as O_3, ozone. We already discussed the harm that ozone in the lower atmosphere may do. Under natural circumstances there is not much ozone at low altitudes, but there is a thin ozone layer at an altitude of 20 to 30 [km]. There it does no harm but, on the contrary, protects living organisms on the earth from harmful solar UV radiation, as it absorbs sunlight with a wavelength shorter than 0.295 [μm] or 295 [nm]. This can be seen on the very left in Figure 1.8, where the absorption is indicated by a grey triangle.

> Keeping sunburn at bay may not be enough for parents wanting to protect their young children from the sun. A study of oppossums has shown that doses of UV-B lower than those needed to cause sunburn in their young offspring, can lead to widespread melanoma in later life.
>
> *New Scientist*, 11 December 1999, 6

A decrease in the ozone layer in the upper atmosphere will not only increase the total amount of UV light of a particular wavelength reaching the earth, but will also allow the transmission of increasingly shorter UV radiation. This has important implications for bio-molecules such as DNA and proteins, which are essential for life. They do absorb sunlight with wavelengths $< 0.3\,[\mu\text{m}]$ when that is not blocked by the ozone layer, with the consequence that they are destroyed or will malfunction. With the thinning of the protecting ozone layer, the bio-molecules will become increasingly vulnerable to UV damage in the future.

1.3.2 *Black body radiation*

The dashed curve in Fig. 1.8 is that of a black body radiating at a temperature of 5900 K. The term '*black body*' originates from experiments with a black cavity held at a certain temperature. It appeared to emit a smooth spectrum with the intensity I_λ as a function of wavelength λ given by the shape of the dashed curve. The position of the peak shifts to lower wavelengths for higher temperatures.

> A physicist in 1912:
> Both Professor Einstein's theory of relativity and Professor Planck's theory of quanta are proclaimed somewhat noisily to be the greatest revolutions in scientific method since the time of Newton. That they are revolutionary there can be no doubt, in so far as they substitute mathematical symbols as the basis of science and deny that any concrete experience underlies these symbols. The question is whether this is a step forward or backward.
>
> [14], p. 74,75

In the beginning of the 20th century, attempts were made to calculate the spectral shape of black body radiation. In these calculations it was assumed that at each wavelength all energies could be emitted. There was no relation between wavelength and energy. However, the resulting spectral composition of the black body radiation did not correspond to experiment at all.

In an historic contribution to quantum mechanics, Planck solved the problem by making the assumption that the energies of the electromagnetic field are restricted to integral multiples of the energy

$$E = h\nu = h\frac{c}{\lambda} \tag{1.3}$$

Here ν is the frequency of the electromagnetic field and λ is the corresponding wavelength while c is the velocity of light. The symbol h denotes Planck's constant, given in Appendix A. The result is that the electromagnetic field consists of *photons* with energies $E = h\nu$. Thus, Planck arrived at the so-called Planck energy distribution for the radiation $I_E\mathrm{d}E$ [Wm^{-2}] emitted by the black body with energies between E and $E + \mathrm{d}E$:

$$I_E \mathrm{d}E = \frac{2\pi\nu^3}{c^2}\frac{1}{e^{h\nu/kT}-1}\mathrm{d}E \tag{1.4}$$

where T is the absolute temperature of the black body [K] and k is Boltzmann's constant [J K^{-1}]. The radiation I_E is also called the radiation *flux* or briefly, the flux, as it refers to the radiation passing a square metre per second. In Equation (1.4) the flux is given as a function of the energy $E = h\nu$. We are also interested in the flux as a function of wavelength λ. We may use Equation (1.3) and find

$$I_\lambda = \frac{2\pi hc^2}{\lambda^5}\frac{1}{e^{hc/\lambda kT}-1} \tag{1.5}$$

Of course, $I_\lambda \mathrm{d}\lambda$ again is the radiation energy with wavelengths between λ and $\lambda + \mathrm{d}\lambda$ passing a square metre per second [W m^{-2}] (Exercise 1.1). For large frequencies ν or small wavelengths λ it is easy to find the peak value λ_{max} of the flux I_λ as (Exercise 1.2)

$$\lambda_{max} T = 2898\,[\mu\text{m K}] \tag{1.6}$$

(The exercises can be found at the end of the chapter.)

This approximate but useful relation is called *Wien's displacement law*. It shows how the maximum of the emitted radiation distribution depends on the temperature of the black body. For $T = 5900$ [K] the maximum of the emitted radiation is at $\lambda \approx 0.5\,[\mu\text{m}]$, which agrees with the dashed curve in Fig. 1.8. By integrating Equation (1.4) over all energies, one finds another well known equation, the *Stefan–Boltzmann law* (Exercise 1.3). This law expresses how the total amount of energy I [W m^{-2}] emitted by a black body depends on its temperature T, given by:

$$I(T) = \sigma T^4 \tag{1.7}$$

Here, σ is the Stefan–Boltzmann constant, which is independent of the material.

In Figure 1.9 we show the emission spectrum of a black body at 5800 [K]. Because of Equation (1.6) this curve is shifted a little bit to the right of the dashed curve of Figure 1.8, but it still gives a good fit and 5800 [K] is the approximation for the sun's black body temperature we will use in this book.

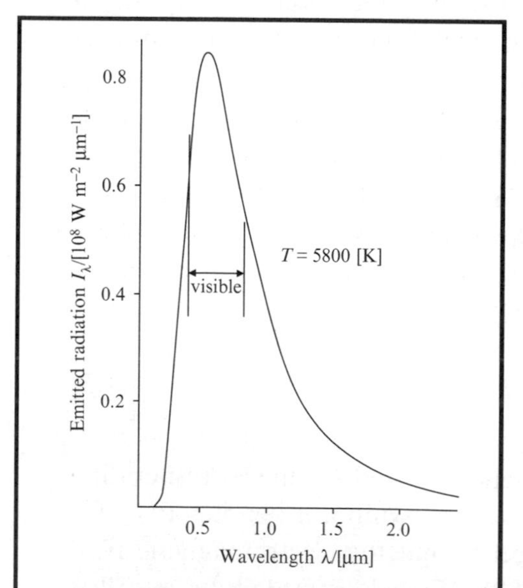

Figure 1.9 Black body spectrum with variables for the sun's surface temperature $T = 5800$ K

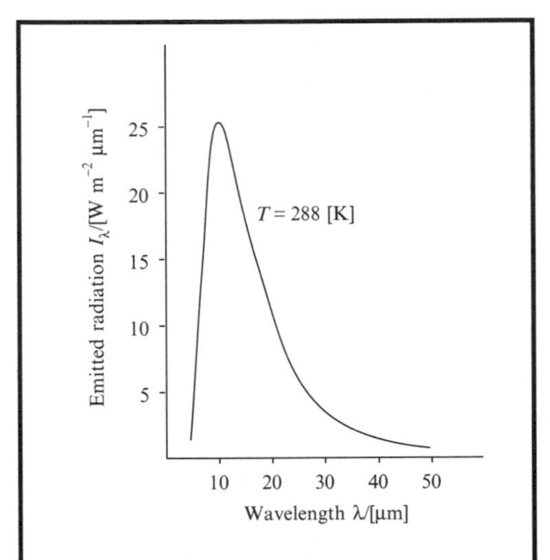

Figure 1.10 Black body spectrum for $T = 288$ [K]. This represents the average emission spectrum of the earth. Note the difference in vertical scale with the solar emission spectrum in Figure 1.9

Note that Figure 1.9 gives the emission per [m^2] from the surface of the black body (the sun), whereas Figure 1.8 shows what is arriving at the earth. This explains the difference in scales of both figures (Exercise 1.4)

The solar emission spectrum still peaks at about 500 [nm], and it has a long tail to the long wavelengths of the infrared (to the right in Figure 1.9). There are photosynthetic organisms that extend their absorption of sunlight up to 1.1 [μm]. These photosynthetic bacteria have a hard time competing with the H_2O molecules in the atmosphere which may absorb the same photons (see Figure 1.8).

In Figure 1.10 we show the black body spectrum for a temperature $T = 288$ [K]. This may be interpreted as the radiation that the earth emits into the atmosphere. It peaks at about 11 [μm]. We note in Figure 1.10 that, again, the spectrum is very broad. We shall see in Chapter 3 that for $\lambda > 4\,[\mu\text{m}]$ all radiation is absorbed by CO_2 molecules in the atmosphere.

We note that, at temperatures such as we experience on earth, the black body radiation has wavelengths far in the Infrared, which is way out of the visible region. Bodies at temperatures of, for example, 100 [°C] = 373 [K] still do not emit in the visible. We cannot see, but 'feel' the IR radiation when we put our hands near a hot water bottle.

Example 1.1: Geothermal heat flow

The heat flow F from the interior of the earth reaching the earth's surface is estimated as about 42×10^{12} [W]. What is the flow per [m^2]. Assume that the solar influx would not be present. Calculate the equilibrium surface temperature of the earth arising from the internal heat flow only. (Data without a reference is taken from [15]).

Answer

The heat flow per [m^2] is found as $F/(4\pi R^2)$ with R the radius of the earth. With the data of Appendix A we find that this becomes 0.082 [W m^{-2}].

If there were no solar input, the internal flow would be the only source. In a steady-state situation the inflow from the interior equals the outflow at the temperature T. Following Stefan-Boltzmann's law [(1.7)] we have $F/(4\pi R^2) = \sigma T^4$, which gives $T = 35\,[\text{K}]$. So, without the influx of solar energy the temperature of the earth surface would be down to a value about 35 [K]. Therefore, as we know well, the present, pleasant temperature on earth is due to the sun.

Example 1.2: Energy of the solar peak

The surface temperature of the sun ($\approx$ 5800 [K]) produces a spectrum with wavelengths shown in Figure 1.9. In accordance with Equation (1.6) it peaks around $\lambda \approx 0.5\,[\mu m]$. What is the peak energy in [eV]?

Answer

The energy of the emitted photons at the wavelength $\lambda \approx 0.5\,[\mu m]$ can be found from the relation (1.3).

$$E = h\nu = \frac{hc}{\lambda}$$

With the data given in Appendix A one finds $E = 2.5$ [eV]. There is a difficulty though, since Figure 1.9 and Equation (1.5) are written as a function of wavelength. From a physics point of view, one should look at Equation (1.4), which describes the flux as a function of energy, as energy is liberated in units of $h\nu$. Because of the inverse relation (1.3) the peak of intensity as a function of energy (1.4) is at a different position. This turns out to be at about 1.6 [eV], as we shall see in Figure 5.7.

We notice that the excitation energies in atoms and molecules are of the order of 1 [eV] or lower. This means that the solar photons have just enough energy to excite higher levels in these molecules. Nature has found a way to store these energies by photosynthesis, as we shall now see.

1.3.3 Photosynthesis: capturing and storing solar energy

When one considers an arbitrary grass meadow left to itself without supply of energy in the form of fertilisers, then it appears that a harvest may recover about 1 percent of the solar input. This number may seem small, but one has to consider that the energy is not only captured but stored as well. This remarkable achievement may possibly be improved and this is the subject of much research.

Photosynthesis produces carbohydrates for which we recall Equation (1.2)

$$6\,CO_2 + 6\,H_2O + 4.66 \times 10^{-18}[J] \rightarrow C_6H_{12}O_6 + 6\,O_2 \qquad (1.8)$$

> The world's most abundant protein is called RuBisCo. It is an enzyme which, in the dark, acts as a catalyst for the inverse of photosynthesis. It almost certainly is the most inefficient enzyme in primary metabolism. Therefore, the quest for a better RuBisCo is the Holy Grail of plant biology.
>
> *Science* **283**, 15 January 1999, 314–16

Humans or animals may consume the carbohydrates on the right-hand side of the Equation. Ultimately, all our food stems from reaction (1.8) with all kinds of subsequent reactions, which form proteins and fats.

The carbohydrates in (1.8) may also be left as residual biomass. In the course of time they then may experience chemical reactions by which they lose CO_2 and CH_4. As a result, the relative carbon content of the residue increases; after a long time it may form fossil coal, which is just carbon, or it may form hydrocarbons, abbreviated as C_xH_y. These will be found in crude oil or natural gas. Burning fossil fuels will always produce CO_2 and H_2O of which the carbon dioxide is one of the greenhouse gases mentioned earlier. As the biomass originates from plants, which contain sulphur and nitrogen, coal and oil still contain a fraction of these elements, whereas natural gas is relatively free of them. So burning coal, oil or biomass produces sulphur oxides and nitrogen oxides, besides CO_2.

1.3.4 Harnessing solar power by technology

Renewable energy is energy of which there is a continuous inflow. Besides heat from the interior of the earth which delivers geothermal energy, all other renewable energy originates from the sun. (Tidal energy is somewhat of an exception as it receives its power from sun and moon together, where the moon is the main contributor.) We already mentioned photosynthesis and will discuss other examples later on. Let us just say a few words about technologies, which use solar power directly.

An example of high technology is the *solar cell*. Here, a solar photon is absorbed by an electron, which results in electric power. In the conventional design a so-called semiconductor is used which absorbs photons above a certain energy $E > E_g$. It so happens that the energy E_g [eV] determines the output of the solar cell.

Perhaps less fancy, but as important, is the *solar collector*. This device absorbs the solar heat and converts it into hot water. This may be used in the tap, for showers and bathing and – possibly – for central heating purposes. The task of technology here is the reduction of heat losses and the fine tuning of the equipment.

A lady scientist from the Florida Solar Energy centre had her house built as one out of a batch of six. She insisted on a sheet of Al foil underneath her roof. The builder reluctantly obeyed what he saw as the whim of a silly, single female. He was converted however when he observed that his workmen had their coffees in just this house. It indeed was protected against the infrared radiation of the roof in the hot Florida sun.

Oral history, told to the authors by the lady herself

When solar cells or solar collectors are added to an existing building, one just has to do the best one can, depending on the local climate. For a new building one may and should optimize, following the requirements of the climate. The architect has to decide whether to get as much sunlight as possible inside (a cool climate) or keep the solar heat outside (a hot climate). A clever design of the windows can do this. Also one may choose the orientations of the building such as to optimize the performance of the installed equipment, e.g. by putting solar cells or collectors perpendicular to the mean incoming sunlight. In short, using solar power should be part of an integrated concept.

1.4 *THE GREENHOUSE EFFECT*

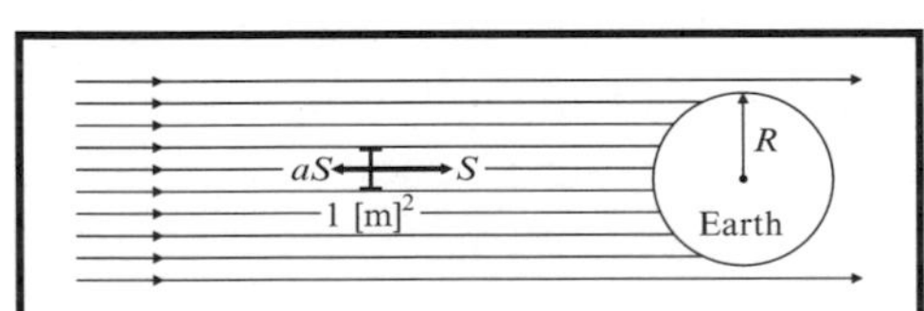

Figure 1.11 Solar radiation is entering the atmosphere from the left with S [W m^{-2}]. A fraction a, called the albedo, is scattered or radiated back.

The mean surface air temperature on earth is around 15 [°C]. In the tropics it is somewhat hotter and in the arctic regions somewhat colder. Except in regions with extreme conditions, most of the earth is habitable; it supports plant and animal life, which in turn provides food for the human population.

Let us make an estimate of the surface temperature of the earth. On average it will be equal to the mean temperature of the air just above the surface with the measured value of 15 [°C]. In a steady-state situation, the incoming energy must then be equal to the outgoing energy, or

$$\textit{energy in} = \textit{energy out} \tag{1.9}$$

We first look at the left-hand side of the energy equation. The amount of heat coming from the interior of the earth is negligible (see Example 1.1), so we restrict ourselves to the bulk: solar radiation. The amount of radiation entering the atmosphere per [m^2] perpendicular to the radiation is called S, the *solar constant*. Its units are [W m^{-2}] = [J s^{-1} m^{-2}]. Looking at the earth from outer space, it appears that a fraction a, called *albedo*, is scattered back or radiated back. As illustrated in Figure 1.11 an amount $(1 - a)\ S$ is

> Most of the glaciers in the Himalayas will vanish within 40 years as a result of global warming. The retreating glaciers have formed dozens of precarious meltwater lakes, accumulating behind an unstable edge of moraine. The Imja glacier lake will burst within 5 years releasing 30 million [m^3] of water and flooding the region downwards.
>
> *New Scientist*, 5 June 1999, 18

penetrating down to the surface. With a radius R the left of Equation (1.9) reads $(1 - a)\, S\pi R^2$.

In order to make an estimate of the right-hand side of Equation (1.9) we approximate the earth as a *black body* with temperature T. According to the Stefan-Boltzmann law (1.7) it produces outgoing radiation σT^4 [W m^{-2}]. The total outgoing radiation from the earth's surface then becomes $\sigma T^4 4\pi R^2$. Substitution in Equation (1.9) gives:

$$(1 - a)S\pi R^2 = \sigma T^4 4\pi R^2 \tag{1.10}$$

or

$$(1 - a)\frac{S}{4} = \sigma T^4 \tag{1.11}$$

Numerical values of σ, R and S are given in Appendix A. For the albedo a the experimental value is $a = 0.30$. Substitution gives $T = 255$ [K], which is way below the true value of 15 [°C] = 288 [K]. The difference of 33 [°C] is due to the *greenhouse effect* for which we should be grateful as it makes life possible. Its origin can be explained by looking at the spectral distribution of the incoming solar radiation and that of the outgoing black body radiation from the earth.

In Figure 1.8 we have already noticed that gases like CO_2 and H_2O absorb the incoming radiation far into the Infrared. As we shall see in Chapter 3 this absorption varies with wavelength but may be considerable up to a wavelength $\lambda \approx 30$ [μm]. From Figure 1.10 we know that the earth, approximated by a black body, emits strongly between 5 [μm] and 30 [μm]. Consequently much of the earth's emission will be absorbed by the atmosphere and radiated back. The net emission from the earth's surface, therefore, is smaller than that of a black body and may be simulated by a factor $f < 1$ on the right of Equation (1.11). This gives

$$(1 - a)\frac{S}{4} = f\sigma T^4 \tag{1.12}$$

With this very rough approximation, a value $f \approx 0.61$ would give the required temperature $T = 288$ [K].

1.4.1 Human induced changes

For ages man has been burning biomass such as wood, peat or dung. Consequently, reaction (1.1) runs to the right, producing CO_2. From the beginning of the industrial revolution around 1760, fossil fuels were used intensively in the form of coal (C) and later oil and natural gas. The latter two consist of varying proportions of carbon and hydrogen, hence its abbreviation $C_x H_y$. Burning fossil fuels is producing CO_2 at an increasing rate. Consequently, the atmosphere is absorbing more and more of the infrared radiation emitted from the earth's surface.

The result of more gases like CO_2 in the atmosphere is that the net emission of the earth becomes smaller. Therefore, the factor f in Equation (1.12) becomes smaller and the earth's surface temperature becomes higher. In that way the left-hand side of Equation (1.12) remains the same.

Figure 1.12 Garden greenhouse. It is kept warm by the glass cover, which radiates back in the infrared and provides shelter against the winds

The mechanism just described is called *radiative forcing*, where the net decrease in outgoing infrared radiation forces the climate system to respond with a temperature increase. In a garden greenhouse (Figure 1.12) something similar happens. During the day the sun passing through the glass windows heats the greenhouse. Its emitted infrared radiation is absorbed by the glass and is partially sent back. It must be said, however, that in a garden greenhouse an additional and important effect is that the heated air cannot be blown away by the wind but is kept inside. Nevertheless, the name *greenhouse effect* has been coined to describe the phenomenon that the earth's surface is kept at a higher temperature (288 [K]) than the 255 [K] expected from Equation (1.11)

1.4.2 Natural changes

> Up to half the global warming we have experienced over the past 130 years may have been caused by an increase in the Sun's output of energy. Since 1970, however, the Sun has been the source of less than a third of the global warming. The build-up of greenhouse gases is to blame for the rest. Summary from *Nature*
>
> in *New Scientist*, 5 June 1999, 5

There are changes in nature not influenced by man. An example is set by the solar 'constant' S, which is not really a constant but varies in time. Another example is given by volcanic eruptions, which introduce small particles into the upper stratosphere. These prevent the sunlight from reaching the surface of the earth and in terms of Equation (1.10) and (1.12) this makes the albedo a larger.

Not all of the natural variations are well understood. They must be separated from human induced changes when testing climate models. By these models we usually just calculate the consequences of man-made emissions on climate.

1.5 TRANSPORT OF POLLUTANTS

The best care for the environment is to use only renewable energies and very few scarce materials; this should be combined with production methods where polluting materials are made harmless at the factory or at the farm. This is not always possible or the price may be considered as unacceptably high. We therefore must accept some unavoidable pollution and understand by which mechanism pollutants disperse through the environment.

Let us start with a point source where the pollutants are emitted from a position, which is localized in space and time. The model here is the *plume* from a chimney as shown in Figure 1.13. In this case a wind blows from the left to the right and the reader will notice that the plume spreads when it is dispersing from the chimney. These two aspects, a flow of air or water along the source and the spreading of the plume are rather general. One calls the geometrical shape in which a particular point source disperses its pollutants the *plume* from the source.

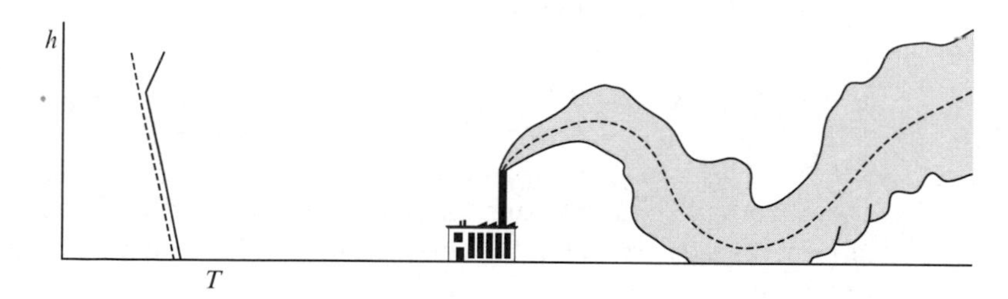

Figure 1.13 Model of a point source in an external flow: the smoking chimney

Without going into the mathematics, it will already be clear that the actual shape of the plume will be difficult to calculate. For the remarkable thing about a smoking chimney is that it changes all the time, sometimes even on a timescale of minutes. So, approximations must be made which allow one to describe the plume in an average way. This is the so-called *Gaussian plume*, described by a Gauss function, which is widely used in practice for many purposes.

1.5.1 The Gauss function

The *Gauss function* is given by the mathematical formula

$$f(x) = \frac{1}{\sigma\sqrt{2\pi}} e^{-x^2/2\sigma^2} \tag{1.13}$$

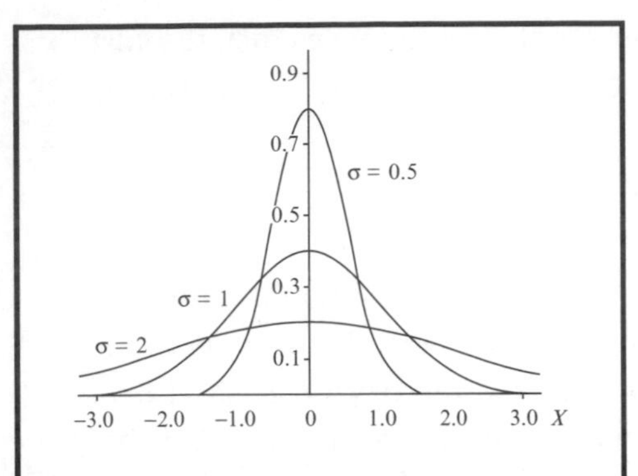

Figure 1.14 The Gauss function (1.13) for three different values of the width parameter σ

The factor in front of the exponential function guarantees that the integral over the function from $(-\infty)$ to $(+\infty)$ equals one. This *normalization* of the Gauss function makes it easy to connect the function with a certain amount of pollutants emitted.

Example 1.3: Normalization of the Gauss function

Check that the Gauss function (1.13) is normalized to one.

Answer

We write down the integral

$$\int_{-\infty}^{\infty} f(x)\mathrm{d}x = \int_{-\infty}^{\infty} \frac{1}{\sigma\sqrt{2\pi}} \mathrm{e}^{-x^2/2\sigma^2} \mathrm{d}x = \frac{1}{\sqrt{\pi}} \int_{-\infty}^{\infty} \mathrm{e}^{-u^2} \mathrm{d}u = 1 \tag{1.14}$$

where we have changed our variables to $u = x/\sigma\sqrt{2}$. The fact that the last integral equals one can be verified by using any mathematical table or software.

In Figure 1.14 the Gauss function is shown for three different values of the parameter σ. It is clear that, for increasing σ, the function goes down in height and spreads out. So, when the Gauss function (1.13) is used to describe the behaviour of the cross-section of a plume perpendicular to the external flow, σ should increase with time t. In this way the known spreading of a plume in the course of time is reproduced. In fact, later on we will find the relation $\sigma = \sqrt{2Dt}$ where D is the diffusion constant, which depends on the properties of the medium.

Banknotes sometimes display pictures of a great national scientist. Here, the German 10 Mark note of the pre-Euro era displays Gauss with his function.

Besides the cross-section perpendicular to the main flow, one needs the dispersion in the direction of the flow. The trick we will use here is to assume that in the x-direction the transport by the flow itself, is the main effect. This process is called advection, or in other words, sticking to the flow.

When one needs to go beyond the approximation of the Gaussian plume, one almost always has to rely on advanced computer methods [16].

> For calculating a calamity release of pollutants in a complex terrain such as industries, surrounded by hills, Gauss models are inadequate. It is possible to solve the appropriate differential equations by inserting parameters to account for atmospheric turbulence.
>
> [16], p 375–384

The physics approach described above should not blind us to the fact that there are many more ways of transporting pollutants than just dispersion in a flow. Animals, for example, may consume pollutants in their food and move to another place. There they may be devoured by other animals or slaughtered for human consumption. These effects are much more difficult to model reliably.

1.6 *THE CARBON CYCLE*

In a steady-state situation, the ecosystem of Figure 1.1 will be stationary. All elements needed for the development of organisms, not only C, H and O but also S, N, Se and Fe will move from one location to another; but the total amount of material in any part of the system will remain constant. In calculations of the effects of human emissions, one takes as a starting point such a steady state, also called a stationary state, and then estimates the resulting changes. These calculations have been done for several of the elements mentioned above, but most extensively for carbon, as CO_2 is one of the most important greenhouse gases.

> Well-documented examples of past imbalances among natural sources and sinks of CO_2 are evident in the record of atmospheric CO_2 increase and climate change at the end of the last ice age about 10000 or 20000 years ago. Climate and the C cycle appear to have been so tightly coupled that they must be considered as parts of the same global system.
>
> *Science* **259**, 12 February 1993, 935, 937

As we do not come back to the carbon cycle in detail, we will spend some time on that cycle now. In Figure 1.15 we show a simplified overview of the global carbon cycle as summarized in the readable and comprehensive ESCOBA report of the European Commission [17]. The boxes in the figure represent the estimated carbon contents in units of [10^{12} kg]. The arrows represent fluxes in [10^{12}kg yr^{-1}]. Full-line arrows represent relatively rapid changes and dashed ones are relatively slow. Similar graphs

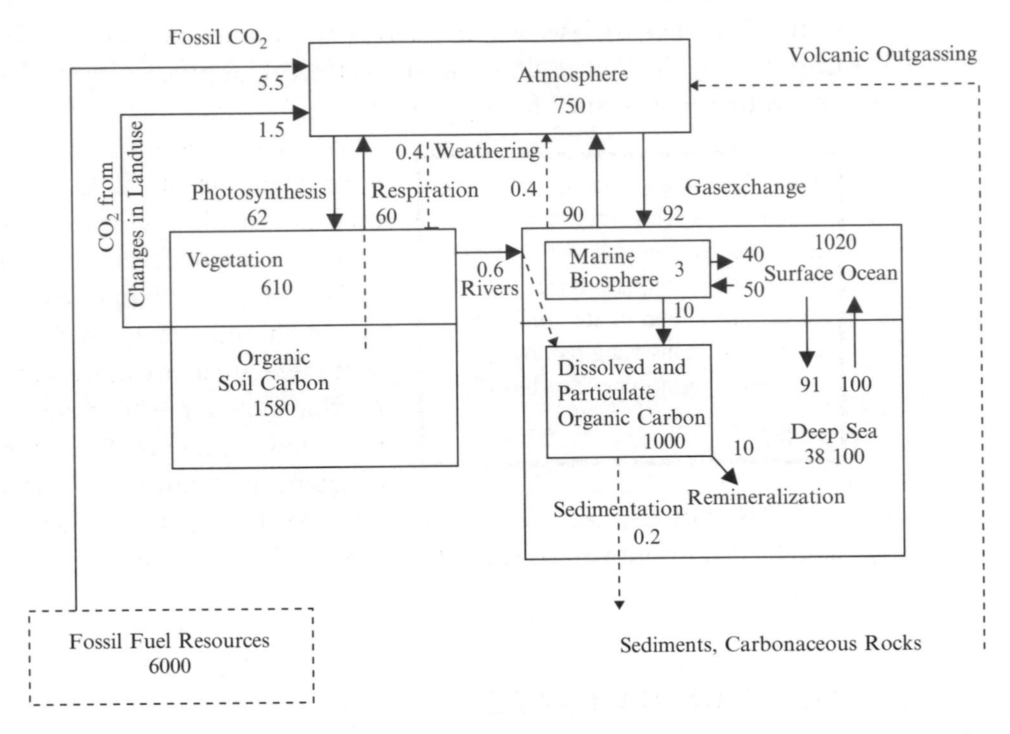

Figure 1.15 Overview of the global carbon cycle. Numbers in boxes refer to the carbon content in units [10^{12} kg]. Numbers near arrows refer to fluxes [10^{12} kg/yr]. Drawn arrows represent fast fluxes and dashed ones represent slow exchanges. From [17], p. 3

may be found elsewhere in the literature with different numbers, which illustrates the difficulty of the estimates. The figure itself also needs some time to digest, so we will discuss some of the data it contains.

We start from the top of the graph, which shows the atmosphere, since the carbon content there directly influences the greenhouse effect as we have seen earlier. Far left, the input from the burning of fossil fuels is shown. This number is easiest to estimate as the use of fossil fuels is well-documented in UN statistics.

The input from changes in land use, a little to the right in the figure, already is a little more complicated. Conversion of forests to agriculture releases the biomass present in the trees (Equation (1.1)). Also, the carbon and nitrogen content of the soil decreases. The crops that replace the trees have their own carbon content, but the amount of carbon is less as the volume of crops is smaller than that of trees.

1.6.1 *The oceans*

We skip the exchange between atmosphere and vegetation for a moment and move directly to the exchange between atmosphere and oceans. The uptake of

CO_2 by the oceans and out-gassing of CO_2 from the oceans could, in principle, be found by measuring the partial pressure of CO_2 gas in the ocean surface layers and comparing it with the pressure in the atmosphere. In this way it is found that in the tropics the out-gassing dominates and at moderate latitudes, it is the uptake. However, estimates are difficult, as one needs to know the relationship between pressure difference and uptake or out-gassing. Even when that relation is a simple proportionality, the constant of proportionality will depend on the roughness and pureness of the sea surface, which is difficult to simulate in a laboratory.

One therefore supplements the measurements of pressure by a study of the circulation of tracers, which are coupled to the carbon cycle, such as the ratio of carbon isotopes to be discussed in Chapter 10. Finally, all knowledge is put into ocean models, which results in the numbers shown in Figure 1.15. Note that we are left with two relatively big numbers (90 and 92 [10^{12} kg/yr]) of which the difference is decisive for the net ocean uptake. So this difference will have a relatively large error.

It is even more difficult to predict what would happen with the atmosphere – ocean interaction when the CO_2 concentration in the atmosphere would double from that of pre-industrial times. To get an idea we look at the mechanism by which the carbon circulates through the oceans. The CO_2 gas in the ocean top layers will react with water according to the equilibrium reaction

$$CO_2\,(\mathrm{gas}) + H_2O \leftrightarrow H_2CO_3\,(\mathrm{aq}) \tag{1.15}$$

Here the addition of (aq) means that the carbonic acid H_2CO_3 is dissolved in water. Next, the acid dissociates into bicarbonate and carbonate according to the two reactions

$$H_2CO_3\,(\mathrm{aq}) + H_2O \leftrightarrow H_3O^+\,(\mathrm{aq}) + HCO_3^-\,(\mathrm{aq}) \tag{1.16}$$

$$HCO_3^-\,(\mathrm{aq}) + H_2O \leftrightarrow H_3O^+\,(\mathrm{aq}) + CO_3^{2-}\,(\mathrm{aq}) \tag{1.17}$$

Both equations are governed by an equilibrium constant. The relative magnitude of the HCO_3^- and CO_3^{2-} concentrations strongly depend on the pH of the water. For sea water with pH ≈ 8 one finds that HCO_3^- dominates, with a relative concentration of 97 percent (cf. Exercise 1.9). As a consequence, the carbon will react with calcium in the marine life according to the reaction

$$Ca^{2+} + 2\,HCO_3^-(\mathrm{aq}) \leftrightarrow CaCO_3 + CO_2 + H_2O \tag{1.18}$$

Therefore the carbon is stored in calcium carbonate (or calcite), which is used for the construction of shells and tissues. Part of it slowly sinks down to the bottom of the oceans.

> If the balance between carbonate chemistry and photosynthesis shifts in the future, certain algae might switch from being a carbon source to a carbon sink.
>
> *Nature* **405**, 25 May 2000, 412

In a steady state this downward flux is compensated by an upward flux of carbon-rich waters near the equator, along the continental margins and in high-latitude waters. The process of transport down and up again is called the 'marine biological pump'. Its magnitude does not only depend on the carbon content in the waters but also on the availability of major nutrients like N, P and micro-nutrients like Fe and Mn.

With a doubled CO_2 concentration in the atmosphere, the climate will change and it is difficult to predict how the availability of nutrients will respond to the change of climate. It may go both ways. If we are lucky, more nutrients will result in more plankton going down, which takes part of the excess of CO_2 down, mitigating the climate change. If we have bad luck and the nutrients are lacking, the plankton will be killed, the 'pump' will stop functioning and the atmospheric CO_2 concentration may increase by another 150 [ppm].

1.6.2 The land

> Old and middle aged forests are net absorbers of CO_2. This net effect is the small difference between two huge numbers representing photosynthetic gains and respiratory losses. Respiration may be more sensitive to long term climate change than photosynthesis. The answer to this question determines the role of the forests as a carbon sink.
>
> *Nature* **404**, 20 April 2000, 819/820

When we return to Figure 1.15 and look somewhat left of the middle at the exchange between atmosphere and vegetation, we notice again that two large numbers (62 and 60 [10^{12} kg/yr]) determine the exchange: photosynthesis and respiration. The net result is an uptake of 2 [10^{12} kg/yr] of carbon.

The net effect of the landmass on the CO_2 budget may also be calculated directly from the carbon budget without reference to large numbers. This is shown in Table 1.1. The input into the atmosphere from industrial emissions and the changes in land use can be estimated rather accurately. It adds up to $7.1 \times [10^{12}\ \text{kg yr}^{-1}]$. The measured accumulation in the atmosphere is much smaller, so the rest goes into the ocean, which is modelled, and in the land uptake, which is inferred. This latter number then appears as

Table 1.1 Global mean CO_2 budget / [10^{12} kg yr^{-1}] for 1980–1989 [18], p. 17

Estimated industrial emissions	5.5 ± 0.5
Estimated emissions from land use changes	1.6 ± 1.0
Total emissions	7.1 ± 1.1
Measured accumulation in the atmosphere	3.3 ± 0.2
Modelled ocean uptake	2.0 ± 0.8
Inferred land uptake	1.8 ± 1.6

$(1.8 \pm 1.6) \times [10^{12}$ kg yr$^{-1}]$. Ref [18] from which Table 1.1 was taken gives a value of (0.5 ± 0.5) [10^{12} kg yr^{-1}] as uptake in the Northern Hemisphere from regrowth of forests. The rest is a 'missing' carbon sink.

> A big carbon sink in North America could lead to the conclusion that there is no need for the USA and Canada to curb emissions. But maturing forests eventually stop storing carbon, so this part of the missing sink won't be with us forever or even much longer.
>
> *Science* **282**, 16 October 1998, 386

An indication that there is a very active carbon sink in the Northern Hemisphere is found from measuring the small differences in CO_2 concentrations in the atmosphere ([18], p. 15). The burning of fossil fuel, for example, chiefly happens in the northern part of the world. Not only because the landmass and populations are bigger in the northern half of the world, but also because it is richer and is using more energy. It is estimated that the resulting CO_2 concentration in the Northern Hemisphere would be 6 [ppmv] larger than in the Southern Hemisphere.

As the mixing between the Northern and Southern Hemispheres is slow because of the strong tropical winds separating them, this difference of 6 [ppm] should be found in measurements. In reality one finds a difference of only 3 [ppmv], which indicates a strong sink in the Northern Hemisphere, perhaps larger than the $(0.5 \pm 0.5)[10^{12}$ kg yr$^{-1}]$ mentioned above.

1.6.3 The final steady state

The numbers in Figure 1.15 refer to the present situation. One could, however, use the graph to calculate the final steady state of the system. For example, assume that in some period of time a large fraction of the fossil fuel resources of 6000 [10^{12} kg] is burned and its CO_2 emitted into the atmosphere. Of course, much of these resources are so diluted that a commercial exploitation would not make sense. A reasonable assumption would be that a quarter of the resources are burnt, or $1500 \times [10^{12}$ kg]. Next one has to make an educated guess about the timescales in which the different parts of

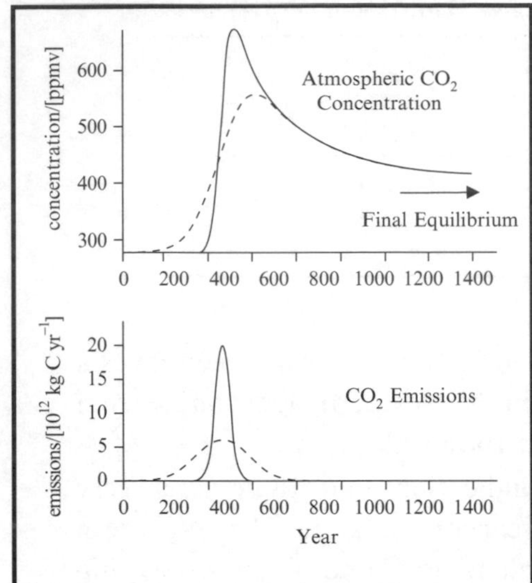

Figure 1.16 The lower graph shows the release of a quarter of the fossil fuel resources according to two timescales: quick (full) and slow (dashed). The upper graph shows the response in terms of CO_2 concentrations. From [17], p. 4

the system would respond to change. This is not easy as several timescales are involved. The shortest ones correspond to a day–night cycle or a summer–winter cycle, where others correspond to glacial–interglacial variations in the order of thousands of years.

The result of a model calculation is shown in Figure 1.16. In the lower graph two types of emissions are drawn, both peaking at 400 years after some zero. The full curve corresponds to a rather quick release of carbon, within 200 years, whereas the dashed one corresponds to a slower exhaustion of resources, in about 400 years. The upper graph shows the resulting atmospheric CO_2 concentration as a function of time. In both cases, a final equilibrium would be reached some 1500 years after zero with a concentration of 400 [ppmv]. When the pulse is distributed more evenly in time (the dashed curve) the peak height is lower. In both cases the climatic consequences will be large. But it stands to reason that in the second case they might be somewhat smaller as the ecological system has more time to adapt.

1.6.4 Conclusion

The student will appreciate that there is much work to do for chemists, biologists, geologists and physicists. There are still many things to measure, to study and to model. Some of them will be discussed in the rest of this book.

EXERCISES

Numerical data to be used in the exercises may be found in Appendix A. Some common data which were not put in Appendix A may be found in tables such as [15].

1.1 Check the dimensions of Equations (1.4) and (1.5). Derive (1.5) from (1.4) using (1.3).

1.2 Find Wien's law (1.6) from (1.5) using the approximation of small wavelengths.

1.3 Find (1.7) by integration of (1.4) using $E = h\nu$. Look for the nasty integral in a book or use mathematics software to find it. Compare the value of σ you find with the value given in Appendix A.

1.4 Deduce from Figure 1.9, the peak value of black body radiation to be expected at the earth. Use for the radius of the sun $R_{sun} = 6.96 \times 10^8$ [m] and the distance from the centre of the sun to the centre of the earth $R_{se} = 1.49 \times 10^{11}$ [m]. Check with Figure 1.8

1.5 The Greenland ice sheet has an area $A = 1.834 \times 10^6$ [km^2] and an average thickness $d = 1790$ [m]. Suppose the sheet breaks off and starts flowing in the North Atlantic Ocean. How big would the rise r in sea level be? If the sheet then melts, how big would the additional rise be (all calculations in first approximation)?

1.6 The home of one of the authors is using $V = 3000$ [m^3] of natural gas a year for central heating and hot water. It has a horizontal roof area of $A = 40$ [m^2]. Calculate the number of thermal Joules used from the gas. Estimate the incoming solar power on the roof during a year, assuming that half of the influx will reach the roof. Ignore any correction for the geographical latitude. Check whether a solar hot water installation would make sense. Repeat the calculation for your own home and energy needs.

1.7 The home of Exercise 1.6 is using $E = 2200$ [kWh] per year of electricity. Assume that the roof is covered with solar cells with an efficiency of 10 percent at the output of the electricity generating system. Calculate the number of electric Joules used and the number of Joules the solar cells would effectively produce. Note that the storage problem is larger than in the previous exercise. In fact, the solar system should be connected to the grid. Also note that a choice between renewable hot water and renewable electricity will be influenced by cost considerations for the consumer.

1.8 According to Figure 1.15, the total pool of carbon in the atmosphere was $M = 750 \times 10^{12}$ [kg]. Show that this corresponds with 347 [ppmv], a value which was reached in 1987.

1.9 For chemistry students: For sea water one may take pH = 8. Use the equilibrium constants $K_1 = [H_3O^+][HCO_3^-]/[H_2CO_3] = 4.3 \times 10^{-7}$ and $K_2 = [H_3O^+][CO_3^{--}]/[HCO_3^-] = 4.8 \times 10^{-11}$ to calculate the fractions in sea water of $[CO_3^{--}]$, $[HCO_3^-]$ and $[H_2CO_3]$. For students proficient with computers: plot the three fractions as a function of pH.

2 Weather and Climate

At moderate latitudes the weather is an inexhaustible subject of conversation [19]. It is changing rapidly and unexpectedly, and in many countries it is wise to take an umbrella with you, even if the weather forecast is 'good'. The word 'weather' encompasses temperature, sunshine, cloudiness, rainfall, wind and wind directions; these are the topics which are given in the daily weather forecast.

> European weather outlook
>
> **Scandinavia:** A cool and wet day in the north. There will also be some showers in southern parts of Norway, especially over the hills. However, Sweden and Denmark will be dry with the best of the sunshine in Denmark, where it will be quite warm. Highs 9[°C] in the north, but 23[°C] in Denmark.
>
> **France:** It will stay unsettled.
>
> **Italy:** Some heavy and thundery storms in the north and over the Alps.
>
> **Greece:** Dry and very hot inland
>
> From: *The Guardian* (European edition), August 7, 1999

'Climate' is more than just the average of all these measurable quantities over one or more years. It also implies the daily and seasonal changes of these variables and the rate at which the changes occur. The difference between the tropical climate in the Caribbean and the maritime climate in North-west Europe is not only described by the difference in temperature. Even more different are the day to day and hour to hour variations of the weather in both places.

Weather and climate are difficult subjects to define and to calculate, the same holds true for climate change. Old people may tell you of particularly hot summers in their youth, and of some harsh winters. The weather columns of many newspapers compare the present weather with that of the past years and discuss the differences. As these sometimes large variations from year to year have always occurred, it will not be easy to deduce climate change from occasional observations. It will be even more difficult to separate 'natural' variations of climate from human induced changes. One will need some physical insights for that purpose.

In this chapter we therefore start, in Sect. 2.1, with some physical concepts of energy transport, which we need in appreciating the discussions on climate and climate change. In Section 2.2 we apply our concepts to vertical

Summer 1548:
This summer was a great drought for lack of rain and in July the plague (the pest) reigned sore in London with great death of people.

October 1548:
In Paris the beginning of the autumn is very dry. One organizes the procession of Ste Geneviève to call for rain. In Ost-Friesland (Germany) strong winds at high tides causes the sea water to submerge the dikes and flood the lands.

November 1548:
Incessant rain in the south of France filled the Rhône, which flooded Avignon.

From ancient chronicles [19]

atmospheric phenomena and in Section 2.3 to horizontal motion in the atmosphere. In this section we need some vector notation, which may be a little hard for some students, but we explain it in the main text and summarize it in Appendix B. We assure these students that it will be rewarding to take the time to understand it, not only for use in the rest of this book but also for their further career in science. In Section 2.4 we will discuss the oceans, whereas in Chapter 3 we will use the subject matter of the present chapter to describe the human influence on climate, resulting in climate change.

2.1 *ENERGY TRANSPORT*

Figure 1.8 shows the solar radiation entering the atmosphere and from the left of equations (1.11) and (1.12), we recall that short-wave radiation is entering from the sun. From the right-hand side of Equations (1.11) and (1.12) we know that long-wave radiation is leaving the earth. For the earth in total, the energy entering should be equal to the energy leaving (Equation (1.9)). One out of many simplifications in our discussion was that a global average was taken, which ignored all variations with latitude ϕ and longitude λ.

We therefore have a look at Figure 2.1, which summarizes experimental data at the top of the atmosphere [20]. The graph shows the annual mean absorbed solar radiation and the emitted long-wave radiation as a function of latitude ϕ, while an average over longitude λ was taken along a circle of constant latitude ϕ. The absorbed radiation displays a peak at the tropics and a rather small value in higher latitudes North and South. From Figure 1.11 we remember that the perpendicular input is taken as $S = 1367\,[\mathrm{W\ m^{-2}}]$, the solar 'constant'. About 30 percent of it is scattered back or reflected, and we have a day-and-night variation as well, so the peak value in the solar absorption of about 350 $[\mathrm{W\ m^{-2}}]$ looks reasonable.

The variation of absorbed sunlight with latitude ϕ is easily understood qualitatively, for at high latitudes the sun stands lower in the sky, which gives a lower influx. Moreover the snow-caps in the Arctic and Antarctic

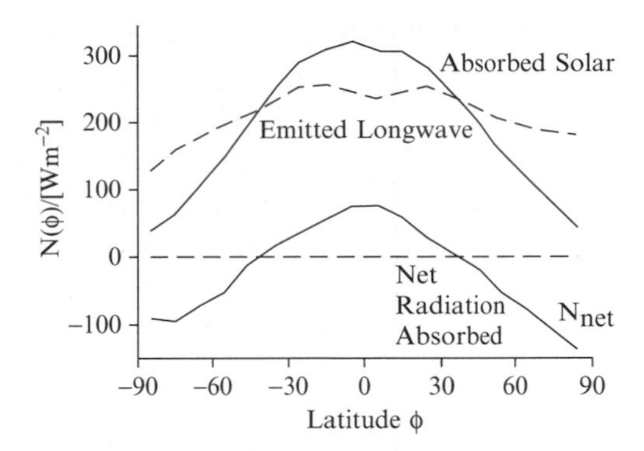

Figure 2.1 Annual mean absorbed solar radiation and emitted long-wave radiation, as a function of latitude and averaged around latitude circles. The net radiation absorbed is negative at the poles, implying energy transport. Reproduced with permission of Academic Press from [20], Figure 2.12

regions reflect much of the incoming radiation. In this way we understand the general behaviour of the solar absorption curve.

The dashed curve in Figure 2.1, representing the emitted long-wave radiation is even more interesting. It also decreases from the tropics to the poles but less strongly than the solar input. From Equation (1.11) we know that the emission is given by the black body radiation, which reads σT^4 [W m^{-2}], where T is the temperature in Kelvin. Expressed in Kelvin, the poles of course are still colder than the tropics but the difference in Kelvin is small. Consequently, the variation of emission with latitude λ is small as well.

The third curve shown in Figure 2.1 is the net absorbed radiation, which is the difference between the two other curves. This net absorbed radiation is positive in the tropics and negative in moderate and higher latitudes. This does not imply that the tropics are getting hotter all the time and the other regions colder, as we know that on average our climate is fairly constant. So, the third curve demonstrates that there must be a net heat transport from the tropics to the poles.

Example 2.1: Energy passing the 35° parallel

From Figure 2.1 make a rough estimate of the total amount of energy [W] passing the circle of latitude $\phi = 35°$ (where the net radiation $N_{\text{net}}(\phi)$ absorbed is zero) to the North.

Answer

The simplest assumption is that all net positive radiation $N_{\text{net}}(\phi)$ north of the equator goes to the North and south of the equator goes to the South.

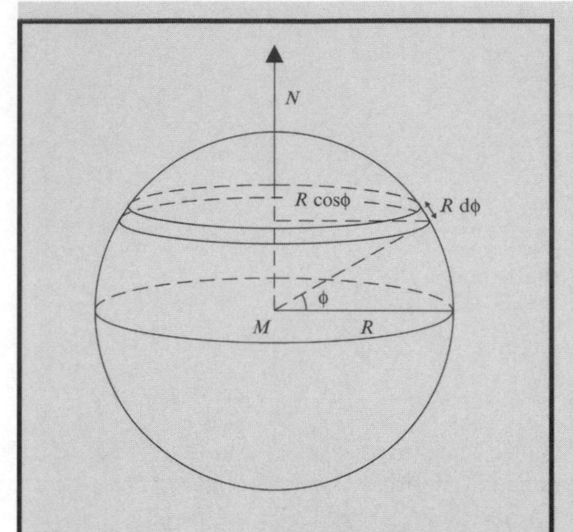

Figure 2.2 The net radiation absorbed in a strip around a latitude circle is found by multiplying its surface area with $N_{net}(\phi)$ from Figure 2.1

Therefore we have to integrate the net incoming energy in the Northern Hemisphere from $\phi = 0°$ to $\phi = 35°$. In mathematics we then consider a thin strip along a sphere between latitude angles ϕ and $\phi + d\phi$. This strip is indicated in Figure 2.2. The width of the strip equals $Rd\phi$, where R is the radius of the sphere. In order to get the length, or circumference of the strip, we notice that its radius becomes $R \cos \phi$. The surface area of the strip therefore equals $2\pi R \cos \phi R d\phi = 2\pi R^2 \cos \phi d\phi$. In mathematical physics, quantities like $d\phi$ are called differentials, or infinitestimal small quantities. The change of ϕ over the strip therefore may be ignored and the area is multiplied by the net radiation $N_{net}(\phi)$. So the total contribution to the North becomes

$$\int_0^{35^0} N_{net}(\phi)\, 2\pi R^2 \cos \phi \, d\phi = F \tag{2.1}$$

In numerical work we always have to express angles in *radians*, or [rad], defined by $180° = 2\pi$ [rad]. Therefore, $\phi = 35°$ corresponds with $\phi = 0.61$ [rad]. To keep it simple we approximate the function $N_{net}(\phi)$ in Figure 2.1 for $\phi > 0$ by a triangle: $N_{net}(\phi) = 75 - 123\phi$ [W m^{-2}]. The reader may check in Figure 2.1 that $\phi = 0$ gives the correct peak value 75 [W m^{-2}]. For $\phi = 0.61$ we indeed find $N_{net}(\phi) = 0$. Substituting the radius R of the earth from Appendix A we find with some simple software that $F = 5.65 \times 10^{15}$ [W]. To appreciate the enormous amount of energy, we may divide by the circumference of the latitude circle and find about 1.7 $\times 10^8$ [Wm^{-1}], which is 170 [MW m^{-1}]. So each metre is crossed each second by the output of a medium sized power station.

Atmosphere and oceans are responsible for the energy transport calculated in Example 2.1. They roughly contribute to the same extent ([20], Figure 2.14). We will now look at the global atmospheric circulation, but leave the oceans to Section 2.2.

2.1.1 The energy transport to the poles

The net heat transport is gigantic (Example 2.1) and it is not easy to see how nature organizes it, let alone to calculate it. For, at first sight (Figure 2.3), the dominant winds at ground level near the equator are easterly (from the east to the west) with a tendency to blow towards the equator. These

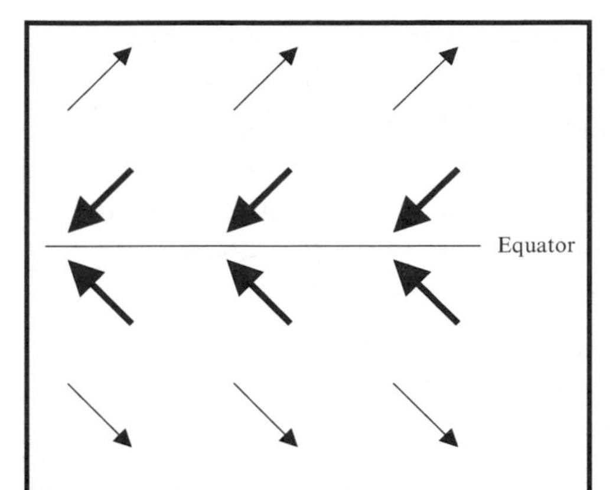

Figure 2.3 Schematic view of the mean airflow near ground level. Near the equator it is easterly, further away it is westerly

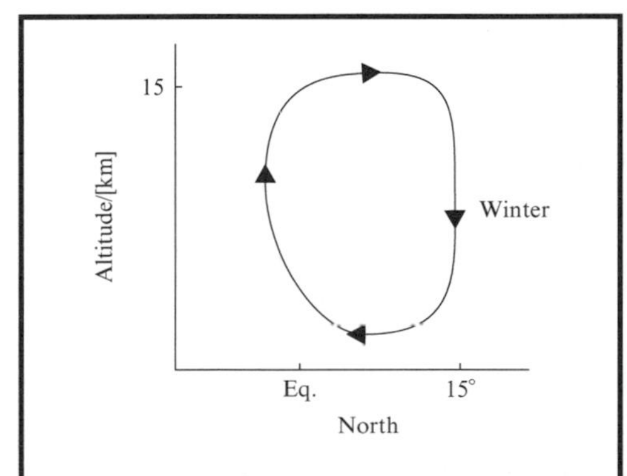

Figure 2.4 The Hadley circulation in the Northern Hemisphere during winter as based on experimental data. See [20], Figure 6.5

The consequences of vertical motion are immediately obvious in the association of ascent with the clouds and rains near the equator and of descent with the arid subtropical highs. The horizontal areas over which the ascent and descent are spread are so large, however, that mean vertical speeds are too small to be observed directly, and have to be deduced by various indirect measures.

From [21], p. 89

winds are traditionally called trade winds as sailing ships could rely on them for navigating in the tropics.

Somewhat further from the equator the winds are westerly, but with many more cyclones, anticyclones and depressions, so for the casual observer the dominant trend is not obvious.

In order to judge the energy transport one should not only look at the atmosphere at ground level, but take into account all layers of the atmosphere from bottom to top. For the oceans, as we shall see later, one similarly has to look at all depths.

A summary of the observations of vertical air mass movement near the equator is shown in Figure 2.4. There is one dominant circulation with its bulk at the winter side of the equator, which here is taken as the Northern Hemisphere. At ground level one notices a flow towards the equator, which is consistent with Figure 2.3. At the equator the air rises; at the same time the sun heats the oceans causing ocean water to evaporate. The rising air therefore takes with it latent heat in the form of water vapour and transports it to the North, where it descends again. In the meantime the air rains out and the latent heat warms up the remaining air as heat of condensation. It can be shown that up to about 15° North this is the dominant mechanism for energy transport ([20], Figure 6.11). At that latitude the cyclones and depressions take over. That type of energy transport is usually named *eddy* transport as this word suggests the whirlpool of motion in the atmosphere.

The dynamics of eddy transport is not simple. We can show, however, that a mean northern flow of air masses in the Northern Hemisphere is consistent with strong westerly winds. To do this we need to go back to *Newton's laws of motion*. His first law says that a particle or a body on which no forces are working will move in a straight line with a constant velocity. In order to describe the particle we need a coordinate system and Newton suggests that there is such a coordinate system, which is connected to the stars and the sun.

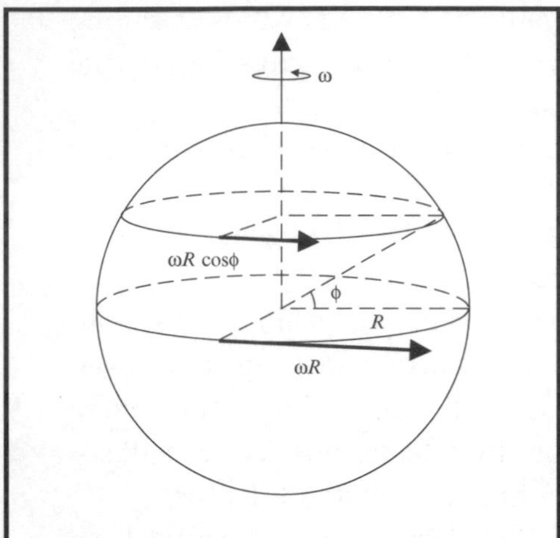

Figure 2.5 An air mass moving north will get a higher eastward (= westerly) velocity with respect to the earth

Let us now look at a small volume of air, which often is called a 'parcel' of air. As shown in Figure 2.5, when the parcel is at rest at the equator, its velocity when viewed from the stars will be ωR. Here ω is the angular velocity of the earth around its axis and the velocity $\omega R = 464\,[\mathrm{m\ s^{-1}}]$ to the East. At a higher latitude ϕ the velocity of a parcel at rest would be $\omega R \cos\phi$. This is a smaller eastward velocity and one therefore expects that a parcel of air, moving to the north would acquire an eastward velocity.

Indeed one finds eastward velocities at an altitude of 15 [km] with a magnitude of about 40 [m s^{-1}] to the East in the so-called *jet stream*. This is smaller than one would expect at first sight. The main reason is that one should take into account that the parcel of air at the equator has a circular motion and is not moving in a straight line. So it performs a rotation which results in an angular momentum around the earth's axis. This angular momentum is preserved ([1], p. 47). Part of it results in the jet stream, but most of the rotation is transferred into the large-scale rotation of the air masses in eddies like cyclones and depressions.

2.2 *THE ATMOSPHERE, VERTICAL STRUCTURE AND MOTION*

Before looking in more detail at the horizontal motions in the atmosphere and the connected physical laws, we study the vertical structure of the atmosphere and the laws that determine the vertical motion of air.

2.2.1 *Vertical structure of the atmosphere*

Figure 2.4 puts the top of the Hadley circulation at 15 [km]. This altitude finds its origin in the vertical structure of the atmosphere, summarized in Figure 2.6. On the left-hand side we note the altitude above sea level in [km] and in the middle a wide full curve, which corresponds with the temperature, shown on the horizontal axis. We note that the temperature goes down with altitude up to about 15 [km], where it remains stable and, at a higher altitude, starts rising again. The lowest part of the atmosphere is called the *troposphere*. The region where the temperature stops dropping is called the

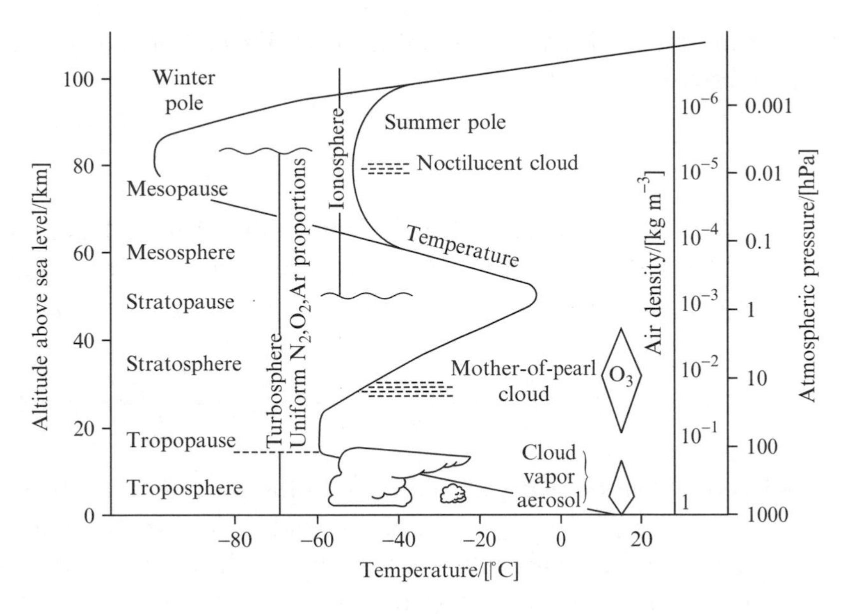

Figure 2.6 The vertical structure of the atmosphere. The altitude on the left axis corresponds to the rapidly decreasing scales of density and pressure on the right axis. The wide full curve indicates the temperature as a function of altitude with a definite dependence on summer and winter at higher altitudes. (Adapted and reproduced by permission of the author from [21], Figure 3.1, p. 48)

tropopause. The majority of the phenomena, which determine weather and climate, happen in the troposphere.

When, for completeness, we look further up in Figure 2.6, we notice that the temperature rises in the *stratosphere* and stops at the *stratopause*. The highest part of the atmosphere is called the *ionosphere* as practically all atoms are ionized, which means that they have lost some of their electrons. Further, notice the clouds, most of which develop in the troposphere, although at higher levels some minor clouds may be formed as well.

On the right-hand side of Figure 2.6, we notice a scale for air density and atmospheric pressure which both go down exponentially, with factors of 10, when the altitude rises on an ordinary, linear scale. To explain this phenomenon we need two relations, the equation of state for air and the hydrostatic equation. The *equation of state* describes the relationship between thermodynamic properties of a certain amount of material. For a homogeneous material these are the pressure p, volume V,

> The peak height of the Mount Everest is close to 10 [km], where the air density is less that 1/3 of the sea level values. As the oxygen fraction is the same as at sea level, a climber inhaling in a normal fashion reduces his intake of oxygen to one third. This cannot sustain the life of even the fittest individual for more than a few weeks nor allow hard exertion for more than a few hours at a stretch.
>
> From [21], p. 82

temperature T and the number of moles n. For an *ideal gas* the equation becomes very simple

$$pV = nRT \tag{2.2}$$

With a mole weight of M it follows that 1 mole has a mass of M grams $= M \times 10^{-3}$ [kg] and n moles have a mass $m = nM10^{-3}$ [kg]. Writing Equation (2.2) in terms of mass and rearranging some terms yields

$$p = \frac{nRT}{V} = \frac{mRT}{VM \times 10^{-3}} = \frac{m}{V} \frac{R}{M \times 10^{-3}} T = \rho R_{air} T \tag{2.3}$$

In equations (2.2) and (2.3) the symbol R indicates the universal gas constant, so $R_{air} = R/(M \times 10^{-3})$ is a gas constant for air. Since air is a mixture of gases also the value of the molar weight M has to be adapted which leads to a value of R_{air} quoted in Appendix A. Note that in eq. (2.3) we introduced the symbol $\rho = m/V$ for the density of the air [kg m^{-3}].

From Equation (2.3) it follows that, for a constant temperature T, the pressure p and the density ρ are proportional. Figure 2.6 shows that in the atmosphere the temperature as a function of altitude varies between–60 [°C] and 15 [°C]. In Equation (2.3) the temperatures have to be expressed in Kelvin as 213 [K] and 288 [K]. The air temperature as a function of height is not really constant, but deviates by some 40 [K] from the average 250 [K], which is less than 20 percent. So, the scales for ρ and p on the right in Figure 2.5 are only proportional in an approximate sense.

Let us now look at the *hydrostatic equation*. The 'hydro' in the name refers to a column of water in which the pressure increases downwards, because of the weight of the water. The hydrostatic Equation (2.5) to be derived below, applies to any *fluid* (liquid or gas) without vertical accelerations.

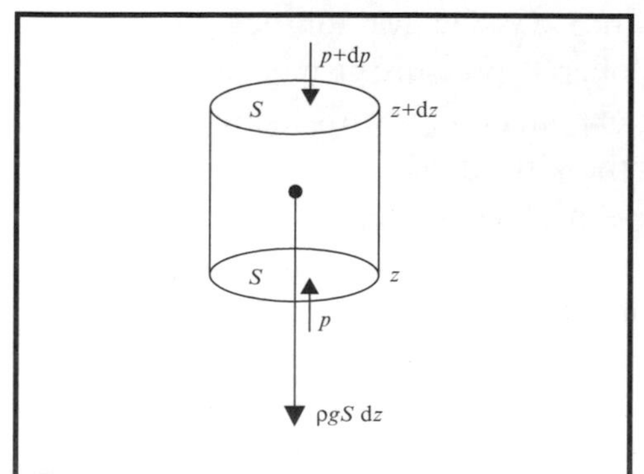

Figure 2.7 The hydrostatic equation. A mass of air $\rho S\mathrm{d}z$ is kept in its place by a pressure $S\mathrm{d}p$

To derive the equation we consider a small disk of the fluid column with, for the bottom, a vertical coordinate z and, for the top, a coordinate $z + \mathrm{d}z$. The pressures at the two positions are p and $p + \mathrm{d}p$ respectively, as shown in Figure 2.7. The surface area of the disk is S. The total downward gravity force for a volume S dz with mass $\rho S\mathrm{d}z$ may be written as $g\rho S\mathrm{d}z$. The downward pressure force equals $(p + \mathrm{d}p)S - pS = (\mathrm{d}p)S$. The sum of the two must be zero, as there are no other forces present and accelerations are assumed to be absent as well:

$$g \rho S \, \mathrm{d}z + \mathrm{d}p \, S = 0 \tag{2.4}$$

or

$$\mathrm{d}p = -g \rho \, \mathrm{d}z \tag{2.5}$$

The negative sign shows that the pressure decreases as one goes up with $\mathrm{d}z > 0$. Equation (2.5) can also be written as

$$\frac{\partial p}{\partial z} = -g\rho = -g \frac{p}{R_{\mathrm{air}} T} = -\frac{g}{R_{\mathrm{air}} T} p = -\frac{1}{H} p \tag{2.6}$$

In the second equality we needed Equation (2.3) in order to get a relation where the pressure p is written on both sides. The last equality becomes interesting if the temperature is taken as roughly constant at $T \approx 250$ [K]. Then the right-hand side may be interpreted as a constant $1/H$ multiplied by pressure p as follows

$$\frac{\partial p}{\partial z} = -\frac{p}{H} \tag{2.7}$$

The Italian scientist Torricelli created a vacuum by pouring mercury into a long glass tube, closed at one side, and then putting the column upside down in mercury again. The French scientist Pascal understood that the mercury was in equilibrium with a column of air. To prove this he had his brother-in-law climb the Puy de Dôme. 19 September 1648. Indeed, as we understand from equations (2.8) and (2.9) the mercury went down (Example 2.2). Incidentally, Pascal could only perform his famous experiments with many varieties of Torricelli's tubes because he had access to the best glass-blower of France in Rouen.

Units and dimensions

On both sides of Equation (2.7) there is a pressure in the numerator. On the left there is a length in the denominator, so the variable H must have the character of a length as well. This argument is based on the fact that in physics the dimensions at both sides of an equation must be equal. A *dimension* is a measurable physical quantity such as mass [M], length [L], time [T] and temperature [t]. In principle, all measurable quantities have a physical dimension.

Up to this point we have used units like [kg], metre [m], second [s] and derived units like Newton [N] and written them between square brackets. We will do the same with dimensions, if we use them. It is often more convenient and quicker to check an equation on the units. If the units agree, the dimensions should agree as well. But you may get confused, when on one side you end up with [N m^{-2}] and on the other side with [Pa]. Then you should know that both are the same, or look at a book where these units are written out in their fundamental dimensions and you will find that they match.

Pressure as a function of height

Let us go back to Equation (2.7). We now understand that H has the dimension of a length and therefore can be expressed in [m]. With $T = 250$ [K] and the values of Appendix A, it follows that $H = 7321$ [m] (Exercise 2.2).

To solve Equation (2.7) for a constant value of H we note that the derivative of p with respect to z should be proportional to p itself. If the derivative of a function is proportional to the function itself we know from mathematics that the function must be exponential:

$$p = p_0 e^{-z/H} \tag{2.8}$$

One may check by substitution that this indeed obeys (2.7). The constant p_0 in front of the exponential is found by substituting $z = 0$, which gives $p = p_0$, so p_0 is the pressure at ground level. In Chapter 5 we shall learn that the exponential form in Equation (2.8) has a deep physical meaning.

It is easy to write (2.8) in terms of powers of 10 (Exercise 2.3). We find

$$p = p_0 10^{-z/H_{10}} \tag{2.9}$$

with $H_{10} = 16.9$ [km]. It follows that for every 16.9 [km] increase in altitude the pressure should go down by a factor of 10 and this indeed corresponds with the scale on the right-hand side of Figure 2.6.

Example 2.2: Pressure decrease on top of a mountain

The Pascal experiment* on atmospheric pressure was made by climbing from a village at 490 [m] height to the summit of the Puy de Dome in France at 1465 [m]. Calculate the number of [cm] the mercury barometer would go down as follows:

(i) check that at sea level the barometer would show a height of 76 [cm] for 1 [atm] pressure.

(ii) Use (2.8) or (2.9) to calculate the pressures at 490 [m] and at 1465 [m] and convert them to barometer heights. Pascal's brother-in-law measured a difference of 8.5 [cm].

Note: In student labs, students are often asked to perform the same experiment to measure the height of their university buildings. If accuracy is required, the effect of the temperature difference on the density of mercury also has to be taken into account; H also has to be adapted to the average local temperature, and finally a difference in humidity and temperature together may influence the vapour pressure of water vapour. This contribution should be subtracted as Equations (2.8) and (2.9) apply to dry air.

Answer

(i) At sea level we have a pressure which used to be called 1 [atm]. Officially this equals $p_0 = 101.3$ [kPa]. A column of h [cm] of Hg and a cross-section of 1 [m^2] has a volume $V = h/100$ and a mass of ρV, which gives a pressure $\rho Vg = \rho hg/100$. This should equal 101.3 [kPa]. Use $\rho = 13545$ [kg m^{-3}] for mercury at 20 [°C] and we find $h = 76.26$ [cm].

(ii) We use Equation (2.8) with $H = 7321$ [m], as given above. Then
p (490 [m]) $= 0.9353\, p_0 = \rho hg/100$, which corresponds with $h = 71.32$ [cm]
p (1465 [m]) $= 0.8186 p_0 = \rho hg/100$, which corresponds with $h = 62.43$ [cm]
so the difference amounts to 8.89 [cm].

*Data on the Pascal experiment may be found in his 'Oeuvres Complètes', in the letter sent to Pascal by his brother-in-law Périer, who performed the experiment, 22 September 1648. The authors acknowledge the help of Dr C. de Pater in finding this source and the conversion to modern SI units.

The hydrostatic Equation (2.5) is a very good approximation to the vertical build-up in the atmosphere. The vertical accelerations usually are small and can be ignored. The further approximation that $T =$ constant, which leads to Equations (2.8) and (2.9) is less general and these equations are only meant to give a general idea.

The pressure at $z = 0$ should be interpreted as the sea level pressure. This pressure p_0 is not a constant, but is dependent on location and time, so it varies and, as we shall see, gives rise to the horizontal winds. Weather maps, for example, depict the lines of constant pressure in units [hPa] (hectopascal), which is close to the old unit of 1 [mbar] (= mean pressure at sea level/1000 = 1.013 [hPa]). Meteorologists also discuss isobaric surfaces. On these surfaces the pressure is constant and one can draw lines of constant altitude within such a surface. In one example, the 500 [hPa] surface shows contour lines varying between 5200 [m] altitude and 5800 [m] altitude ([20], Figure 6.7).

2.2.2 *Vertical motion of (humid) air*

So far, we have neglected the vertical flow of air. Rising warm air, however, is a typical example of vertical transport of energy and mass. Often clouds are formed as well. Let us first describe a cloud by quoting [22] (pp 82,83)

'A cloud is an assembly of tiny droplets usually numbering several hundred per [cm^3] and having radii of about 10 [μm]. This structure is extremely stable. The droplets show little tendency to come together or to change their sizes except by general growth of the whole population. Precipitation develops when the cloud population becomes unstable and some drops grow at the expense of others. On their way down, water drops will start evaporating, which means that they need a minimum size of about 0.1 [mm] radius to reach the ground. A typical raindrop has a radius of 1 [mm] and may reach the ground with a velocity of 6.5 [m s^{-1}].'

> Liquid water contents of stratus clouds are usually in the range from 0.05 to 0.25 [g m^{-3}]. In stratocumulus and deep nimbostratus the values can reach nearly 1 [g m^{-3}]. When a cloud of water droplets is cooled to temperatures below 0 [°C] there is a chance that ice crystals will begin to appear. But because the water is highly dispersed and the atmosphere has a relatively short supply of particles that can serve as centers for ice formation (ice nuclei), ice crystals may not be observed until the cloud is cooled to −10 [°C] or colder.
>
> [22], pp 74, 75

The importance of vertical energy transport may become apparent in the following example from daily life.

Example 2.3: Latent heat and sensible heat in air

Consider 1 [m^3] at ground level in the morning with a temperature of 10 [°C] and a humidity of 50 percent. (a) How much dry air does it contain and how much water vapour?

In the course of the day the air is heated to 20 [°C] and, because of evaporation of water in the neighbourhood, the humidity increases to 80 percent. The air is still at ground level. (b) How much dry air does it contain and how much water vapour? (c) What is the increase in latent heat of the sample and what is the increase in sensible heat?

Answer

(a) The percentage of humidity of 50 percent means that the air contains half of the maximum possible amount of water vapour at that temperature and pressure. From tables (e.g. [15], p. 26, 38) one may find that it contains 1.23 [kg] of dry air and 4.7 [g] of water vapour.

(b) The volume of 1 [m^3] expands a little, which we will ignore, but the mass of dry air remains the same anyway. The air now contains 13.9 [g] of water vapour ([15], p. 38). The origin of the water vapour is evaporation of surface water by the absorption of solar energy. Thus the vapour contains what is called *latent heat*.
(c) The latent heat of water vapour is tabulated ([15], p. 346) in [J g^{-1}] as a function of temperature. It increases from 11.7 [kJ] to 34.1 [kJ]. Of course solar heating also causes a temperature increase of the dry air itself. This is found by multiplying mass [kg] times specific heat at constant pressure ($c_p = 1004$[J $kg^{-1}K^{-1}$]) times the temperature increase [K]. In this example the increase in sensible heat turns out to be 12.4 [kJ].

The remarkable fact from this simple calculation is that the latent heat of only a few grams of water vapour is comparable to or even greater than the sensible heat of a [kg] of air. So heat transport by motion of air containing water vapour can be very effective.

The air in Example 2.3 most probably will rise. Then it will cool with a rate to be determined below, and finally it will start condensing. Ignoring the fact that the pressure will drop at higher altitudes and that the volume will expand, we again look at a table and find at what temperature the 13.9 [g] of water vapour in 1 [m^3] will saturate the air. That turns out to be 16 [°C], corresponding with an altitude of 400 [m], as we will show below. So, from this height onward, cloud formation could start. The latent heat will be liberated again, when the vapour condenses. The sensible heat also can be returned to the surroundings at a place different from its origin.

The adiabats

In a few simple cases it is not difficult to calculate the temperature drop of an ascending parcel of air. For a parcel of air, which moves vertically without exchange of heat with its surroundings this is called an *adiabatic change*. When the air is completely dry, one finds the so-called *dry adiabat* with the property

$$\frac{\partial T}{\partial z} = -\frac{g}{c_p} = -\Gamma_d \tag{2.10}$$

Here g is the gravity acceleration [m s^{-2}] and c_p is the specific heat at constant pressure [J $kg^{-1}K^{-1}$]. The decrease of temperature T with height z is defined as Γ_d and is called the *dry adiabatic lapse rate*. It follows from

Equation (2.10) that the lapse rate Γ_d is a positive number, which means that the temperature goes down with height. With the data of Appendix A one finds $\Gamma_d \approx 0.01[\mathrm{K\ m^{-1}}]$ (Exercise 2.4). Consequently, for every hundred metre rise, a parcel of dry air will cool by 1 [K] = 1 [°C]. Also for humid air this result is approximately correct, as long as the water vapour does not condense. This result we used above, when stating that at 16 [°C] the air, which had a temperature of 20 [°C] at ground level would have risen by 400 [m].

If air is saturated and still rises, it will start condensing, not all at once, but in such a way that the remaining air remains saturated with water vapour at the lower temperature. The latent heat present in the vapour is slowly liberated with the consequence that the decrease of temperature occurs less quickly than for dry air.

Consequently, for saturated air the *saturated adiabatic lapse rate* Γ_s is smaller than the dry one. So, $\Gamma_s < \Gamma_d$. This is illustrated on the left in Figure 2.8. One has to be careful in reading the figure, as the axes are drawn such that z is vertical. But the derivative (2.10) has the horizontal variable in the numerator and the vertical variable z in the denominator, which is contrary to usual mathematical convention and therefore may be a little confusing. The easiest procedure is to take a certain $\mathrm{d}z > 0$, look at the figure and note that the corresponding $\mathrm{d}T < 0$.

Figure 2.8 (left) also shows that the saturated adiabatic lapse rate is dependent on the altitude and that, for higher values of z, both the rates tend to become the same. This happens when most of the water vapour has condensed and disappeared.

The adiabats are also interesting when we remember the Hadley circulation. At the tropics, wet air ascends, it will then follow the saturated curve, and cool, as shown on the right of Figure 2.8. At high altitudes the air moves

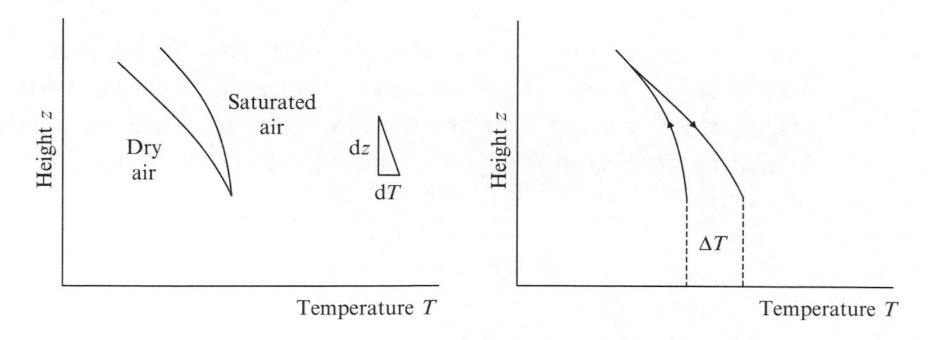

Figure 2.8 On the left is plotted the adiabatic rise of a parcel of air, either dry or saturated. The slope is called the lapse rate. For high z the dry and saturated adiabatic lapse rates become the same. On the right one finds the history of a saturated parcel going up by the saturated adiabat, while after rain-out it returns down by the dry adiabat and obtains a temperature higher by ΔT

northwords in the Northern Hemisphere or southwards in the southern and then descends. Most water vapour will have rained out during the horizontal transport and the descending air will be dry and will follow the dry curve. So, the descending air will heat up more then be cooled when going up. Consequently, the descending air is hotter by a temperature ΔT and one would expect hot regions at geographical latitudes of 20 to 30 degrees. This indeed is the region where most deserts are found! So, the air is not hot, because there is a desert, but there is a desert because the air is hot and dry.

Cumulus cloud formation

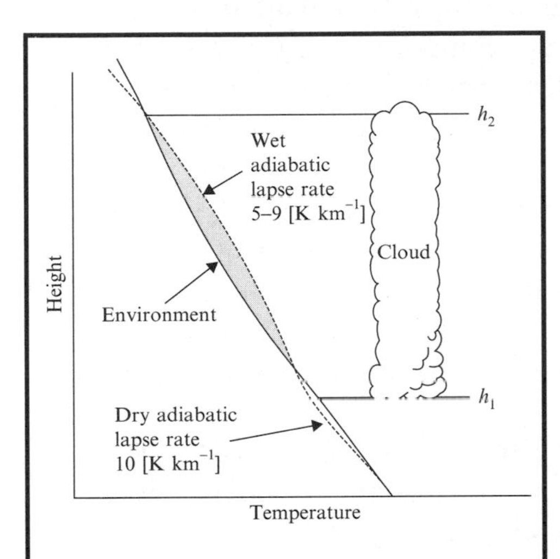

Figure 2.9 Cumulus cloud formation. Below altitude h_1 a rising parcel of air follows the dry adiabat but gravity is stronger than buoyancy (stable atmosphere). When condensation occurs, it follows the wet adiabat and buoyancy wins (unstable atmosphere). Above $z = h_2$ there is stability again

In a first approximation one may regard the vertical temperature distribution of the atmosphere as given. In Figure 2.9 this temperature distribution of the local environment is indicated by the full curve. At the bottom of the graph ($z < h_1$) one notices the broken curve, indicating the dry adiabatic lapse rate. It is close to the environmental curve, but shifted a little to the left, i.e. to lower temperatures. This means that a parcel of air with a small upward velocity will become a little cooler than its environment, its density then becomes a little higher than the environment and, according to Archimedes' law, the downward gravity force is larger than the buoyancy and the parcel goes-down again. This not only happens for dry air, but also for humid, but not yet saturated, air. In other words: the atmosphere is *stable*.

When the parcel contains water vapour and, for some reason, moves a little higher, then a point will be reached ($z = h_1$) where it starts condensing and following the wet adiabatic curve. In the graph it moves to the right of the environmental curve, so it becomes warmer than the surrounding and less dense and, in this case, buoyancy is stronger than gravity and the parcel stays on the move up: the atmosphere is *unstable*. In the meantime the small condensing drops are forming clouds, as indicated in Figure 2.9. At still higher altitude ($z = h_2$) the wet adiabatic curve crosses the environmental curve again. At that point almost all water vapour has condensed, as one may see by looking at the slope: it is again that of the dry adiabat. Therefore, at higher altitudes ($z > h_2$) the atmosphere is again stable and the vertical motion of air will stop (Exercise 2.5).

2.3 HORIZONTAL MOTION OF AIR

Although vertical motions of air are often ignored, we cannot disregard horizontal motions. We know from experience that the wind velocity is changing all the time in magnitude and direction, so there must be forces acting on parcels of air. Let us consider a small volume or parcel of air indicated by $\mathrm{d}V$. As we want to keep the letter V for volume, we will denote the velocity of air (and of other fluids) by a vector $\mathbf{u}$.

A vector may be indicated by an arrow with a length and a direction, or alternatively by its three components in an orthogonal coordinate system. The simplest vector is the position vector $\mathbf{r}(t)$ as sketched in Figure 2.10. One will note both the arrow and the three components x, y and z, which are time dependent. So the change of position in time may be written as $\mathbf{r}(t)$ with bold $\mathbf{r}$ or alternatively as $x(t)$, $y(t)$ and $z(t)$. The strength of the vector notation is that it is more concise than the notation with components in a coordinate system. For readers who are not acquainted with the vector notation or who want to refresh their memories, the relations, used in this book are summarized in Appendix B.

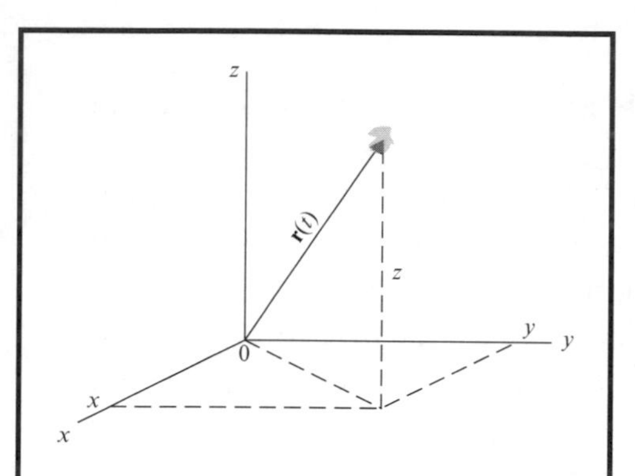

Figure 2.10 The position of a parcel of air at time t is defined by the arrow $\mathbf{r}(t)$ from a fixed point O. Alternatively, it is defined by the three components (x, y, z) of the position in an orthogonal coordinate system

The velocity vector $\mathbf{u}(t)$ is defined as the time derivative of the position vector. In mathematical terms this can be written as

$$\mathbf{u}(t) = \frac{\mathrm{d}\mathbf{r}}{\mathrm{d}t} = \mathrm{Lim}_{\Delta t \to 0} \frac{\Delta \mathbf{r}}{\Delta t} = \mathrm{Lim}_{\Delta t \to 0} \frac{\mathbf{r}(t + \Delta t) - \mathbf{r}(t)}{\Delta t} \tag{2.11}$$

This equation summarizes three similar equations for the x, y, and z components separately. For completeness we write out the x component of the first and the last terms of Equation (2.11), which gives

$$u_x(t) = \mathrm{Lim}_{\Delta t \to 0} \frac{x(t + \Delta t) - x(t)}{\Delta t} \tag{2.12}$$

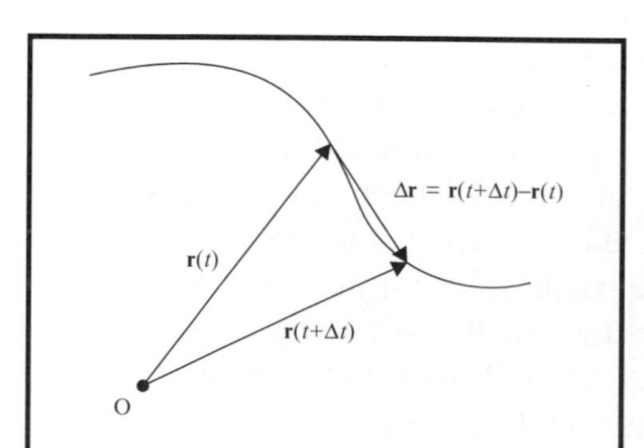

Figure 2.11 The trajectory of a parcel of air with two positions at times t and $t + \Delta t$. The velocity is defined as the difference vector in the limit $\Delta t \to 0$ and follows the tangential to the curve

Figure 2.11 shows the trajectory, which a parcel of air may follow. On this curve the position vectors at two times t and $t + \Delta t$ are indicated and the vector $\Delta \mathbf{r} = \mathbf{r}(t + \Delta t) - \mathbf{r}(t)$ is drawn. It is clear that the direction is approximately along the tangent line or the tangential of the curve and indeed for the limit that $\Delta t \to 0$ it precisely follows the tangential.

If we need to specify the x, y and z components of the velocity $\mathbf{u}$ we will use one of the two conventions (u_x, u_y, u_z) or (u_1, u_2, u_3). The convention (u, v, w) is used in meteorology but we will avoid it here.

The acceleration $\mathbf{a} = \mathrm{d}\mathbf{u}/\mathrm{d}t$ is defined analogously to Equation (2.11) as

$$\mathbf{a} = \frac{\mathrm{d}\mathbf{u}}{\mathrm{d}t} = \mathrm{Lim}_{t \to 0} \frac{\Delta \mathbf{u}}{\Delta t} = \mathrm{Lim}_{t \to 0} \frac{\mathbf{u}(t + \Delta t) - \mathbf{u}(t)}{\Delta t} \tag{2.13}$$

To find the direction of the acceleration $\mathbf{a}$, imagine that in Figure 2.11 the velocity has the direction of the tangential. As the acceleration is the change of the velocity, the acceleration vector will make an angle with the trajectory; in fact it always points to the convex side (the local 'inside' of the curve) (Exercises 2.6 and 2.7).

2.3.1 Newton's Equations of Motion

The physics problem now is to find the equations which determine the acceleration of a particle. We have already mentioned Newton's first law in connection with Figures 2.4 and 2.5. In a system fixed to the stars, we recall that a body without forces is in uniform motion along a straight line. In the numerator of Equation (2.13) $\mathbf{u}(t + \Delta t) = \mathbf{u}(t)$ and at all times the acceleration is zero.

Newton's second law expresses mass × acceleration of a particle as the sum of all the forces acting on it, assuming again a coordinate system fixed in space and tied to the stars. We apply this to a parcel of air with volume $\mathrm{d}V$. Its mass may be written as $\rho\, \mathrm{d}V$ where ρ indicates the density of the air. For convenience we write acceleration × mass and find

$$\frac{\mathrm{d}\mathbf{u}}{\mathrm{d}t} \rho\, \mathrm{d}V = \mathbf{F}_{\text{total}} = \mathbf{F}_{\text{gravity}} + \mathbf{F}_{\text{pressure}} + \mathbf{F}_{\text{Coriolis}} + \mathbf{F}_{\text{friction}} \tag{2.14}$$

We want to apply Newton's laws to motions on earth and we know that the earth is moving through space around the sun and rotates daily around its axis. For all intents and purposes the annual motion can be ignored, but the daily motion may not. Therefore, we need correction terms of which only $\mathbf{F}_{\text{Coriolis}}$ is important enough to write down explicitly. We come back to that force below, but discuss the forces in Equation (2.14) one by one, beginning on the left.

The first force on the right-hand side in Equation (2.14), the *gravity* force, acts vertically. As discussed above, its main effect is shown in the hydrostatic

Equation (2.5), where it takes care of the increasing pressure from top to bottom of the atmosphere. As vertical accelerations are usually small, we will ignore the vertical components of Equation (2.14) in most cases. The horizontal components of Newton's law (2.14) may be written as

$$\frac{\mathrm{d}\mathbf{u}}{\mathrm{d}t}\rho\,\mathrm{d}V = \mathbf{F}_{\text{total, horizontal}} = \mathbf{F}_{\text{pressure horizontal}} + \mathbf{F}_{\text{Coriolis, horizontal}} + \mathbf{F}_{\text{friction, horizontal}} \tag{2.15}$$

It is indicated that, on the right, the horizontal components are taken.

The pressure force

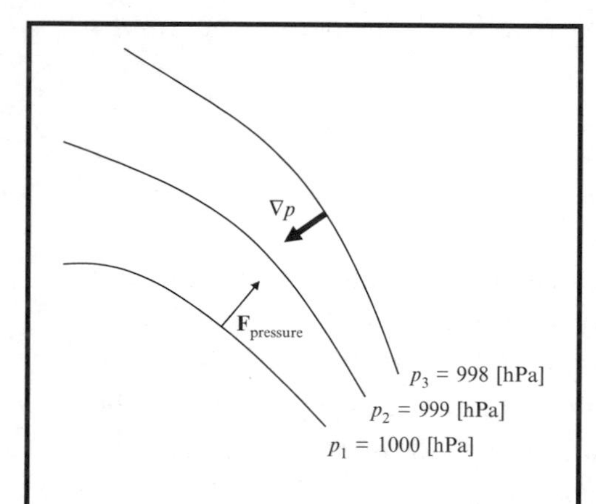

Figure 2.12 Pressure force. The pressure force will act perpendicularly to the isobars and therefore in the direction of $-\nabla p$

The horizontal components of the first force on the right of Equation (2.15), the *pressure* force, are found from the curves of constant pressure, the *isobars*, as in Figure 2.12. At a sufficiently small scale, the isobars are almost parallel since they cannot cross, as a location cannot have two pressures at the same time.

In Figure 2.12 one would guess that the air would move from high pressure to low pressure in order to reach a situation of equal pressure everywhere (which of course will never happen). One would also guess that it follows the shortest route, that is that it will move perpendicular to the isobars. In Appendix B (Figure (B.6)) we introduced a vector called gradient ∇p, which had the direction perpendicular to the isobars, pointing from a low value of p to a high value and with magnitude

$$|\nabla p| = |\text{grad}\ p| = \frac{\partial p}{\partial s} \tag{2.16}$$

Here the derivative on the right, $\partial p/\partial s$, is the increase of pressure p per metre [Pa m^{-1}] along the perpendicular line. So, we expect that the pressure force would have a direction precisely opposite from the gradient ∇p. It will also be proportional to the volume of the parcel, as a parcel which is twice as big will experience a pressure force, which is twice as big as well. We would guess

$$\mathbf{F}_{\text{pressure}} = -\nabla p\,\mathrm{d}V \tag{2.17}$$

In order to check this guess, we look whether the left and the right of Equation (2.17) have the same dimensions, as they should. The left has the dimension of Newtons [N]. The pressure itself is expressed in [N m^{-2}]. The gradient gives an extra length in the denominator, so ∇p has dimension [N m^{-3}]. The volume dV has the dimension [m^3], so the right of Equation (2.17) has dimension [N m^{-3}] $\times$ [m^3] = [N] like the left, which fits.

Of course, there could be an extra numerical factor in front of the gradient in (2.17), which does not show up in a check on dimensions. However, if one derives the pressure force from first principles ([1], pp 44–45), it turns out that Equation (2.17) is indeed correct.

The Coriolis force

The next term in Equation (2.15) is called the *Coriolis force*. It originates from the fact that Newton's laws of motion apply only to coordinate frames, which are not accelerating. The earth, however, moves around its axis with an angular velocity $\mathbf{\Omega}$. This vector points from the South Pole to the North Pole along the earth's axis. Its magnitude corresponds to one rotation of the earth in 24 hours and is given in Appendix A. For the magnitude of $\mathbf{\Omega}$ we will use either upper case Ω or lower case ω whichever is convenient.

Example 2.4: The angular velocity of the earth

Check the value for the earth's angular velocity $\mathbf{\Omega} = 7.292 \times 10^{-5}[s^{-1}]$ given in Appendix A. Take into account the annual motion around the sun.

Answer

Make a sketch of the annual motion around the sun and take two subsequent days, exaggerate the difference, as in Figure 2.13. Note that the axis of the daily and annual motion both point in the same direction. The earth is back at its first position after one year = 365.24 days. In one day we see from the graph that the earth has rotated from noon to noon a little more than 360° = 2π [rad]. In a complete year it has made one extra rotation, so per day it rotates $2\pi \times 366.24/365.24$ [rad]. Divide by $24 \times 60 \times 60$ seconds to find the angular velocity.

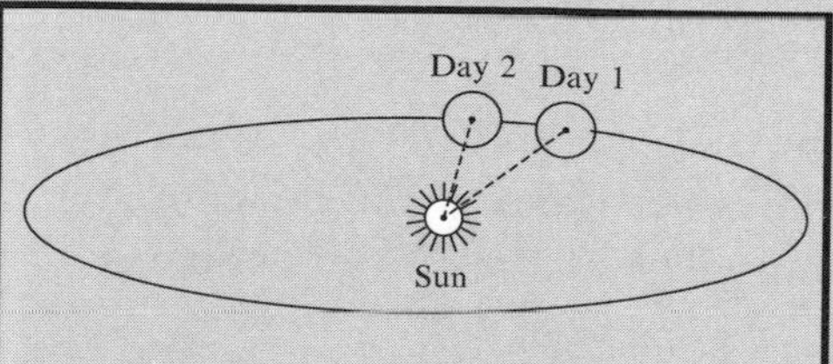

Figure 2.13 Calculation of the angular velocity Ω of the earth around its axis

In a rotating coordinate system, one may still apply Newton's laws if one adds correction terms. In environmental applications the most important correction is the Coriolis force, given by

$$\mathbf{F}_{\text{Coriolis}} = -2\mathbf{\Omega} \times \mathbf{u}\,\rho\,\mathrm{d}V \tag{2.18}$$

On the right we note the mass $\rho \mathrm{d}V$ of the air parcel. This means that the term $2\mathbf{\Omega} \times \mathbf{u}$ represents an acceleration. This is understandable since a correction for moving frames will introduce extra velocities and accelerations. A dimension check shows that $\mathbf{\Omega}$ has the dimension [s^{-1}] and the velocity $\mathbf{u}$ has the dimension [$\mathrm{m\ s}^{-1}$]. Multiplied this gives [$\mathrm{m\ s}^{-2}$], which is indeed an acceleration.

The Coriolis force will become less mysterious if one works out Exercise 2.8. For the present we just take the Coriolis force at its face value and will see later that the force is needed to describe the observations, at least qualitatively.

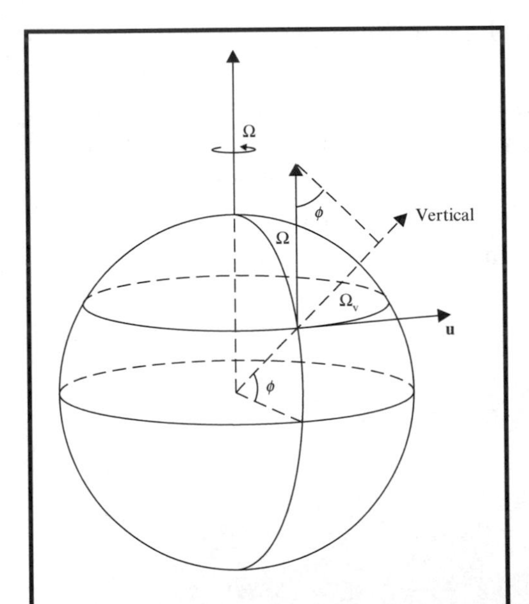

Figure 2.14 The horizontal Coriolis force at latitude ϕ is determined by the vertical component $\Omega_{\rm v}$ of $\mathbf{\Omega}$

Let us have a closer look at Equation (2.18). The first thing to note is that it is written as a vector product, a product of two vectors, which by itself again is a vector (Appendix B, Equation (B.6)). Let us look at the vector part of Equation (2.18), which gives with (B.8) $-\mathbf{\Omega} \times \mathbf{u} = \mathbf{u} \times \mathbf{\Omega}$. We separate the angular velocity vector $\mathbf{\Omega}$ of the earth in a horizontal part with component $\Omega_{\rm h}$ and a vertical part with component $\Omega_{\rm v} = \Omega \sin\phi$, where ϕ is the geographical latitude. Note that the vertical direction at any place on earth points from the centre of the earth outwards, toward the zenith. It therefore depends on location. The components of $\mathbf{\Omega}$ are indicated in Figure 2.14. We indicate the unit vector along the horizontal component by $\mathbf{e}_h$ and along the vertical by $\mathbf{e}_{\rm v}$ and may write the Coriolis force (2.18) as

$$\mathbf{F}_{\text{Coriolis}} = -2\mathbf{\Omega} \times \mathbf{u}\,\rho\,\mathrm{d}V = 2\mathbf{u} \times (\Omega_{\rm h}\mathbf{e}_{\rm h} + \Omega_{\rm v}\mathbf{e}_{\rm v})\,\rho\,\mathrm{d}V \tag{2.19}$$

This is the sum of two vector products. In the first one both vectors $\mathbf{e}_{\rm h}$ and $\mathbf{u}$ are horizontal. Their vector product therefore is vertical and can be ignored with respect to gravity. The horizontal component of the Coriolis force is meant in Equation (2.15). It becomes

$$\mathbf{F}_{\text{Coriolis, horizontal}} = 2\mathbf{u} \times \Omega_{\rm v}\mathbf{e}_{\rm v}\,\rho\,\mathrm{d}V = 2\Omega\,\sin\phi\,u(\mathbf{e}_{\rm u} \times \mathbf{e}_{\rm v})\,\rho\,\mathrm{d}V \tag{2.20}$$

where $\mathbf{e_u}$ is a unit vector in the direction of the horizontal velocity $\mathbf{u}$. The vector product $\mathbf{e_u} \times \mathbf{e_v}$ is horizontal, perpendicular on $\mathbf{u}$ and has the length one, as the angle between the two unit vectors equals 90°. So, the magnitude of the horizontal Coriolis force becomes

$$2\Omega \sin \phi u \rho \, \mathrm{d}V = fu\rho \, \mathrm{d}V \tag{2.21}$$

which defines the *Coriolis parameter* $f = 2\Omega \sin \phi$. We see that at the equator, where $\phi = 0$ the horizontal component of the Coriolis force vanishes. The vertical component, as noted before, is neglected with respect to the large gravity acceleration.

Friction

> Wind is a flowing wave of air, moving hither and thither indefinitely. It is produced when heat meets moisture, the rush of heat generating a mighty current of air. This we may learn from hollow bronze balls, with a very small opening through which water is poured into them. Set before a fire, not a breath issues from them before they get warm. But as soon as they begin to boil, out comes a strong blast due to the fire.
>
> Vitruvius (25 BC), On Architecture, I, VI, 2

The third force on the right-hand side of Equation (2.15) is the friction force. The major source of friction in the atmosphere is the presence of the ground surface with its hills, mountains, trees, buildings etc. In fact, very close to the ground (in the order of [mm]) the air velocity goes to zero. As we know that at altitudes of metres and higher the wind velocities may be considerable, there must be internal friction between the air layers as well.

The friction between air and ground is difficult to describe, so we will not even try to do so, but just mention that it may be taken into account empirically by a so-called *roughness parameter* z_0. In a first approximation, which holds in the lowest few hundred meters of the atmosphere, the increase of the horizontal velocity u as a function of altitude z is written as

$$u(z) = \frac{u*}{k} \ln \frac{z}{z_0} \tag{2.22}$$

Here, u^* is the so-called *friction velocity* [m s^{-1}] and $k \approx 0.4$ is called the Von Karman constant. The friction velocity u^* takes account of the tangential stress of the air along the surface. The roughness z_0 is taken as some 10 percent of the height of the grasses or plants that determine the terrain. If one accepts a logarithmic relationship between $u(z)$ and z, a relation like (2.22) follows directly, since a logarithm must be taken of a dimensionless quantity, hence z/z_0. If one wants a velocity u on the left, one also needs one on the right, but apparently one that is independent of height and takes into account the friction at the boundary, like u^*. As u^* is defined in a very specific way, one needs an additional constant k in Equation (2.22). Note that Equation (2.22) may not be used for $z < z_0$ where the logarithm becomes negative.

In the logarithmic relation (2.22) the ground situation dominates the velocities way up. This can only be right in an average way, as the internal friction between the layers of air is not taken into account. Indeed, for accurate calculations and for higher layers it is the friction between the layers that connects them with the ground level down.

Geostrophic flow

If we go to high enough altitude, above 5 [km] or so, the influence of the ground disappears and the wind velocity changes very little with altitude. This means that the left-hand side of Equation (2.15) vanishes and also the last term on the right. What remains is the sum of the horizontal pressure force and Coriolis force $\mathbf{F}_{\text{pressure, horizontal}} + \mathbf{F}_{\text{Coriolis, horizontal}} = 0$. Thus it follows from equations. (2.15), (2.17) and (2.18)

$$0 = (-\nabla p - 2\rho\,\mathbf{\Omega} \times \mathbf{u})_{\text{horizontal}} \tag{2.23}$$

where the mass $\rho\,\mathrm{d}V$ of the parcel of air was omitted. We omit the subscript horizontal at ∇p, as it acts almost horizontally anyway and find from Equation (2.20) and (2.21)

$$0 = -\nabla p - 2\rho\,\Omega \sin\phi\, \mathbf{e}_\mathrm{v} \times \mathbf{u} = -\nabla p - \rho f\, \mathbf{e}_\mathrm{v} \times \mathbf{u} \tag{2.24}$$

Equation (2.24) says that the sum of two vectors must be zero. Since ∇p points perpendicular to the isobars from low to high (Figure B.4), $\mathbf{e}_\mathrm{v} \times \mathbf{u}$ should point perpendicular to the isobars as well. With $\mathbf{e}_\mathrm{v} \times \mathbf{u}$ perpendicular to both $\mathbf{e}_\mathrm{v}$ and $\mathbf{u}$ (Equation (B.5)) it follows that $\mathbf{u}$ blows parallel to the isobars. This is illustrated in Figure 2.15. There are, of course, two possibilities parallel to the isobars. The easiest procedure is to take a direction and check, whether Equation (2.24) is satisfied. If not, it should be the other direction.

The resulting velocity is called the geostrophic wind or more general, the *geostrophic flow* with magnitude u_G. The magnitude of the last term in Equation (2.24) is $\rho f u_G$, as $\mathbf{e}_\mathrm{v}$ and $\mathbf{u}_G$ are perpendicular and the sine of the angle of 90° equals one. So, (2.24) gives, for the magnitude u_G:

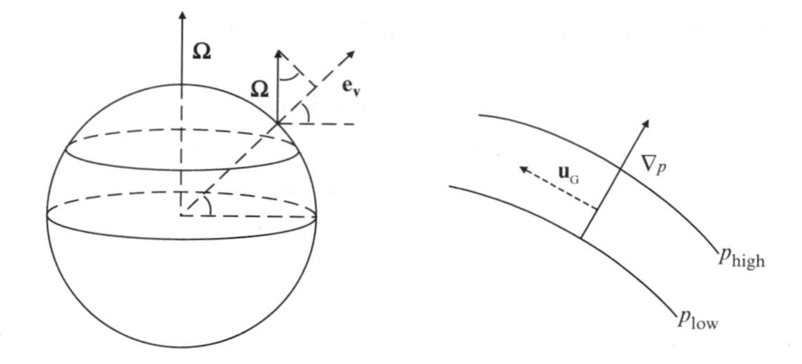

Figure 2.15 Coriolis force and the geostrophic flow. On the left, one observes that the vertical component of the earth's angular velocity determines the horizontal Coriolis component. On the right, the vertical unit vector $\mathbf{e}_v$ points upwards out of the paper. One should check that the dashed arrow indeed obeys the direction, required by (2.24)

$$u_G = \frac{|\nabla p|}{f\rho} \tag{2.25}$$

where we put the magnitude of ∇p in the numerator. The weather maps at 5 [km] to 10 [km] of altitude indeed show that the wind blows parallel to the isobars with deviations that are usually smaller than 10°. Of course, one should not apply (2.24) or (2.25) near the equator where the Coriolis parameter f becomes zero. The observed wind speeds at high altitudes correspond to Equation (2.25) with deviations that seldomly exceed 10 percent ([21], p. 198).

The Dutch pioneer meteorologist of the 19th century Buys Ballot stated as an empirical law: 'low pressure in the Northern Hemisphere is on the left hand when facing down wind'. The law was recognized from observations of surface winds blowing around extra-tropical and tropical cyclones, before it was formally linked to the Earth's rotation through the equation of motion.

Near sea or ground level, the friction terms in (2.15) play an important role. Consequently the wind speeds are lower than high up, and the Coriolis force is correspondingly smaller. The pressure force becomes more dominant in determining the wind direction. Without Coriolis forces the wind would blow from high pressure to low pressure, which is perpendicular to the isobars. It would not blow parallel as the geostrophic winds do. The Coriolis force does not vanish, however, so in practice one has an intermediary situation, which one may express by saying that the wind blows from high to low pressure with a deviation to the right.

Figure 2.15 applies to the Northern Hemisphere where the angular velocity of the earth points upward. On the Southern Hemisphere it points into the earth, and the geostrophic flow would be in the opposite direction. On the Southern Hemisphere near ground level, the wind will blow from high pressure to low pressure with a deviation to the left.

2.3.2 *Origin of pressure differences*

Up to this point we have assumed pressure differences. We know of course, that they originate from solar heating. We already discussed the Hadley circulation in Figure 2.4, where air was heated at the equator and, by a detour via higher altitudes, returned to the equator. Thus, because of the rise of air, the equatorial regions experience low pressure and the returning air from higher pressures at higher latitudes fills the gap again. This was shown in Figure 2.3.

The pressure differences at higher latitudes than the tropics, in the North and South, must originate from solar heating as well. The complicated pressure distributions, which are changing all the time, reflect the uneven distribution of landmasses over the planet. We also know that the temperature of the seawater lags behind that of the surface air over land. The sea is warmest at the end of the summer when the temperatures on land are already dropping, and coolest at the end of the winter. Also, the day and night cycle gives different temperatures on sea and land. Finally, the gaps from rising warm air cannot simply be filled in from high pressure to low pressure, as the Coriolis force twists the air into another direction. All together this results in the heat transport to higher latitude, deduced from Figure 2.1, which makes temperature differences on earth less extreme than otherwise would have been the case.

The general conclusion so far is that there are many variables determining weather and climate. It is therefore not surprising that weather forecasts are phrased in global terms, such as 'there will be occasional showers'. It is not feasible yet to forecast a shower in Berlin's Zoo between 4 o'clock and 4.15 o'clock for the next day, after which the sun will be shining again. In view of the complexity of the phenomena it is doubtful whether such a precise forecast could ever be made.

For climate, the same will hold. Calculating a climate from scratch, given physical parameters, will be complicated. It will be easier to calculate climate *change*, the perturbing effect of certain natural or human induced changes. These topics are discussed in Chapter 3. First, however, we look at the oceans, which occupy most of the world and therefore must have an important influence on climate.

2.4 *THE OCEANS*

All realistic models take the effect of the oceans into account. The reasons are clear. The oceans cover some 70 percent of the surface area of the world (which suggests that it is more appropriate to talk about Planet Ocean instead of Planet Earth!). They reflect little sunlight and therefore look 'dark' to the observer! So, the oceans absorb much of the solar energy entering. Their waters evaporate easily and form the clouds, which transport latent heat and sensible heat to other places. They contain all kinds of salts; indeed the salinity is so high that sea water is not suitable for consumption. Also, the oceans may absorb many gases, among which are large quantities of CO_2, as is illustrated in Table 1.1. Finally, as mentioned in our discussion of Figure 2.1, the oceans transport about half of the surplus energy at the tropics to the higher latitudes (Example 2.1).

The present section will discuss the structure of the oceans and the big ocean streams, which transport their waters. It finishes by describing the El Niño phenomenon, to appreciate the difficulties climate modellers are facing.

2.4.1 *The big oceans*

In a simplified picture one may distinguish four layers in the oceans: a mixed layer at the top, upper waters underneath, deep waters in the middle and a bottom layer deep down in the oceans. This structure is sketched in Figure 2.16, together with an indication of the depths involved.

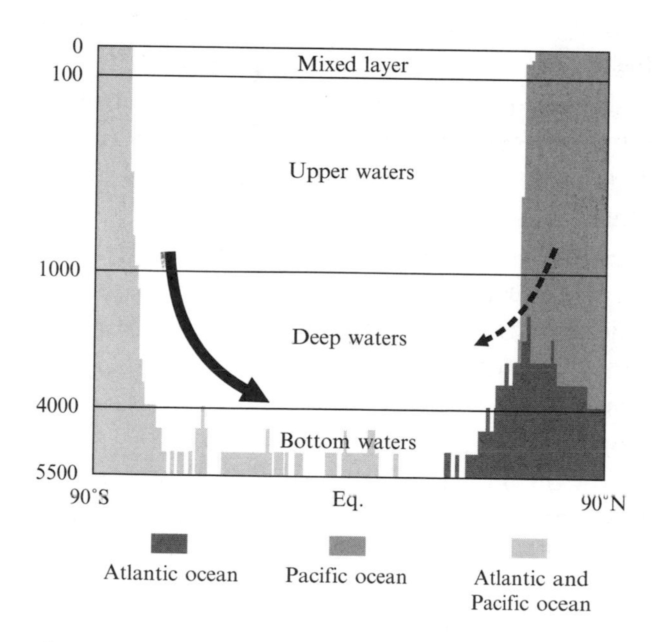

Figure 2.16 Layered structure of the oceans. Data for the depths only give an indication of the geography involved and change scale at 1000 [m]. The grey blocks at the bottom represent the structure of the ocean floor. In the South (left) they are similar for the Pacific and the Atlantic; in the North (right) the Atlantic floor is coloured dark grey and the Pacific land mass is medium grey. In the South, bottom water is formed in both oceans (full arrow), in the North in the Atlantic only (dashed arrow)

The structure of the ocean floor is depicted in Figure 2.16 as well. In the North (on the right) the Atlantic has the Arctic Sea, where water of high density is formed. The surface water is reasonably warm because of the Gulf Stream, to be discussed below. This water evaporates and cools, as it has to provide the heat of evaporation. It also becomes saltier, as the evaporated water leaves its salt behind. For both reasons the water becomes denser than the water underneath and it sinks to the bottom of the oceans, as indicated. In this way the density variations drive the *thermohaline* (= heat + salinity) circulation. In the Pacific one has landmass in the North and other climatic conditions, so no bottom water is formed.

In the Antarctic regions (on the left in Figure 2.16) both oceans are similar and all around the continent, water sinks to the bottom of the oceans. The amount of deep bottom water formed is about 10^{10} [kg s^{-1}] in the Antarctic Ocean and about double that amount in the North Atlantic ([20], p. 201).

> As a consequence of global warming, increased stratification of the oceans would occur, reducing the nitrogen flux to the upper ocean. At the same time enhanced precipitation could lead to a reduction in the flux of Fe, blown in by the winds. As Fe is essential to the fixation of CO_2 by plankton, less CO_2 is bound. This would decrease the effectiveness of the biological pump, contribute to a net efflux of CO_2 and lead to a positive feedback (= more warming).
>
> *Science* **281**, 10 July 1998, 200–5

The top layer of the oceans is in direct contact with the atmosphere, it 'feels' the winds and absorbs the sunlight. These processes are indicated in Figure 2.17. The penetration of solar radiation is shown on the left and the temperature profile resulting from irradiation, turbulence and heat exchange is found on the right and at the bottom.

In Figure 2.17 one will note that the temperature first decreases only slowly, because of the turbulent mixing in the top layer. This temperature profile defines the mixed layer (hence the name). It is thin in warm seasons (with top water of low density) or when there is not much stirring by the winds. In cold, stormy weather it may increase to a few hundred metres. On average it is about 70 [m] thick.

Below the mixed layer the temperature decreases to about 4 [°C] or 5 [°C] at a depth of 1000 [m]. The temperature decrease is called the *thermocline*, the same word is often used for the beginning of the temperature decrease below the mixed layer. A depth of 1000 [m] is usually taken as the point where the deep waters begin, as indicated in Figure 2.16.

The downward thermohaline motion, indicated in Figure 2.16 induces horizontal motion as well. Since the mass of water that the descending waters are replacing has to go somewhere. The horizontal motion is determined by other factors also. In the first place the wind stress at the ocean surface will

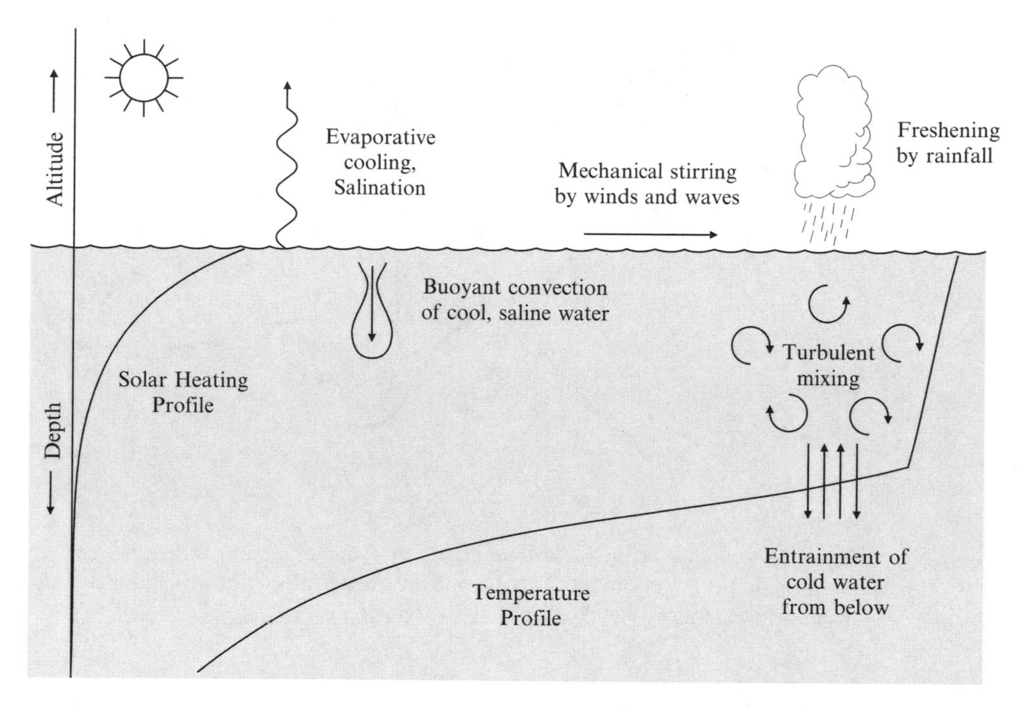

Figure 2.17 Processes in the mixed top layer of the oceans. In this layer the atmosphere causes rapid and turbulent mixing and almost constant temperature. Somewhat lower, the temperature goes down (curve right and bottom). The thickness of the mixed top layer varies between a few metres and a few hundred metres and averages about 70 [m]. Reproduced with permission of Academic Press from [20], p. 179

> The moon is receding from the Earth at about 4 [cm] a year, inserting 4×10^{12} [W] of energy in the form of dissipation of tidal energy. This happens partly in the deep sea and the resulting mixing is controlling the overall ocean circulation.
>
> In fact, half of the power required to return the deep waters to the surface is coming from mixing, driven by dissipation of tidal energy.
>
> *Nature* **405**, 15 June 2000, 743–44

move the waters, and when there is a dominant wind direction, such as the easterly trade winds in the tropics, the wind stress always acts in roughly the same direction. Other factors influencing the horizontal motion are the pressure differences in the atmosphere, which induce similar pressure differences in the oceans and the Coriolis force, defined in Equation (2.18). A factor which is usually ignored, and we only mention in passing, is the influence of the tides in pushing the circulation.

We restrict ourselves to three dominant forces: wind stress, pressure differences and the Coriolis force. Besides, there are density differences and boundary conditions as the oceans can only flow where there is no land, so geographic details pose a strong constraint to the big ocean currents. The final result, of course, is three-dimensional; it depends on location and depth and is difficult to represent. Figure 2.18 gives an impression, nevertheless.

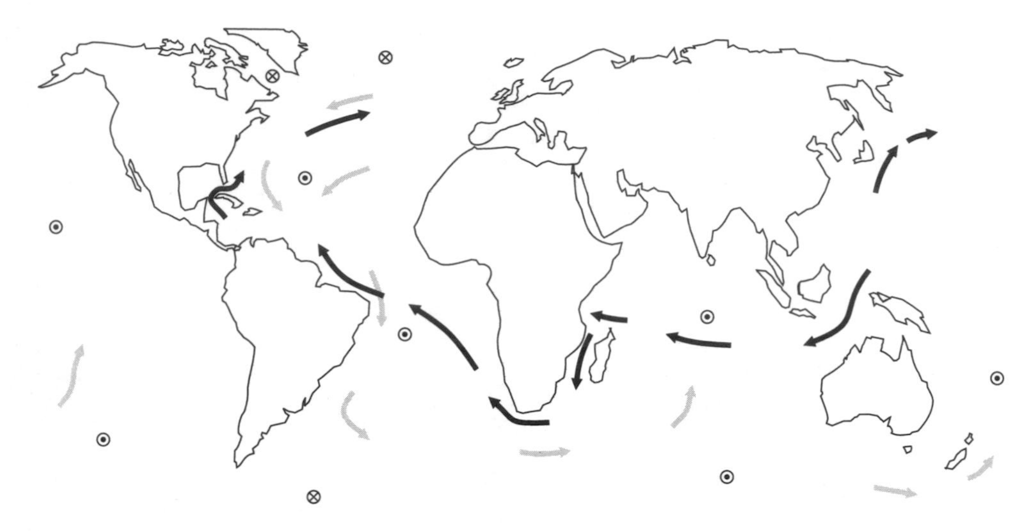

Figure 2.18 Global ocean circulation. Dark arrows indicate warm surface currents and grey arrows represent deeper return currents. A circle with a cross indicates down-welling and a circle with a dot up-welling of deep waters. Based on [20] with the help of [23]

In Figure 2.18 dark arrows indicate warm surface currents and grey arrows the circulation in the deeper waters. We notice the Gulf Stream in the North Atlantic with its origin in the Gulf of Mexico, hence its name. Near the East Coast of North America it has a width of 60 [km] only, is 500 [m] deep and has a velocity of 1 [m s^{-1}]. Its importance becomes apparent when we note that the Gulf Stream alone transports 3×10^{10} [kg s^{-1}] of water as compared with a down-welling of 10^{10} [kg s^{-1}] in the polar regions. Near Japan one may notice the so-called Kuroshio for which a similar mass transport is assumed (Exercise 2.10).

> In a model calculation the atmospheric CO_2 concentration was assumed to grow steadily. The North Atlantic thermohaline circulation weakens in all global warming simulations and collapses at high levels of carbon dioxide
>
> *Science* **284**, 16 April 1999, 464–67

In Figure 2.18, circles with a cross indicate locations where deep waters are formed (in the Arctic and Antarctic) and circles with a dot denote locations where deeper waters come to the surface again. It is difficult to measure the deep currents directly. One may measure the concentration of radioactive ^{14}C from atomic bomb explosions in the 50's and 60's as an indicator of the movement of the waters during the last 50 years (Chapter 10). These explosions gave pulses of ^{14}C during a short period of time at accurately known intervals; by studying this 'fingerprint' in deep waters one may deduce the time since the water left the surface. For waters which have left the surface longer ago, one measures the oxygen concentration and uses a modelled rate for oxygen depletion by plants and animals, that need oxygen for their existence. The turnover time of the ocean is in the order of 1000 years ([20], p. 197).

As mentioned above, the big currents are a result of many different prevailing forces. Together the currents act as a conveyor belt transporting half of the excess energy at the tropics to middle and high latitudes. One result is that the climate in Europe is much milder than at comparable latitudes in the USA and Canada. One prediction is that a significant global warming may cause the belt to break down, so that Northern Europe will become colder instead of warmer. The unanswered question again is how stable the conveyor belt is against climate change.

2.4.2 *El Niño and NAO*

Weather and climate are the result of interaction between atmosphere and oceans. This becomes clear from two well known phenomena: ENSO, better known as *El Niño Southern Oscillation* and NAO, the *North Atlantic Oscillation*.

El Niño is Spanish for 'the Christ Child' and is the name given by Peruvian fishermen to the occasional warming of coastal waters around Christmas time. This happens with an irregular period of 2 to 7 years. The warm waters push down the usual up-welling of cold water that brings nutritious water, rich of fish, to the surface. We first describe the 'normal' situation in the Pacific and then the deviation from it in El Niño [24].

The 'normal' situation at the equator is sketched in Figure 2.19. The wind is blowing from South America towards the west, introducing a westward surface ocean current. This pushes up the water about 40 [cm] at the land barriers of Indonesia and Australia. At those places the mixed layer thickens and pushes down the thermocline where the water gets colder.

> The rise and fall of Canada's lynx population is tied to the climate cycle of El Niño's Atlantic cousin, researchers have discovered. It is believed that changes in temperature and snow cover affect how easily lynx catch their prey.
>
> *New Scientist*, 21 August 1999, 19

In the east, near South America, water must flow in to supply water for the westward surface current. This happens by a subsurface flow just underneath the mixed layer, as is indicated in Figure 2.19. This undercurrent is like a cold tongue of perhaps 100 [m] thick and 300 [km] wide, which comes to the surface near South America. The result is cold water in the east and consequently a relatively cold atmosphere, while in the west both are warmer. The air pressure in the east (South America) then becomes higher than in the west, which is consistent with our initial assumption of a westward airflow.

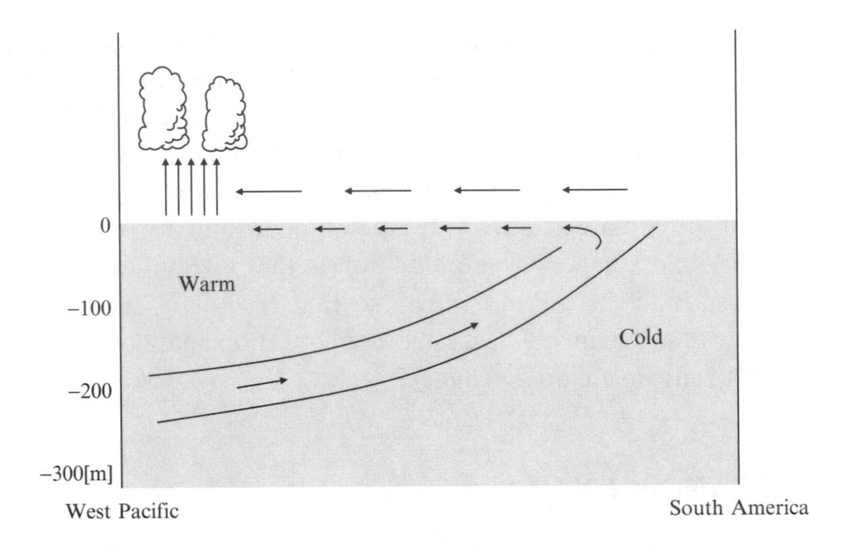

Figure 2.19 The Pacific Ocean at the Equator in its normal mode. Westward winds induce westward surface currents, which push down the thermocline. An eastward undercurrent (a cold tongue) supplies the necessary water. Evaporation in the west (Indonesia and Australia) produces clouds and precipitation

Let us now look at the undercurrent. Although it meanders a little to the North and the South, it must remain close to the equator. For when it would go too much to the North the Coriolis acceleration (2.20), which is zero at the equator, gives the current a correction to the right. Similarly, when it roams too much to the South, it will experience a Coriolis acceleration to the North.

Finally, we notice in Figure 2.19 that the warm ocean waters in the west easily evaporate, causing high clouds and precipitation in that region.

The 'normal mode' of the Pacific described above, results from a complicated interplay between forces. So, perturbations may easily happen and cause the El-Niño. Suppose that for any reason the cold tongue does not surface at South America. Then the thermocline is lower and the mixed layer relatively warm. Consequently, the pressure difference in the atmosphere between east and west decreases, causing weaker westward winds, weaker westward surface currents, less build-up of warm waters in the west, a decreasing undercurrent and a still lower thermocline in the east. This is a positive feedback: the perturbation has effects, which strengthen its importance and consequently maintain the El-Niño phenomenon.

The ocean, however, contains an enormous amount of water with a high heat capacity. That must be the explanation why further interactions let the ocean circulation return to its normal mode.

The North Atlantic Oscillation, NAO is another example of atmosphere-ocean interaction. Here, especially in winter, the airflow is determined by an 'Icelandic low' pressure and an 'Azores high' pressure. The pressure difference points North, but because of the Coriolis force (2.20) the resulting winds are Southwesterly. The word 'oscillation' in North Atlantic Oscillation refers to the fact that the magnitude of the pressure differences change irregularly from winter to winter. Stronger winds transport more heat and will result in milder winters, up to Siberia. So prediction of the next stage of the North Atlantic Oscillation has importance for agriculture and living conditions in Europe.

It is well known that sea surface temperatures and the atmospheric circulation are related. The mechanism and the predictability of the next stage of the North Atlantic Oscillation are still under discussion [25].

EXERCISES

2.1 In Figure 2.1 it looks as if the net radiation, integrated over latitude ϕ becomes negative. This cannot be true. Explain.

2.2 Calculate $H = R_{\text{air}}T/g$ and show that it has the dimensions of length.

2.3 Derive Equation (2.9) and calculate H_{10}.

2.4 Check the numerical value and the dimension of the dry adiabatic lapse rate from Equation (2.10)

2.5 Argue from Figure 2.9 that a parcel of air going down at $z > h_2$ would experience an upward force. So that the atmosphere is really stable there: a change from an equilibrium situation is counteracted.

2.6 Show that for a curved trajectory the acceleration $\mathbf{a} = d\mathbf{u}/dt$ of a particle points inwards.

2.7 Consider a uniform motion in a circle with $u = \Omega R$, where R is the radius and Ω the angular velocity. Prove that the acceleration obeys $a = u^2/R$, inwards.

2.8 For students interested in mathematical physics: Consider a mass at rest in a non-accelerating frame. Describe the mass from a frame rotating at a constant angular velocity Ω around a constant axis through an origin O. In this frame the mass performs a uniform circular motion with velocity $u = \Omega R$. It experiences two 'correction accelerations', the centrifugal, which may be written as u^2/R outward, and the Coriolis acceleration. Show that together they result in the centripetal acceleration, which drives the circular motion observed in the rotating frame.

2.9 Check the dimensions in Equation (2.20).

2.10 Calculate the energy transport of the Gulf Stream, assuming that the return currents are on average $\Delta T = 12$ [°C] cooler. Assume that the Kuroshio near

Japan transports a similar amount of Watts. Notice whether this compares well with the result of Example 2.1 on Northward energy transport.

2.11 Consult an atlas, which includes climate data. Compare the mean temperature in your town with locations at the same latitude elsewhere. Conclude that there are differences of a few degrees, often more in the winter than in the summer.

3 Climate Change

In this chapter we discuss the natural changes in climate, that have occurred in the past (Section 3.1), the changes that are induced by man (Section 3.2), their consequences (Section 3.3) and the possible reactions to change (Section 3.4). As the climate does not indicate which changes are 'natural' and which are human induced, one of the problems will be to disentangle both.

3.1 *NATURAL CHANGES OF CLIMATE*

Changes of climate are an intrinsic part of nature. In Figure 3.1 we show the surface air temperature, deduced from several sources, among which are ice core drillings in Greenland, with methods to be discussed in Chapter 10. Notice that the scale becomes a factor of ten coarser when one moves up one frame. In the graphs a horizontal line indicates the average temperature. One observes in the lowest of the three pictures that the average turns out to be 15 [°C], in the one next upward it is 14 [°C] and in the top one, 13 [°C]. So, one may conclude that at present we are experiencing a rather warm overall climate. However, we note at the same time that the period indicated by W2 in the top graph was warm as well, and also earlier, some 300 000 years ago it was rather warm. Obviously, this cannot be caused by man, who was at the time only present in small numbers with a low level of consumption. So, in order to understand the human-induced changes we first have to look at the natural fluctuations.

Figure 3.1 is based on measurements made in the 1970's and 1980's. Data were compiled from several sources and give an indication of the surface air temperature on earth. These experiments are still being improved. So, one should not deduce the temperature 235 000 years ago from the graph, but use it to get an idea about the fluctuations of temperature with time.

One then notices a rather irregular behaviour in the lower three graphs. But a more regular behaviour is observed if one looks at the top graph of the last million years, where all minor fluctuations had to be omitted. In that graph the lower temperatures correspond with glacial periods, in which ice was covering large parts of the earth. The Beograd scientist Milankovitch

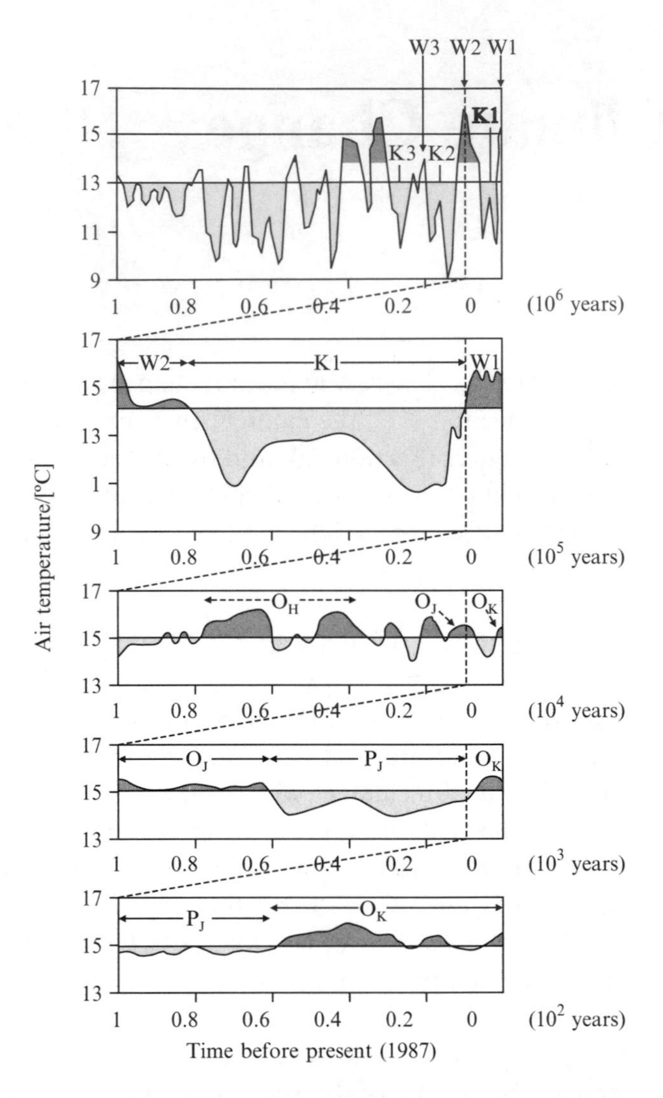

Figure 3.1 Surface air temperatures during the last million years. Note that the horizontal scale jumps a factor of ten between figures. Also note that the average temperature goes down from bottom to top. Reproduced by permission of the authors from [26], Figure 9, p. 44

recognized this regular behaviour at large timescales in the period 1915–1940 [27]. He connected the glacial periods with changing insolation, which in turn was caused by a changing orientation of the earth.

Below, we will discuss the, now widely accepted, Milankovitch theory, then move on to changes in insolation which originate in the sun itself and end this section with some other natural influences on climate. In science new experiments force one to rethink accepted theories. Some difficulties with Milankovitch's theory are pointed out in [28].

3.1.1 *The orientation of the earth (Milankovitch effect)*

The earth moves around the sun in an elliptic orbit, of which the sun itself occupies one of the two focal points. If the long axis of the ellipse is called a, then the short axis becomes $a\sqrt{(1-e^2)}$, where e is called the *eccentricity* of the ellipse. For $e = 0$ we have a circular orbit as a special case. The eccentricity e of the orbit is not large; at present it amounts to $e = 0.0167$. This means that the orbit is almost a circle and in many calculations it is indeed treated as one. The orbit, however, is an ellipse and its eccentricity e changes with time between the values of 0.002 and 0.055 with a period of 100 000 years ([20], p. 308).

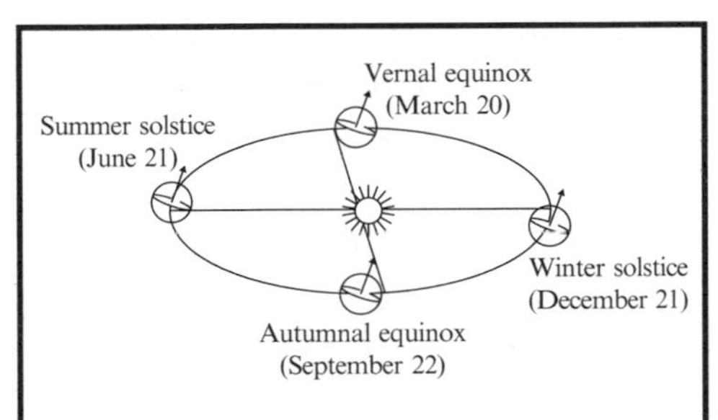

Figure 3.2 The orbit of the earth around the sun. The sun is in one of the focal points of the ellipse, a little to the right of its centre. The dates in the figure correspond with the year 2001, in which year the earth is closest to the sun (the perihelion) on January 4th

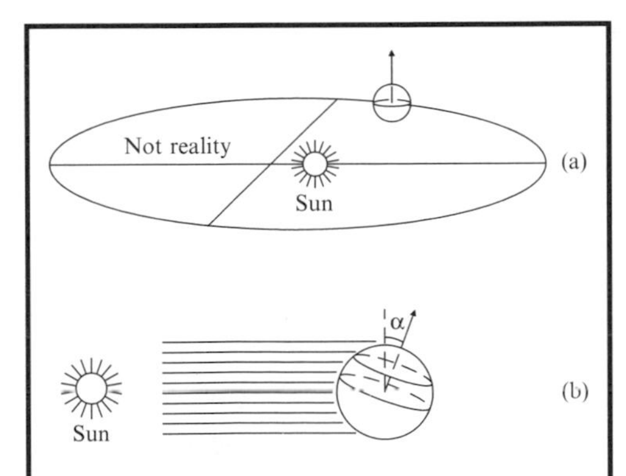

Figure 3.3 (a) When there is no tilt between the axis of the earth and the orbital motion there is no difference between the length of day and night or between summer and winter. (b) With a tilt there is, in (b) It is summer in the South and winter up North

The present orbit around the sun is depicted in Figure 3.2, together with some dates of the position of the earth in the year 2001. The position at the long axis, closest to the sun is called the *perihelion*. According to Kepler's laws the earth passes its perihelion at a relatively high speed, so although the solar irradiation around that location is large, it only lasts for a relatively short time.

Notice in Figure 3.2, that the axis of rotation of the earth points upward and to the right in the figure. The arrow on the earth's axis points from South to North. It corresponds with the rotation of the earth according to the rule of thumb of the vector product (B.6). When putting the four fingers of the right hand in a rotation from West to East, the thumb will point upwards, to the North. This is the direction of the *internal angular momentum*, or *spin*, of the earth. For the motion of the earth around the sun, the thumb will also point upwards when the fingers follow the dates March, June September. This is called the *orbital angular momentum* of the earth. As shown in Figure 3.3 there is an angle between the directions of the two angular moments, called the *tilt*.

The tilt forms the basic explanation for the seasons on earth. In Figure 3.3 (a) we show the situation without a tilt. The Northern Hemisphere (N) and the Southern Hemisphere (S) receive the same amount of sunshine, and at all locations on earth, day and night are equally long (12 hours).

In Figure 3.3 (b) we show a tilt with an angle α between the earth's axis and the orbital angular momentum. Even by eye we notice that in the figure the Southern Hemisphere receives most of the solar radiation. Moreover, the day in the South will be longer. These combined effects result in summer in the South and winter in the North. When the tilt remains constant the situation half a year later will be the other way round, which is precisely what happens. The tilt does change however, albeit very slowly. With a period of about 41 000 years it varies between 22.5° and 24.5° with, at present, a value of 23.5° ([20], p. 308)

The effect of a change of tilt is best understood when we imagine in Figure 3.3 (b) an extreme tilt of 60° or so. We notice that the summer in the Southern Hemisphere in that case will be very hot, while at the same moment the winter in the North will be very cold for lack of sunshine. Half a year later the effect is reversed, so an increase in tilt makes the seasons more extreme.

> Milankovitch frequencies really are deep in the climate data set. Although cycles of 23,000 and 41,000 years in the climate record do match up precisely with Earth's wobbling and nodding, the ice ages don't keep to the 100,000 year schedules that should be set by the periodic elongation of Earth's orbit.
>
> *Science* **285**, 23 July 1999, 503–4

Figure 3.2 indicates the beginning of the seasons, where the precise dates refer to the year 2001. It is accidental that at present the winter starts at a position close to the perihelion. The winter solstice is defined as the point in time where the sun is lowest in the skies and the days are shortest. That is the beginning of winter. The summer solstice is defined as just the opposite. The sun is at its highest and the day lasts longest. The beginning of the spring and the beginning of the autumn are in between when day and night last equally long. The latter situations are easiest to explain.

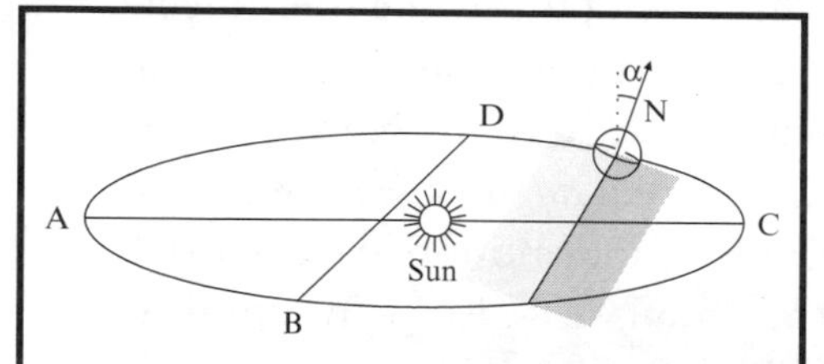

Figure 3.4 Precession and the change of seasons. The earth moves in an ellipse with long axis AC and short axis BD. The sun is a little beside the centre at a focal point. The (grey) equator plane of the earth and its orbital plane around the sun cut in the drawn intersection line. For a position on the orbit where the sun lies on that line (Figure 3.5), day and night are equally long: the autumn point or vernal point (spring). The precession of the earth's axis around the 'vertical' causes a change in the position on the orbit where the seasons occur

Assume for a moment that the earth's axis has a constant direction in space. Look then at Figure 3.4, where the earth is still close to the perihelion (point C). In the picture we have drawn the intersection between the equator plane of the earth and the orbital plane. This line is to the right of the sun. When we would follow the earth to the left in its orbit, the intersection follows until at a certain date the sun lies on the intersection. That situation is shown in Figure 3.5. All points on each meridian* will see sunrise at the same moment in the morning and sunset at the same moment at night. Figure 3.5 shows that precisely half of the earth's surface is lit by the sun. So the length of the day is the same

*A meridian connects all points on earth with the same longitude.

everywhere and half of 24 hours, which is 12 hours. This defines the beginning of spring or autumn. The position at the beginning of spring is called the *vernal point*.

The dates in Figure 3.2 approximately represent the present situation. The axis of the earth, however, performs a *precession* around the normal on the orbital plane with a period of about 23 000 years. Assume for the sake of argument that the tilt would stay constant in the figure and the earth would stay at the same position in the orbit. Then the only (slow) change is precession and the corresponding change of the earth equator plane in space. Consequently, the intersection line of the (grey) equator plane and the orbital plane will rotate. If we now allow the earth to move along its orbit, we understand that the location of the earth, where the intersection line crosses the sun, slowly changes from year to year. So, the quarter piece of the earth's orbit where a certain season occurs, moves slowly around the orbit. Still, the beginning of spring is always around March 22nd. The reason is that the definition of the year is the time the earth needs to go from vernal point to vernal point (see Exercise 3.1). Astronomers, since times long past, look to the skies from the earth. So, they talk about the orbit of the sun along the ecliptic around the sky. The vernal point then is the position of the sun at the moment when day and night are equally long. Because of precession, this vernal point moves along the ecliptic, crossing the 12 signs of the zodiac in 23 000 years.

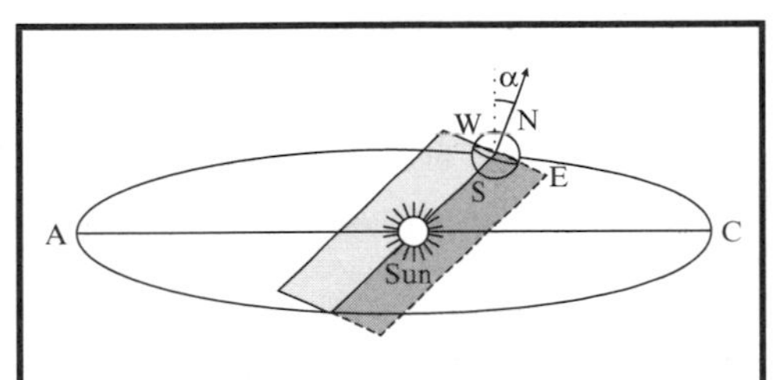

Figure 3.5 The sun lies on the intersection line of earth's equator plane and its orbital plane. Day and night are equally long at the autumn point or vernal point (spring)

We now have three effects on the seasons, which may reinforce or weaken each other. For example, a glacial period in the Northern Hemisphere could start when we have cold winters and cool summers in the North. So, we should have midwinter in the North, when the earth is in the aphelion, combined with a high eccentricity (then the summer will be cool) and a large tilt. The Milankovitch periods and their explanation are established well. Only the 100 000 year period is the subject of many studies.

A physical scientist, more than other natural scientists, is inclined to think in linear relations between cause and effect. This holds true for many cases where one can reduce a complicated problem to a simple one, involving only a few variables. However, for complex systems, like the climate system, one may run into difficulties. Small changes in a parameter like insolation may have big consequences, for example when some threshold is crossed.

Even when the external changes are slow, the reaction of the climate may have sudden jumps. A large ice cover will melt slowly, but suddenly a large piece may break off, go into the oceans, and melt there. This would change the salinity and temperature and might influence the course of big ocean currents such as the Gulf Stream, which at present gives Northwest Europe its moderate climate.

Another possibility is that small changes in insolation will influence the kind of plant species that grow and, consequently, the type of gases that the plants give off. This may influence the greenhouse effect and have a bigger influence on the surface air temperatures than just the insolation itself ([20], p. 287).

We should mention that the change in tilt of the earth's axis and its precession do not change the total insolation of the earth. They change the distribution over our planet. The insolation then interacts with the complex climate system, giving rise to glacial periods, warm interglacial periods and many other effects. The change of eccentricity does change the total radiation received, as mentioned above, but the effect is small. The influence of eccentricity changes on total insolation is smaller than 0.18 percent ([20], p. 306), and if it has an effect at all, it is by interaction with the climate system.

3.1.2 *The changing sun*

Milankovitch variations mainly influence the distribution of insolation over the globe. The luminosity of the sun itself may change as well – with an effect on the insolation all over the earth. Indeed there are two causes for variation of the luminosity of the sun.

The first is a continuous change due to the lifecycle of the sun itself. This goes very slowly. From models of stellar evolution, one estimates that some 2.7 billion years ago the suns luminosity was 30 percent less than at present ([20], p. 287). So there must have been compensating factors to maintain a temperature not far from the present one. We will come back to that later.

> In addition to the climatic changes associated with the linear response to the orbital forcing, we propose that a nonlinear response at sub-Milankovitch frequencies may have been responsible for the onset of the Northern Hemisphere glaciation, 2.75 [MY] ago.
>
> *Science* **285**, 23 July 1999, 568–71

The second change in luminosity is connected with the fluctuation in the number of *sunspots*. These spots are darker parts of the sun with a temperature between 4000 [K] and 5000 [K], as compared with 5800 [K] for the rest of the solar surface. Associated with the sunspots are brighter areas of the surface, called *faculae* with a temperature of perhaps 6800 [K]. In Equation (1.7) we discussed that for a black body the emitted radiation follows the relation $\sigma T^4[\mathrm{Wm}^{-2}]$. Because of the fourth power of the temperature T the emission of the hot faculae per [m^2] more than compensates for the lower emission of the sunspots. As the total area of the faculae is bigger than that of the spots a high number of sunspots corresponds with a higher total emission from the sun.

The number of sunspots varies irregularly with an average period of around 11 years. The number of sunspots has been measured since 1848; accurate measurements of the solar irradiance have only been possible since satellite measurements began around 1980. Both measurements indeed correlate but the observed change in luminosity is a small fluctuation of about 1.5 [W m^{-2}], which is only 1 percent of the solar constant S. In the model of Equation (1.11) such a small change in S hardly influences the earth's temperature T.

There are indications of a stronger correlation between temperature and the sunspot cycle [29], so it may be too simple to represent the effect of solar radiation by a single number S. In Figure 1.9 we gave a picture of the solar spectrum, approximated by that of a black body. Variations in solar constant S refer to fluctuations in the surface area of the curve shown there. The fluctuations of solar irradiation with wavelength, are at small wavelength much larger than the 1 percent mentioned above. They amount to 30 percent or more – albeit for a small part of the spectrum of Fig 1.9 or more precisely Figure 1.8 [30]. For the Lyman α line at a wavelength of 0.1216 [μm] the variations may be as large as a factor of 2. These small wavelengths correspond to high energies. That high energy, Ultra Violet (UV) part of the solar spectrum is responsible for sunburn and has many other effects in the biosphere. The living part of the climate system therefore might respond more strongly to the sunspot cycle than the small change in average irradiation of 1 percent suggests. By influencing the concentrations of greenhouse gases in the atmosphere, the climate system may reinforce the effect of solar variations. Once again this indicates a warning sign when applying simple models – although we cannot do without them as a first-order approach.

> The record of globally averaged sea surface temperatures over the past 130 years shows a highly significant correlation with the envelope of the 11-year cycle of solar activity over the same period. The total range of temperatures was about 1 [°C].
>
> From [29]

3.1.3 *Catastrophes*

Even more difficult to predict than changes of solar radiation is the frequency of catastrophes. They are known to have influenced the climate system for an appreciable time. Eruptions of volcanoes bring huge amounts of SO_2 gas and dust into the atmosphere. The SO_2 gas attracts water vapour, which develop into small particles, called *aerosols* (defined as particles with radii smaller

> The seemingly implausibility of a big bolide (meteorite) crashing into a planet was dramatically refuted in July of 1994, when 20 or so pieces of the disintegrating comet Shoemaker-Levy 9 slammed into Jupiter over a period of several days. The major disruptions produced in the atmosphere of the solar system's largest gaseous planet left little doubt that a similar collision with the earth could indeed occur and that the effects could indeed occur and that the effects could easily be catastrophic.
>
> [2], p. 74.

than 20 [μm]). Aerosols in the troposphere will form the nuclei of larger water drops, which soon precipitate. Aerosols in the stratosphere may remain there for a long time and may cut off the sunlight from entering. After a big volcanic eruption, cooling during a few years of 0.1 [°C] up to 0.5 [°C] has been observed ([20], p. 300; see also [18], p. 116, Figure 2.15).

Another example is the impact of a large meteorite on earth. It is believed that a meteorite entering the Gulf region near Mexico, some 65 million years ago, created an enormous amount of dust which remained in the atmosphere for several years. Plant life reduced and the dinosaurs did not get enough food, so they died from starvation ([2], [31], [32]).

3.2 *HUMAN INDUCED CHANGES*

Since humans started to develop cities and began to cut trees for firewood and building and our ancestors began to irrigate for agriculture, climate was influenced. This, however, happened on a local scale only and the influence on global climate most probably was negligible. This changed with the start of the industrial revolution, which we may put around 1780. Coal was used on a large scale to fuel steam engines or to heat the dwellings of the growing population. The burning of coal is equivalent to oxidation of carbon to CO_2 and consequently the concentration of CO_2 in the air rose steadily. This effect is illustrated in Figure 3.6, where the dotted curve indicates the CO_2 emissions ([10^{12} kg yr^{-1}], right scale) and the full/dashed curve denotes the CO_2 concentrations ([ppmv], left scale). The dashed part of the concentration curve refers to very accurate direct measurements at Hawaii. The little squares represents values deduced from the analysis of old ice cores, of which one knows the age.

3.2.1 *The global radiation balance in more detail*

In order to understand the effect of the rising CO_2 concentration on climate, we look at Figure 3.7. This is a very schematic representation of the incoming

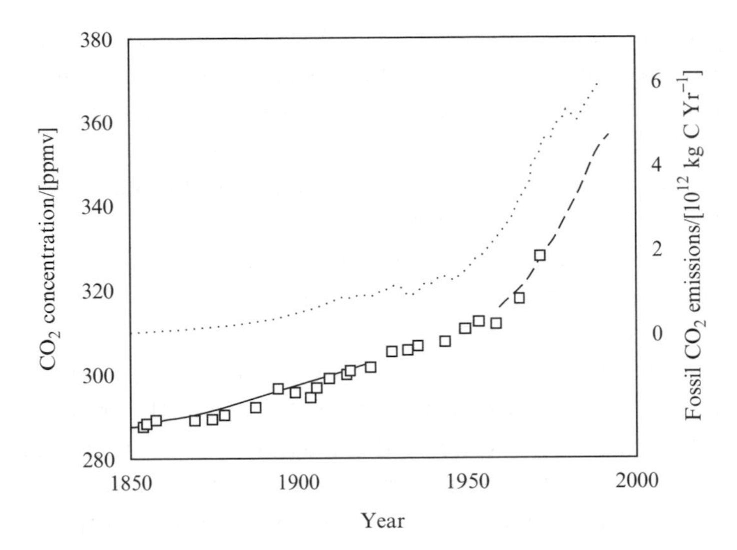

Figure 3.6 Atmospheric CO_2 concentrations (drawn/dashed curve, left scale) and CO_2 emissions (dotted curve, right scale) since 1850. The little squares are data from old ice cores. The dashed curve represents direct measurements. Reproduced with permission of IPCC from [18], Figure 1, p. 16

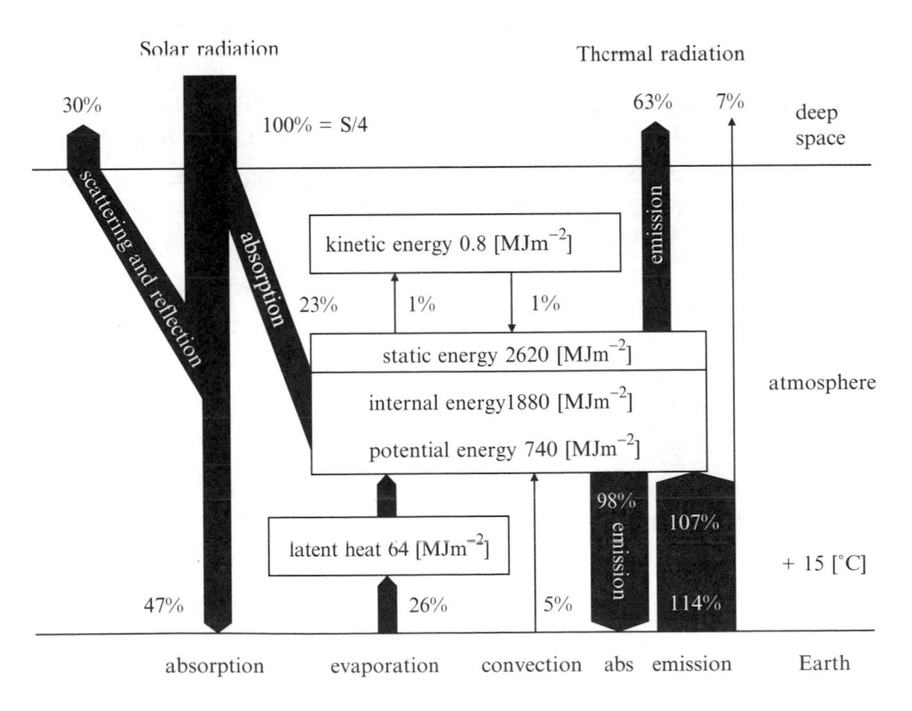

Figure 3.7 Radiation balance for the earth and the atmosphere. The solar radiation on the left is put as 100 percent and its destiny displayed. On the right, the long-wave radiation by earth and atmosphere is presented. In the middle one finds the energy contents of the atmosphere. Reproduced by permission of the Department of Geography, ETH Zürich, from [33], Figure 12, p. 28

and outgoing radiation as deduced from measurements [33]*. On the left the incoming solar radiation is shown, represented by $S/4$ and put at 100 percent to simplify the units. The factor 4 in $S/4$ is due to the average over the globe and over day and night. It is the same factor 4 that was found in Equation (1.11). As noted below Equation (1.11) about 30 percent returns to outer space. Almost half (47%) is absorbed by the surface and the remainder, 23 percent, is absorbed in the atmosphere. 'Surface' here indicates the surface of land or sea and the air just above it. Actually, it is the surface air temperature which is measured.

In the middle of Figure 3.7 one observes that evaporation of water brings latent heat into the atmosphere. Note also the much larger number for the potential energy of the earth's gravitation† (mgh) and the internal heat energy (mc_vT). These data show that small changes in temperature or height distribution may change the static energy of the atmosphere significantly. Finally, one notices the rather small number for the kinetic energy. This represents the wind energy, which is in exchange with the large static energy. So, tapping the winds will not disturb the climate system too much.

On the right of Figure 3.7, a summary is given of the emission and absorption of the long-wave, infrared radiation. Of the total emission of the earth (114%) much is absorbed in the atmosphere (107%). A small fraction disappears into outer space. The atmosphere itself is responsible for emission up and down. If we approximate the atmosphere as a single black body, the emission up and down should be equal at 63 percent, so the remaining part of the downward flux of 98% – 63% = 35% may be interpreted as reflection or back scattering of earth is radiation by the atmosphere.

Figure 3.7 represents a steady state situation, where for each of the compartments (deep space, atmosphere and surface) the incoming energy equals the outgoing energy (Exercise 3.2 and Example 3.1).

> A biologist has trekked the Arizona Mountains for 31 years studying the rites of spring of jays. He found that the jays are laying their eggs earlier and earlier each season. By 1998 the first eggs of the season arrived 10 days earlier than in 1971. He is blaming global warming: the local minimum temperatures have nudged up 2.7 [°C] in 27 years.
>
> *Science* **287**, 4 February 2000, 793–95

*Ref [20], Figure 2.4, p. 28 presents a more detailed model where the atmosphere is divided into the troposphere and the stratosphere. Also the numbers are slightly different from the ones given in Figure 3.7. For qualitative arguments these differences are not important.

†The mass m in this and the next formula is in units [kg m^{-2}] and similarly the energy is in units [MJ m^{-2}] = [10^6 J m^{-2}]. This means that an integral over the vertical atmosphere has been taken.

Example 3.1: Radiation balance

The steady state in Figure 3.7 requires that the rule (1.9): *energy in* = *energy out* holds for the surface of the earth. Check this and deduce the earth's temperature implied in Figure 3.7.

Answer

Energy in = 47% + 98% = 145%
Energy out = 26% + 5% + 114% = 145% (OK)
$\sigma T^4 = (114/100) \times (S/4) = 389.6$ [W m^{-2}]. It follows that $T = 287.9$ [K] = 14.9 [°C]. This agrees with the 15 [°C] of Figure 3.7.

Changes in the steady state may arise from a change in any of the physical quantities shown. If, for example, the 30 percent attributed to scattering and reflection of sunlight became larger (more aerosol particles in the stratosphere) less energy would reach the earth's surface and it would cool. This argument may be made more quantitative, if we just look at deep space only and apply the rule (1.9)

$$\textit{energy out} = \textit{energy in}$$

on 1 [m^2] of surface at the outer atmosphere. This gives

$$a\frac{S}{4} + \sigma T_{\mathrm{a}}^4 + t\sigma T_{\mathrm{s}}^4 = \frac{S}{4} \tag{3.1}$$

On the left we put the energy entering into deep space. The first term represents the total reflection and back scattering of the earth's atmosphere. This is indicated by the albedo a of the atmosphere, multiplied by the average incoming solar radiation $S/4$. The second term represents the thermal black body radiation of the atmosphere, which has the temperature T_{a}. The third term represents the thermal radiation from the earth's surface with temperature T_{s} multiplied by the transmission t through the atmosphere. On the right we put the energy leaving deep space to the earth, which is just the incoming solar radiation. All quantities are in units of [W m^{-2}]. From Figure 3.7 one may deduce (Exercise 3.3) $a = 0.30$; $t = 0.061$; $T_{\mathrm{s}} = 287.9$ [K]; $T_{\mathrm{a}} = 248.2$ [K]. Rearranging Equation (3.1) yields

> At the end of the 19th century the Swedish scientist Svante Arrhenius calculated how changes in CO_2 content would affect the temperature at the earth's surface. He estimated that a doubling of CO_2 would produce a global warming of about 4 [°C] to 6 [°C], not too far from modern calculations.
>
> *Scientific American*, July 1990, 37

$$(1-a)\frac{S}{4} = \sigma T_a^4 + t\sigma T_s^4 \tag{3.2}$$

It would be interesting to see how a change of albedo a or transmission t would influence the temperatures of the earth and the atmosphere. However, we only have one equation with two variables, which cannot be solved without more knowledge. The simplest assumption would be that the coupling between earth and atmosphere would make them both move in the same direction. So, we put

$$T_s = 287.9 + x$$
$$T_a = 248.2 + x \tag{3.3}$$

Global warming is accelerating faster than climate modellers predicted. Back in 1995 IPCC predicted that global temperature would rise between 1 [°C] and 3.5 [°C] in the 21st century. But it has been calculated that the rate of warming is already equivalent to 3 [°C] increase per century.

New Scientist, 4 March 2000, 21

in order to construct a single equation in x. Then a reduction of transmission t from 0.061 to 0.05 increases both temperatures by 1.1 [°C]; another decrease of transmission t to 0.04 increases the temperatures by another 1.0 [°C] (Exercise 3.4). More realistic calculations with troposphere and stratosphere separated, usually result in a cooling of the stratosphere and a warming of the troposphere and the surface ([34], p. 39).

The discussion so far is summarized by Figure 3.8. As the wavelengths of the solar spectrum are relatively 'short', they are indicated by S and as the infrared (IR) wavelengths of the earth spectrum are relatively 'long' they are indicated by L.

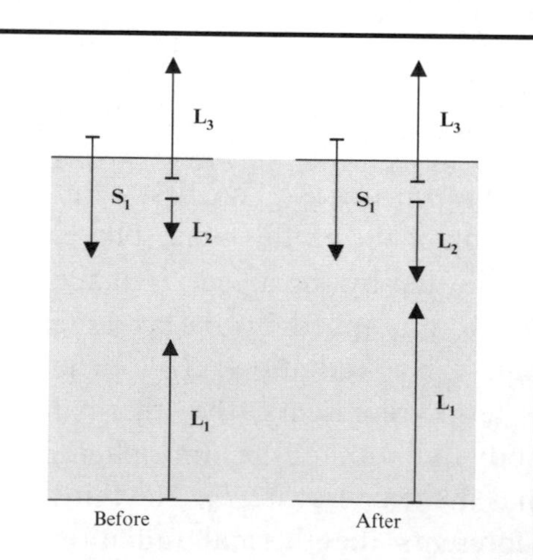

Figure 3.8 Greenhouse effect. Radiation balance between incoming solar radiation (**S**) and infrared radiation from the earth and the atmosphere (**L**). On the left, returning radiation makes a higher surface temperature possible than deduced from equation (1.11). On the right, increase in greenhouse gases cause more returning radiation and therefore a higher earth emission and earth temperature

The gases in the atmosphere absorb the incoming solar radiation (S) at specific wavelengths, which add significant structure to the spectrum as shown earlier in Figure 1.8. About half of the solar energy reaches the surface and this percentage (47%) is not very sensitive to the specific atmospheric composition.

For the infrared radiation emitted by the earth's surface, the situation is different. Trace gases in the atmosphere, in particular CO_2 and H_2O strongly absorb radiation (L) with wavelengths above 1 [μm]. Also here there is a wavelength dependence, but only about 6 percent of the emitted radiation is able to reach outer space. The remainder of it is absorbed and radiated back to the earth.

> Traces of methane-eating bacteria that lived in a pond over 2.7 billion years ago may help to explain the puzzle of how the early earth kept warm. The discovery suggests that methane, a powerful greenhouse gas, was abundant at that time and could have kept the earth from freezing.
>
> Later the carbon cycle began shifting away from a methane rich atmosphere after photosynthesis evolved. Methane concentrations gradually fell, eventually triggering the first widespread glaciation 2 to 3 billion years ago.
>
> *New Scientist*, 17 June 2000, 19

The resulting radiation balance was shown in Figure 3.7. It is summarized in the left part of Figure 3.8. The incoming solar radiation is represented by a vector $\mathbf{S_1}$. The radiation emitted by the earth is represented by a longer vector $\mathbf{L_1}$. The radiation returning from the atmosphere is indicated by $\mathbf{L_2}$, and the long wavelength radiation which goes to outer space by $\mathbf{L_3}$. The latter ($\mathbf{L_3}$) represents the summed contributions of radiation from the atmosphere and that which is transmitted by the atmosphere originating from the earth's surface. As we are discussing a steady-state situation, conservation of energy (Equation (1.9)) requires $L_3 = S_1$ from the point of view of outer space. This was written out more explicitly in Equation (3.2). The IR radiation originates from earth and atmosphere with a correction Δ for absorption, evaporation and convection $S_1 = L_3 = L_1 - L_2 + \Delta$. The returning radiation L_2 makes it possible that $L_1 > S_1$ which explains why the true surface temperature is higher than the temperature deduced from Equation (1.11).

As a next step, the steady state situation is changed to a new steady state. For example, the transmission t in Equation (3.2) may become smaller. For outer space a steady state still requires $S_1 = L_3$. More absorption implies more returning radiation L_2. Therefore L_1 must also be bigger and a higher surface temperature T_s, is implied. This is displayed on the right of Figure 3.8.

The mechanism behind this temperature rise is called *radiative forcing*, since the temperature increase is forced in order to put the balance (3.2) in order again. The radiative forcing is described by ΔI, in the same units [W m^{-2}] as in equations (3.1) and (3.2).

Example 3.2: Change in radiative forcing

Find the radiative forcing ΔI for a change in transmission in Equation (3.2) from $t = 0.061$ to $t = 0.05$.

Answer

The right-hand side of Equation (3.2) becomes smaller by an amount $(0.061 - 0.05)\sigma\, T_s^4$. This is the radiative forcing, ΔI. Assuming that the temperature change is an effect of radiative forcing we take as a first approximation the 'old' value for $T_s = 287.9$ [K] to obtain an estimate, which gives 4.29 [W m^{-2}]. Note that this radiative forcing results in a value of ΔT_s which, below Equation (3.3), was estimated as $\Delta T_s = 1.1$[°C]=1.1[*K*].

A general expression for the relation between radiative forcing and the temperature change ΔT_s is given by

$$\Delta T_s = G\Delta I \tag{3.4}$$

The factor G in between is the so-called *gain factor*, the response of the output (ΔT_s) to the input (ΔI). From our rough estimate above we deduce $G \approx 1.1\,[\mathrm{K}]/4.29\,[\mathrm{W\ m^{-2}}] \approx 0.26\,[\mathrm{K\ W^{-1}\ m^2}]$. This lies within the wide range of values used, of which we will adopt $G = 0.3\,[\mathrm{K\ W^{-1}\ m^2}]$ ([34], p. 43).

As we shall see below, warming of the earth surface will cause changes in the climate system, which may increase the temperature even further. This feedback gives an improvement in Equation (3.4), which reads

$$\Delta T_s = G_f \Delta I \tag{3.5}$$

This corrected value is sometimes taken as high as $G_f \approx 1.0\,[\mathrm{K\ W^{-1}\ m^2}]$ [34]. With ΔT_s from eq. (3.5) one may calculate the temperature of the atmosphere T_a from eq. (3.2) and will find that approximation (3.3) then not longer holds (Exercise 3.6).

3.2.2 *Infrared absorption by greenhouse gases*

The physics responsible for the decreasing transmission with increasing concentration, is the absorption of long-wavelength, infrared radiation by some of the gases in the atmosphere. If the concentration of these *greenhouse gases* increases, the IR absorption will increase as well. More precisely, this may be seen as follows. From Figure 1.10 we know that the black body spectrum of the earth peaks at about $\lambda = 10[\mu\mathrm{m}]$. Figure 3.9 therefore shows the absorption of atmospheric gases in that wavelength region. The vertical ordinate is the strength of the absorption of the earth's emission at the wavelengths,

> Changes in soil structure in response to CO_2 enrichment should be incorporated into global research. For soil structure has a strong effect on soil processes and organisms. On a global scale, the extent of soil degradation and erosion is severe and is accelerated by changes in climate and land use.
>
> *Nature* **400**, 12 August 1999, 628

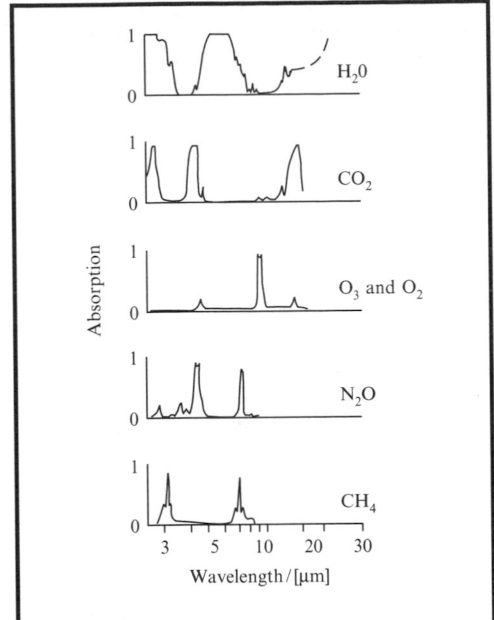

Figure 3.9 Absorption of the earth's emission spectrum by atmospheric gases. Note that the horizontal scale is not linear. Reproduced with permission from the authors from [26], Figure 27, p. 127

shown horizontally. Notice that water vapour strongly absorbs at wavelengths somewhat higher or lower than the peak value of $\lambda = 10\,[\mu m]$. The same holds for CO_2. The absorption by the other three gases shown, oxygen, methane and N_2O, takes place in a narrower wavelength region than for CO_2 and H_2O.

Our conclusion is that, besides CO_2, other gases known as *greenhouse gases*, also contribute to the absorption of the infrared radiation from the earth. With increasing concentration one has to calculate the absorption (which is the lack of transmission) for every gas separately, taking into account their wavelength dependence. For example, the absorption near $15\,[\mu m]$ is practically complete and due to CO_2 only. A higher concentration of CO_2 will not affect the absorption here; it will, however, increase the absorption in the 'tails' of the peak. By careful computation one then finds a global radiative forcing for a certain increase in the concentration of CO_2 and similarly for the other greenhouse gases.

Present and future greenhouse effects

The absorption displayed in Figure 3.9 is considerable. So we understand why at present the greenhouse effect, as discussed in Section 1.4, is responsible for a temperature increase at the surface of about 33 [°C]. In Table 3.1 we show in the 4th column how the different greenhouse gases contribute. We notice that the effect of water vapour is largest at 20.6 [°C], what we already suspect from Figure 3.9. Of the others, CO_2 is responsible for about 7.2 [°C].

Column 6 of Table 3.1 demonstrates that for most greenhouse gases, the concentrations are increasing, leading to an *enhanced greenhouse effect*. It is this effect that is subject to many model calculations and political discussions.

In the last two columns of Table 3.1 one finds the *global warming potentials* (GWP's) of various gases. GWP is the effect on global temperature that addition of one [kg] of gas has relative to the addition of 1 [kg] of CO_2. These GWP's are much larger than 1 and they depend on the time horizon one is considering. This is related to the time after which the amount of added gas is reduced to half of its initial value due to chemical reactions or photo dissociation by sunlight.

Table 3.1 Greenhouse gases and their effects From [26], p. 132 and [18]†, pp. 15, 22, 92

K	Gas	Conc/ [ppmv]	Warming/ [°C]	Lifetime/ [year]	Increase/ [% yr^{-1}]	GWP (20 yr)	GWP (100 yr)
	H_2O vapour	5×10^3	20.6				
*	CO_2	358	7.2	50–200	0.45	1	1
	O_3 troposphere	0.03	2.4				
*	N_2O	0.3	1.4	120	0.25	280	310
*	CH_4	1.7	0.8	12 ± 3	0.47	56	21
*	HFC's			1–200	5?	400–9000	100–12000
*	PFC's		0.6	−50 000	2?	4000 − 6000	6000 − 24 000
*	SF_6			3200	0.6	16 300	23 900

†Ref. [18] is part of the assessment of IPCC, the Intergovernmental Panel on Climate Change of 1995. The next complete assessment is expected in September 2001, after completion of this textbook. We do not expect large changes in the data of Table 3.1 or Figure 3.10. Curious readers may consult the IPCC website (see Appendix C) for the latest results.

In the bottom three rows one finds man-made gases of which the concentrations are still very small. Because of their large GWP's and their long lifetimes in the atmosphere, any increase in their small concentrations may have a large effect.

Although Table 3.1 is already rather compressed, it is still too complicated for international negotiations. Politicians negotiating international agreements on reduction of greenhouse gases prefer a single number or entity about which to negotiate. Therefore the GWP's are used to reduce the addition of greenhouse gases other than CO_2 to the addition of CO_2. In this way one may define an *equivalent CO_2 concentration*, which is the effect of all human induced greenhouse gases reduced to the effect of CO_2 concentration only. As shown in the last two columns of Table 3.1 this will depend on the time horizon one chooses, which again is a political decision.

There are international efforts under way to reduce the emissions of greenhouse gases to the atmosphere and in 1995 the EU Council of Ministers adopted the point of view that the long-term enhanced greenhouse effect should not be larger than 2 [°C]. The so-called *Kyoto protocol* [35] refers to the gases indicated by an asterisk in

> Global warming will cause forests to grow faster in the next 50 years. From about 2050 however, the warming will kill many trees, returning the carbon to the atmosphere and cause a 'runaway' warming. Climate change is not going slowly, like turning a dial on an oven, it is more like turning a light switch.
>
> *New Scientist*, 14 November 1998, 15

the first column of Table 3.1. In Chapter 4 we will discuss technological ways to reduce the emissions of CO_2 and N_2O. Social and political measures required to back up those changes will be discussed in Chapter 11.

Feedback

In going from Equation (3.4) to (3.5) and replacing the gain factor G by the feedback gain G_f, we referred to feedback, which would increase the gain factor of the climate system, perhaps by as much as a factor of three. Qualitatively it is understandable that most of the feedback will reinforce the enhanced greenhouse effect. We mention that:

- A higher temperature will cause more evaporation and a higher water vapour concentration in the air. Because of the enormous effect of water vapour (Table 3.1 and Figure 3.9) this effect may be huge; for the same reason it is difficult to calculate.

- The increase in cloud cover will decrease the sunlight entering, but also decrease the infrared leaving. The last effect is believed to dominate.

- Melting of snow and ice will lower the reflection of the earth and increase the absorption (the 47 percent in Figure 3.7 may become larger)

- A higher temperature will cause a more rapid decay and rotting of plants and increase the concentrations of CO_2 and CH_4. The latter effect might dominate.

- Warming of the oceans may reduce their CO_2 uptake (Figure 1.15) and leave more of CO_2 in the atmosphere.

There are (only) two significant counteracting effects ([34], [18]):

- Water evaporating from the seas will cool the seas and warm the higher layers of the atmosphere where the vapour condenses into small droplets. The effect is a cooling of the lower atmosphere

- The burning of fossil fuels will bring aerosols into the atmosphere, which increase the back-scattering and the albedo a in Equation (3.2). The overall effect is negative.

The influences of aerosols on climate are much more complex than those of the greenhouse gases. The reason is the short atmospheric residence times from less than a day to more than a month. Recent studies demonstrate both the importance of aerosol effects on climate and the complexity of aerosol-cloud interactions.

Science **288**, 12 May 2000, 989, 900

One may bring all these effects together in so-called feedback loops. The resulting gain G_f, defined in Equation (3.5) has a wide range of values, representing the uncertainties in the calculations. In all calculations the reinforcing effects dominate, resulting in values of $G_f > G \approx 0.3$.

IPCC results

The radiative forcing [W m^{-2}] arising from all effects together has been calculated by the IPCC, the Intergovernmental Panel on Climate Change, working under the auspices of the United Nations*. The results are displayed in Figure 3.10 and refer to the anthropogenic radiative forcing since 1850.

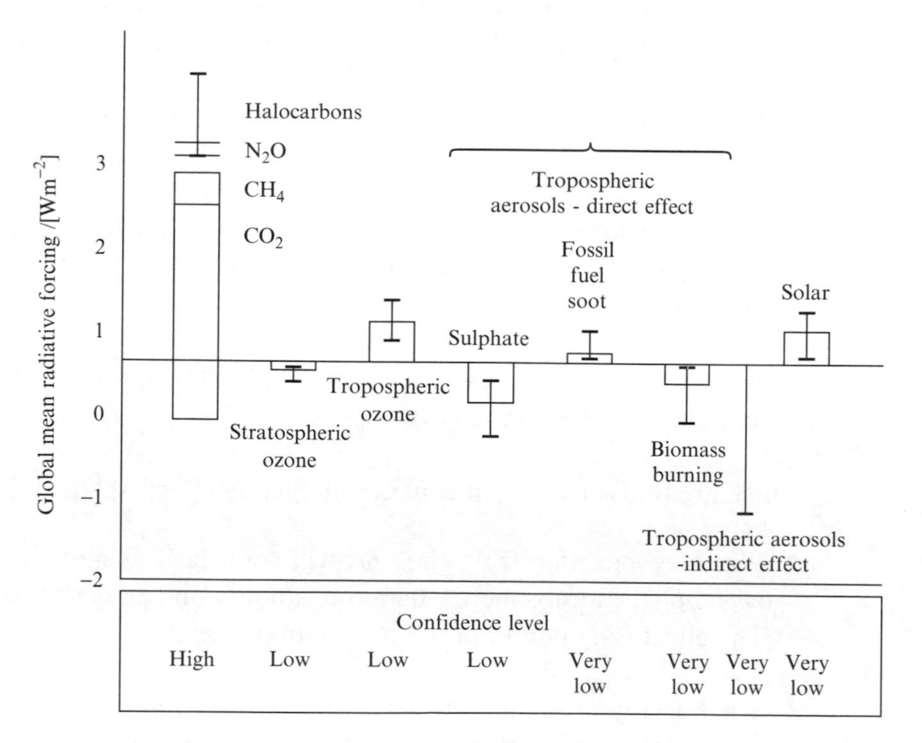

Figure 3.10 Human induced radiative forcing between 1850 and 1990. Note the confidence level in the bottom row. Reproduced with permission from IPCC from [18], Figure 2, p. 17

*Consult the IPCC website (see Appendix C) for more recent information.

Changes in solar input are also taken into account on the far right of the figure.

On the far left one notices the contribution from greenhouse gases, with a high confidence level (bottom line). The stratospheric ozone level is decreasing and the tropospheric ozone level is increasing (Chapter 9) resulting in a negative, respectively positive effect on the forcing. The aerosols have four effects, partly positive, partly negative. Volcanic eruptions are not taken into account.

Most of the calculations refer to radiative forcing, since that quantity is easier to calculate than the resulting temperature increase. It is not only the difficulty in estimating the feedback gain G_f, but also the delaying effects of the oceans (see below). In a cautious analysis IPCC puts the temperature increase from 1850 to 1990, only due to greenhouse gases, at 0.8 [°C]. Including sulphate aerosols one finds a lower number, 0.5 [°C], roughly the experimental value ([18], Figure 15, p. 33).

> Between 1955 and 1995 the world oceans warmed an average of 0.06 [°C] between the surface and 3000 meters. That is about 20×10^{22} [J] added in 40 years, comparable to the loss and gain with the seasons. So, rising ocean temperatures have delayed part of the surface warming. The climate sensitivity may even be close to 3 [°C].
>
> *Science* **287**, 24 March 2000, 2126

In order to discuss long-term effects, one often calculates what happens when there is a sudden jump in CO_2 concentration from its pre-industrial value of 285 [ppmv] to twice this value. Of course, the various models will give different answers to the resulting temperature increase, taking into account all possible delay times. The resulting increase in steady state temperature is called the *climate sensitivity* of the model. It ranges between 1.5 [°C] and 4.5 [°C]. The radiative forcing for a doubling of equivalent CO_2 concentration is usually taken around $\Delta I \approx 4\,[\mathrm{W\ m^{-2}}]$ and the range for the gain function G_f including feedback is put as $0.3\,[\mathrm{K\ W^{-1}\ m^2}] < G_f < 1\,[\mathrm{K\ W^{-1}\ m^2}]$. Much of the effort in computer modelling goes into narrowing the range for G_f.

3.2.3 *Time delays in climate change*

From Figures 2.16 and 2.17 it will be apparent that the surface layer of the oceans is interacting slowly with the layer underneath. This implies that even a sudden doubling of the greenhouse gases concentration today, would not immediately result in a temperature increase ΔT_s given by equations (3.4) and (3.5). On the contrary, the returning infrared radiation depicted in Figure 3.8 as L_2 has to do a lot of warming.

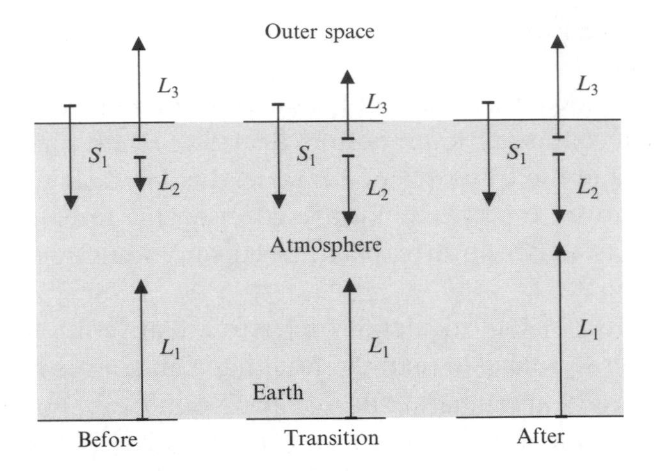

Figure 3.11 Time delays in climate change. Left and right are identical to Figure 3.8, but in the middle the IR emission into outer space is blocked by greenhouse gases. Consequently, the earth has to warm until on the right, the steady state situation again is reached

The beginning of the transition period is sketched in the middle of Figure 3.11. The left and right are identical to Figure 3.8. However, during the transition, L_3 is smaller than S_1. Therefore there is no steady state and the temperature of the earth and the emission L_1 have to increase. This takes time, as we understand from the physics, because the returning radiation first has to heat the first few metres of the ocean surface, where the sun easily penetrates. This water is in turbulent interaction with the rest of the water mass of the mixed layer. So effectively that complete layer has to be warmed. In the meantime the mixed layer exchanges heat with the water in the thermocline, albeit more slowly–and the down-welling waters at the poles will be less cold. So, if they still descend, they will warm the deep oceans. If not, there will be other, very slow exchange mechanisms of heat with the oceanic abyss. The total circulation time within the oceans of at least 1000 years will give an indication of the timescales involved.

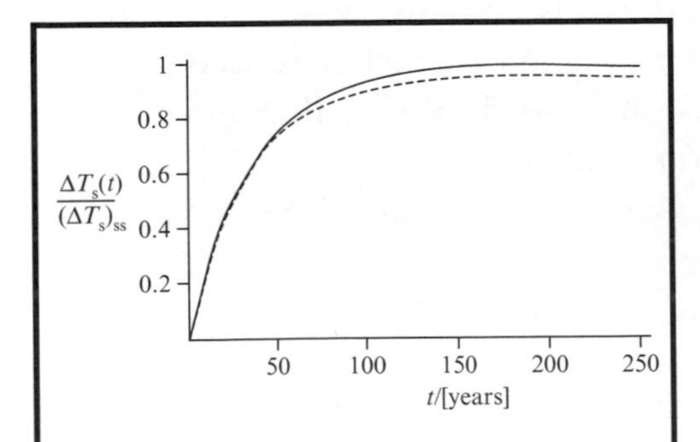

Figure 3.12 Heating of the top layer of the ocean. The top full curve shows the situation when there is no transport of heat to lower layers and the bottom (dashed) curve shows the situation where there is heat exchange

Let us first look at the situation with only one delay time, τ_1, corresponding to heat uptake by the mixed layer only. Suppose that at $t = 0$ we have a sudden doubling of CO_2 concentration for which Equation (3.5) would give a steady state warming of ΔT_s, which we now will call $(\Delta T_s)_{ss}$. Remember that for this value of warming the extra incoming infrared radiation would be compensated by increased outgoing infrared radiation. (The vectors

L_2 and L_1 in Figure 3.11 are both longer on the right (the new steady state) than on the left (the old steady state), but their difference remains the same). When there is a time delay (the middle situation in Figure 3.11) the factual temperature increase will be a function of time and will be indicated by $\Delta T_s(t)$. The simplest way in which the temperature rise goes to its steady state value is given by ([34], p. 45)

$$\Delta T_s(t) = (\Delta T_s)_{ss}(1 - e^{-t/\tau_1}) \tag{3.6}$$

For $t = 0$ the temperature increase (ΔT_s) is zero and for $t \to \infty$ it goes to $(\Delta T_s)_{ss}$. For a value of $\tau_1 = 35$ years* the function $\Delta T_s(t)/(\Delta T_s)_{ss}$ is given as the full (top) curve in Figure 3.12. For $t = \tau_1$ a fraction $f = 0.63$ of the steady state temperature rise is already realized.

Example 3.3: Slow increase in greenhouse gas concentration

The concentration of greenhouse gases is constant at C_0 up to $t = 0$, then increases with 1 percent [yr^{-1}] up to double the concentration 2 C_0 at time $t = t_2$. (a) Calculate t_2.
(b) Assume, not realistically (!) that cause (increase in concentration of greenhouse gases) and effect (temperature rise) are proportional. Takc $\tau_1 = 35$ years in Equation (3.6). Calculate the fraction f of the temperature rise at time $t = t_2$, when the concentration has doubled.

Answer

(a) The time dependent concentration is written as $C(t)$ with $C(0) = C_0$ and $C(t_2) = 2C_0$. We know that the increase $\mathrm{d}C = 0.01C\mathrm{d}t$ when the time $\mathrm{d}t$ is expressed in [yr]. So,

$$\frac{\mathrm{d}C}{\mathrm{d}t} = (0.01)\ C(t) \tag{3.7}$$

When a derivative of a function $C(t)$ is proportional to the function itself the result is an exponential:

$$C(t) = Ae^{t/\tau} \tag{3.8}$$

Substitution of (3.8) in (3.7) gives $\tau = 100$ [yr]. As $C(0) = C_0$ one has $A = C_0$ and consequently $C(t) = C_0e^{t/100}$ Also $C(t_2) = 2C_0$ which gives $C_0e^{t2/100} = 2C_0$. So $t_2/100 = \ln(2)$ and $t_2 = 69$ [yr].

*This is a reasonable value. Ref [20], p. 338 prefers a value of 10 [yr].

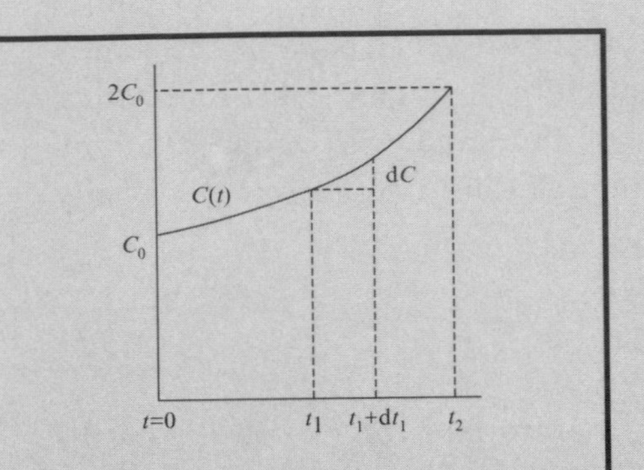

Figure 3.13 Exponential increase (3.8) of a concentration. At each time t_1 an increase of concentration is followed by an exponential growth (3.6)

(b) At any time t_1 a small increase $\mathrm{d}C$ will happen in a time d t_1, as shown in Figure 3.13. The corresponding rise in steady state temperature will be the fraction $(\mathrm{d}C/C_0)$ times the steady state rise for doubling, which is the climate sensitivity $(\Delta T_s)_{ss}$. This can be written as

$$\frac{\mathrm{d}C}{C_0}(\Delta T_s)_{ss} = \left.\frac{\mathrm{d}C}{\mathrm{d}t}\right|_{t_1} \mathrm{d}t_1 \frac{(\Delta T_s)_{ss}}{C_0} \tag{3.9}$$

This is the steady-state increase after a long time. But because of Equation (3.6) the steady state is reached slowly and at a time $t > t_1$ the increase $\mathrm{d}(\Delta T_s(t))$ becomes

$$\mathrm{d}(\Delta T_s(t)) = \left.\frac{\mathrm{d}C}{\mathrm{d}t}\right|_{t_1} \mathrm{d}t_1 \frac{(\Delta T_s)_{ss}}{C_0}\left(1 - \mathrm{e}^{-(t-t_1)/\tau_1}\right) \tag{3.10}$$

At a time t_2, the temperature increase from all preceding steps will be found by integrating

$$\Delta T_s(t_2) = \int_0^{t_2} \mathrm{d}(\Delta T_s(t)) = \int_0^{t_2} \frac{C_0}{100}\mathrm{e}^{t_1/100}\mathrm{d}t_1 \frac{(\Delta T_s)_{ss}}{C_0}\left(1 - \mathrm{e}^{-(t_2-t_1)/\tau_1}\right) \tag{3.11}$$

where we used Equation (3.8) for $C(t)$ with $A = C_0$. In the integral on the right-hand side the climate sensitivity $(\Delta T_s)_{ss}$ can be taken out, so we have $\Delta T_s(t_2) = f(\Delta T_s)_{ss}$. Then it can be simplified and calculated numerically or analytically. For $t_2 = 69$ [yr] and $\tau = 35$ [yr] one finds $f = 0.51$. When you have a computer handy, it is easy to verify that for small $\tau = 0.01$ [yr] the fraction $f = 0.99$. This is understandable, as there is no real time delay in the model then.

Let us now take the case of two ocean layers one with a delay time of $\tau_1 = 35$ [yr] and the other one with a delay time of $\tau_2 = 1000$ [yr]. Heat is flowing from the hotter one to the colder. Using some reasonable parameters, which are not discussed here*, results in the dashed curve of Figure 3.12. Again for $t \to \infty$ the temperature increase $\Delta T_s(t)$ goes to $(\Delta T_s)_{ss}$. The temperature increase of the top layer slows down after approximately 70 [yr];

*The parameters were chosen such that the two curves start to deviate after 50 [yr].

after 100 years at 95 percent of its maximum value, the temperature increase goes very slowly. Apparently, the bottom layer then starts to pick up heat.

What would happen when, for any reason, the concentration of greenhouse gases goes down? Assume, for simplicity that this happens after the oceans have warmed completely, after some 1000 years or so. In that case one has to read Figure 3.11 from the right to the left. The temperature will go down eventually, but in the transition period, the oceans will return their surplus heat to the atmosphere and the reduction in temperature will occur very slowly as well.

3.3 *CONSEQUENCES OF CLIMATE CHANGE*

> The question is not whether global change is occurring, but rather whether it is occurring on such a scale or at such a rate that terrestrial life, including humankind, will find it difficult to adapt [2], p. 173.

The publications of the Intergovernmental Panel on Climate Change give an authoritative overview of the current literature. The present models couple atmosphere and oceans, and include ocean currents in a way. There are many models and many parameters, but to be specific, we use as our reference an economic model (the IS92a, [36], p. 11) which assumes that no measures are taken to reduce emissions and we evaluate the consequences of this assumption. For its calculations it uses current economic and demographic assumptions. As it is a model it may easily underestimate the effects of climate change. A 'worst' case scenario might be deduced from the literature or by multiplying all estimates by a factor of two or more.

3.3.1 *The IS92a IPCC model*

The steady state response to a doubling of (equivalent) CO_2 concentration (the climate sensitivity) for the models discussed by IPCC is between 2.1 [°C] and 4.6 [°C] ([18], p. 298). Scenario IS92a describes the following[†]:

- the climate sensitivity amounts to 2.5 [°C] ([18], p. 40, Figure 19)
- the CO_2 concentration doubles with respect to the pre-industrial value of 280 [ppmv] to 560 [ppmv] in the year 2065 ([18], p. 23, Figure 5b)

[†]For IS92a the economic growth from 1990–2025 is taken as 2.9%; for 1990–2100 as 2.3%; solar electricity costs go down to $0.075/[kWh]; biomass is phasing in ([36], Table 1).

- the corresponding radiative forcing in that year is 4.2 [W m^{-2}] ([18], p. 24, Figure 6a)
- the temperature increase in the year 2065 is 1.2 [°C] ([18], p. 39, Figure 18)
- sea level rise in the year 2065 is 27 [cm].

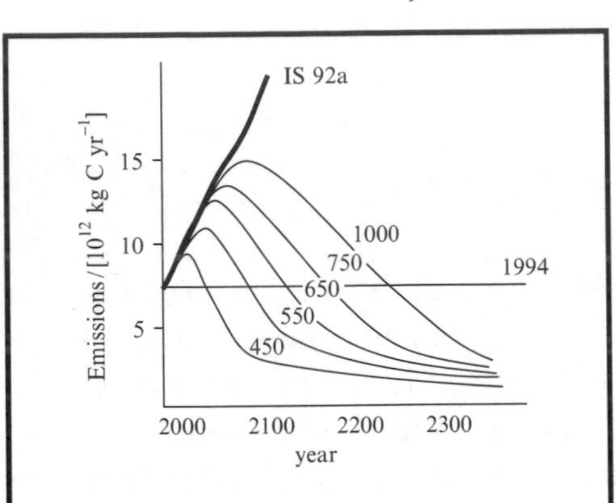

Figure 3.14 Scenarios for anthropogenic CO_2 emissions (vertical axis). The horizontal straight line represents the 1994 (human induced) emissions. The different curves correspond to different starting points in reduction of emissions. Even with drastic reductions with respect to the reference scenario (IS92a) one ends up with high final, constant concentrations, indicated near the curves in [ppmv]. Taken with permission of IPCC from [18], p. 25, Figure 7b

An unchecked increase in emissions would kill many of the world's forests by late next century, including the entire Amazon rain forest. A 750 [ppmv] ceiling would still destroy the Amazon, but delay its loss until the 22nd century. A limit of 550 [ppmv] would probably save it. In all cases the sea level will rise by at least 2 [m] over the next few hundred years.

New Scientist, 30 October 1990, 5

From the discussion at the end of Section 3.2 it follows that even when the increase in CO_2 concentration comes to an end in 2065, the temperature rise and sea level rise would continue. Breaking off the increase of CO_2 concentration is not easy, for even with constant emissions the concentration will keep rising. So, breaking off the increase of CO_2 concentration would require a drastic reduction of CO_2 emissions below the level of the year 1994.

The relation between emissions and concentration is shown in Figure 3.14 for the case when reductions with respect to scenario IS92a are starting in the 21st century. The emissions are put on the vertical axis. The curves correspond to the different times at which, on a global scale, the emissions are starting to decrease. The numbers near the curves indicate the final equilibrium CO_2 concentration in [ppmv]. One will notice the obvious: the later one starts with reductions, the higher the final concentration will be. It is interesting to see that to end up with a concentration of 550 [ppmv], which is double the pre-industrial value, the emissions must go down to 20 percent of the 1994 value. It is worth noting that to calculate Figure 3.14, one needs a carbon cycle model (Section 1.6) in order to calculate the impact of emissions on concentrations.

Assuming stabilization at an equivalent CO_2 concentration of 550 [ppmv] the temperature increase and sea level rise will continue, albeit slowly. The results for the year 2500 are the following:

- a temperature rise of 1.6 [°C] ([18], p. 45, Fig 24, interpolated)
- a sea level rise of 100 [cm] ([18], p. 45, Fig 25, interpolated)

Finally, we mention that the values given above are averages over the globe. The models show regional differences for all variables. The temperature increase, for example, in most places is bigger in the winter than in the summer, having consequences for the formation and melting of ice. The sea level rise may be higher in South East Asia than in Western Europe.

We now give a qualitative discussion of the presumed consequences of climate change ([26], chapter 7; [37]).

3.3.2 *Sea level rise*

The sea level is not constant. In nature the level rises and falls with the tides. In stormy weather the winds may blow the water against the coast and heighten the water level. In places where the coastal regions are protected with dikes, their height is calculated such that the water will not come higher than the dikes for once in the 10 000 years, say. This number holds for the dikes in the West of the Netherlands. With a sea level rise of 1 [m] they will flood every 500 years. A slow rise in average sea level implies that the frequency of flooding of the dikes will increase, unless of course one heightens and broadens the dikes simultaneously.

In many coastal areas, rivers are flowing into the seas. A slow rise in average sea level implies that the sea will penetrate the rivers at high tides more than at present. This increases the slow seepage of salt into the surrounding lands, which is bad for agriculture. Besides, one has to increase the height of the dikes along the rivers as well. Of course, one can build reinforced locks just at the end of a river and leave the dikes as they are, but in that case one needs a mechanism to pump out the river water over the locks, which is also expensive. The Rhine river, for example, has an average outflow of 2200 [m^3s^{-1}], but occasionally it may be as high as 13 000 [m^3s^{-1}].

> Around the world the gages that measure rainfall and stream height are slowly disappearing, victims of a slow erosion in funding. This happens at a time when global warming may be exacerbating weather extremes and water shortages. Now scientists are less able to monitor water supplies, predict droughts, and forecast floods than they were 30 years ago.
>
> *Science* **285**, 20 August 1999, 1199

In delta regions, where big rivers split into many smaller streams, protection will require great expense and adaptations of the eco-structure, or one will lose fertile agricultural lands.

In conclusion, the economic consequences of sea level rise are negative: either one has to apply expensive technological countermeasures, or one has to surrender costly lands, harbours, buildings and infrastructure to the seas.

3.3.3 *Agriculture*

At first sight, the combination of higher temperatures and a higher concentration of CO_2 in the air may be favourable for photosynthesis, see Equation (1.8). A doubling of CO_2 concentration may lead to an increase in the production of biomass of some 40 percent, provided there is enough water available (see Equation (1.8)) and also enough nutrients as solutes in water. This is the bottleneck: although at higher temperatures, more water will evaporate and it will rain more, the problem is to get clean water to the right agricultural place and the addition of nutrients is expensive.

Another point is that changes in temperature and precipitation will result in a movement of fertile, agricultural regions away from the equator to the North (Northern Hemisphere) or South (Southern Hemisphere). At the same time dry regions in countries like Italy or Spain will expand. So, even when the agricultural production in total may increase, the movement of the regions of high production will have its social and financial costs.

Finally, not only do consumer crops profit from a higher CO_2 concentration; but weeds and leaf-eating pests will also flourish. To control them, one will need pesticides, which in many cases are environment-unfriendly, and they will cost money anyway. So it is questionable whether the balance for agriculture will be favourable.

3.3.4 *Extreme weather conditions*

> The Munich reinsurance company anticipates that the number of climate related disasters will rise in the coming years. But the reinsurers will not close shops. The premiums for insuring against damage caused by natural disasters will increase.
>
> [39], p. 14

It is believed that higher temperatures will cause more extreme weather conditions. We have discussed El Niño in connection with a rather warm ocean surface near South America, resulting in heavy rainfalls and storms or hurricanes far away from its origin. It is also intuitively understandable that an adaptation of the ocean-atmosphere system to a higher infrared absorption will occur spasmodically, and not at all smoothly. Indeed, it may be that the increase in the frequency and the magnitude of El Niño is a way in which the greenhouse effect is communicated to the climate system ([18], p. 216).

Hurricanes can cause a lot of damage. Hurricane Mitch, for example, which hit Central America in November 1998 caused damage exceeding $5 billion and killed 11 000 people [38]. In general, the insurance companies measure an overall increase in storm damage over the last 50 years. For all

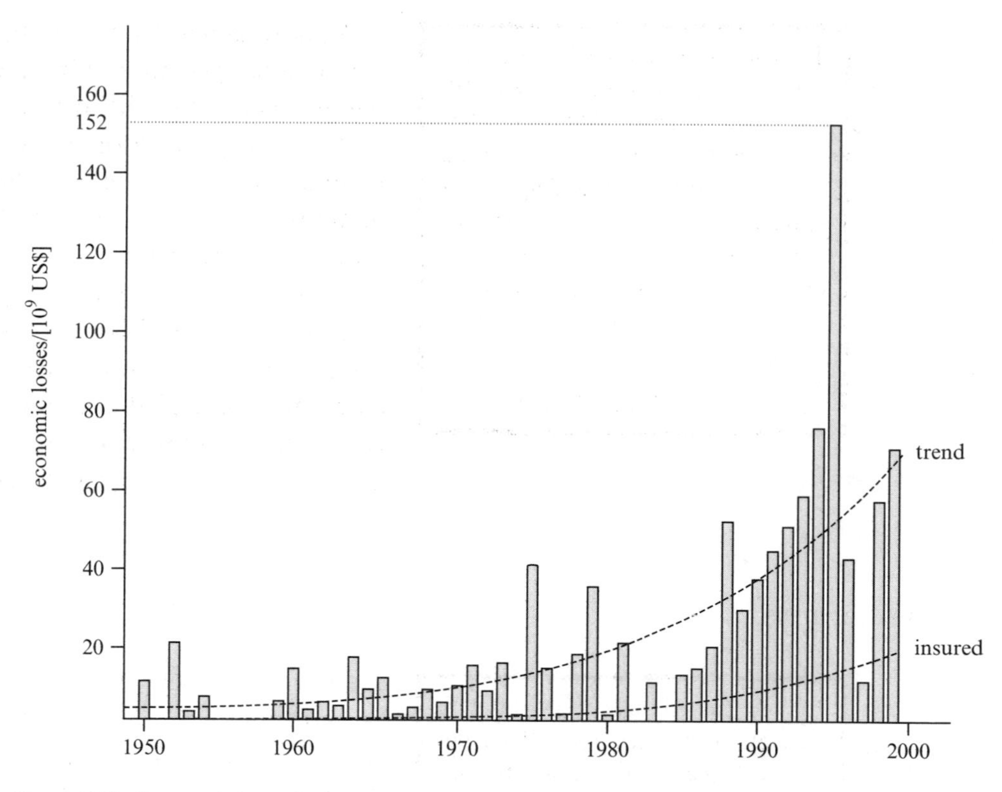

Figure 3.15 Economic losses from great natural disasters. The upper dashed curve shows the trend and the lower curve the part which was insured. With permission of 'Münchener Rück' from [39]

natural disasters together this is shown in Figure 3.15, which was prepared by a world-wide reinsurance company [39]. The trend is a steady increase since 1950 of which only part was insured.

3.3.5 *Health*

Hurricanes are often associated with intense precipitation and flooding. The water excess spawns a cluster of disease outbreaks, including cholera and fevers transmitted by mosquitoes who like these weather conditions [38].

There is a glimmer of hope, however. Extreme weather events are probably connected with a specific pattern of sea surface temperatures some time before the disasters occur. That the correlation is real, we have seen in our discussion of El Niño. With the right kind of models, one hopefully may predict extreme events some months before they occur. Preventive measures like vaccinations can be taken – and perhaps even material damage can be diminished [38].

The rising CO_2 levels will cause a decrease in coral growth between 1880 and 2065 of about 40 percent. The reason is that coral polyps use Ca and CO_3^{--} from surrounding waters to build their skeletons, which become the framework of coral reefs. But CO_2, a weak acid, reacts with water and CO_3^{--} to produce HCO_3^-, which corals cannot use.

New Scientist, 27 May 2000, 8

A more general point is that an increase in temperature of middle latitude regions will bring with it infectious diseases. A disproportionate warming at night and during winter is unhealthy for humans, but insects like it. Mosquito screens can protect one against some of the worst enemies, but not against all. Consequently, vaccination of the general population against common tropical diseases may be called for. So, for some cost, the consequences for health of global warming may be reduced.

3.3.6 *Surprises*

El Niño started some 5800 years ago, as may be deduced from flood deposits off the Peruvian coast. It coincides with the cultural leap around 5800 years ago. Our ancestors had to innovate in order to survive under the new conditions. They started to build temples in that same period. Perhaps to thank their Gods for their survival.

New Scientist, 22 May 1999, 39–41

A surprise usually is something unexpected but nice, as anybody knows who has participated in a surprise party. As the present mode of the climate system is rather pleasant and favourable for human life, surprises will soon be more unpleasant than enjoyable. Although surprises by definition cannot be known beforehand, there are two which come to mind ([18], p. 45, 46).

The first is that the Gulf Stream, which transports heat to Northern latitudes (Section 2.4) breaks down, or stops at lower latitudes. A second surprise is a sudden surge of the West Antarctic ice sheet, causing a rapid rise in sea level. After all, our scientific knowledge is incomplete although research is continuously contributing to more understanding.

3.3.7 *Conclusion*

The most foreseeable consequences of a general rise in temperature are negative or costly to repair. In the one example that may work out positively, the agricultural level of production, it is questionable. So, there is reason enough to try to reduce climate change.

3.4 REACTIONS TO CLIMATE CHANGE

> Global warming is no more than a presumption. Climate is dependent on many more factors than greenhouse gases, for example, on changes in solar activity and aerosols. A reduction of emissions to achieve stabilisation with a final concentration of 550 ppmv requires such drastic policy measures that one cannot take it seriously.
>
> *Essoscope*, 4 Nr. 4 (1996) 7–15, published by Esso oil company.

Governments are indeed trying to reduce greenhouse gas emissions and are reaching international agreements with this aim. We will discuss this in the final chapter of this book. Below, we will look at some other reactions.

3.4.1 Criticisms of the IPCC analyses

There are scientists who point out that the scientific theories and computer models used to calculate the consequences of greenhouse gas emissions are imperfect. They refer to natural fluctuations of climate and ascribe the changes we have experienced over the last decades to that effect. They are specifically warning that no expensive or drastic measures should be taken before climate change has been proven beyond doubt.

The last point is part of a more general problem: how to deal with scientific uncertainties. We will come back to that in Chapter 11. Let us say now, that criticism is an essential element of the scientific method and that scepticism with respect to scientific statements is forcing the proponents of these ideas to argue their case more convincingly – or sometimes, to revise their ideas.

> There is no convincing scientific evidence that human release of carbon dioxide, methane, or other greenhouse gases is causing, or will, in the foreseeable future, cause catastrophic heating of the Earth's atmosphere and disruption of the Earth's climate. Moreover, there is substantial scientific evidence that increases in atmospheric carbon dioxide produce many beneficial effects upon the natural plant and animal environments of the Earth.
>
> Petition against the Kyoto Protocol, 1998, http://www.sepp.org

The student will have to take his or her own position in the debate. Our view is that the greenhouse effect is real enough. The uncertainties are in the magnitude of the effect, in the time delays involved and in the stability of our climate system against major surprises of which we gave a few examples above. If the time delays in the system are in the order of 35 years or more, then, at the moment that the warming has been proven 'beyond doubt' it is too late for reductions to have effect within a human life span. That is the message of Figure 3.14 and the consequence of the rather long time that greenhouse gases remain active in the atmosphere (Table 3.1).

3.4.2 *Improved and continued modelling*

All scientists will agree with the need for more modelling to complete our understanding of the climate system and its response to external influences. Let us therefore summarize how modelling works. One has to divide the atmosphere into a number of layers, as one obviously cannot follow every single molecule of the atmosphere along its path. Looking at Figure 2.6, one would say that three or four layers are the minimum: one for the stratosphere, one for the troposphere above 500 [m] and one for the bottom layers where the influence of the ground surface is felt heavily.

As a second step, one will define ocean layers. With a glance at Figure 2.17 one should take at least three or four ocean layers. Finally, one has to lay a grid over the earth's surface as the vertical composition of the atmosphere will vary over the globe and the vertical composition within the oceans is a function of position also. One needs to describe the global energy transport, such as depicted for the oceans in Figure 2.18. Within the atmosphere, the main depressions and monsoons follow from the modelling.

An example of a grid for the earth's geography is given in Figure 3.16. One may check that the Mediterranean is represented by one box, which means

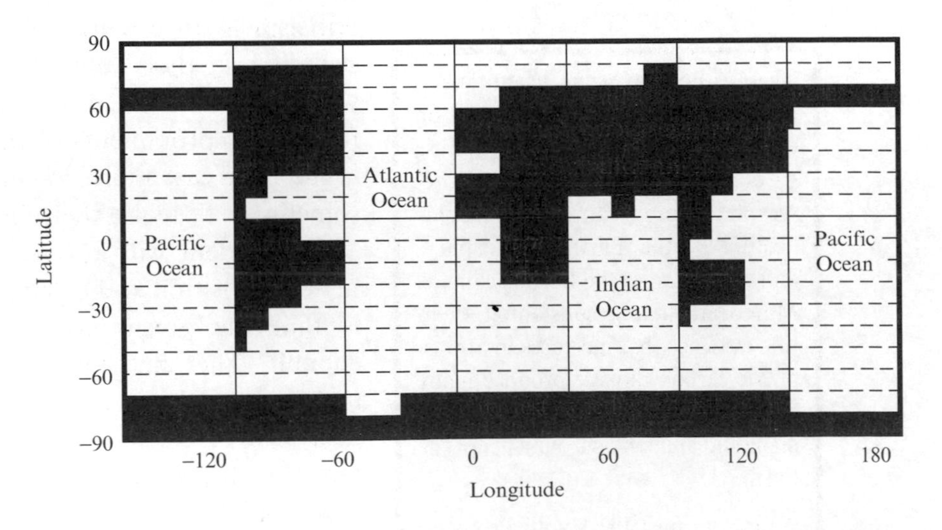

Figure 3.16 Representation of the earth's geography. Dashed lines show the atmospheric grid, solid lines separate the three ocean basins. Reproduced by permission from Springer Verlag from [40], Figure 1, p. 6

that all variables like temperature, humidity, rainfall and cloudiness are averaged over this wide region. From the figure, one may deduce that the Himalayan Mountains are represented by one box also. So a model, by necessity, is only an approximation of reality.

The processes in the boxes of Figure 3.16 (which will be layered vertically) have to follow the laws of nature as described in Section 2.3. But obviously, one has to make assumptions about the thickness of the layers, the interaction between the boxes, the method of averaging etc. One will need parameters to take these unknowns into account. These parameters are then changed until the results of the model agree with a data set of known temperatures, summer–winter differences and other climate variables in the past.

> Much of the period from 1300 to the start of widespread instrumental records may have been relatively cool. This would potentially exaggerate the long-term significance of 20th century warming.
>
> There are unfortunately only a small number of long, well-dated, high-resolution proxy records. Still, it is concluded that the 20th century is indeed anomalously warm.
>
> *Science* **284**, 7 May 1999, 926–27

Advanced and detailed models are usually able to reproduce the known climate and the changes from year to year. They become convincing if they are able to predict the climate from year to year in the future, including regional and continental differences. The student may verify in the newspapers whether the long-term predictions of weather and climate are that advanced.

Even if they are not detailed enough for absolute predictions, they may be good enough to calculate the influence of human induced greenhouse gases since then one calculates the difference between two model outcomes. In such a case inaccuracies and even mistakes, often cancel. So, the relative predictions of IPCC will be more accurate then absolute forecasts.

Figure 3.16 was taken from a so-called *intermediate model* with a limited number of layers and grid points. With such a model one is able to change conditions a little bit, and let the computer simulate the climate thousands of years into the future or in the past. If such a model is able to reproduce new data on past climates without changing its parameters, it adds to its credibility. The ability to look into the far future is necessary to take into account slowly changing processes such as the melting of icecaps.

> By feeding cows and sheep a daily dose of methane munching bacteria, researchers in Scotland hope to limit the animal's contribution to global warming. Ruminants are responsible for almost a third of all methane pollution that passes into the atmosphere. The bacteria should convert methane to the less harmful CO_2.
>
> *New Scientist*, 15 April 2000, 6

Other models have many more layers and divide the earth into smaller boxes than shown in Figure 3.16. That adds to the level of detail one may calculate, but costs so much computer time that it is difficult to look into the distant future or to change many parameters.

The conclusion is that modelling is always incomplete, and that 'we', the society, have to take decisions with incomplete knowledge. The point, of course, is not whether knowledge is complete but whether it is complete *enough*. Also this is a matter of scientific and political judgement.

3.4.3 *Countermeasures*

> With respect to the reduction of greenhouse gases one has to distinguish:
>
> - the technical potential, where costs are not taken into account
> - the economic potential, which can be achieved in the absence of market barriers
> - the market potential, which is possible with present market conditions.

How could one react to the arrival of global warming and the accompanying climate changes? It has been suggested to introduce small particles of SO_2 into the atmosphere, which form aerosols or soot ([2], p. 164). The practical difficulty will be to organise it and pay for it, but the more fundamental question is whether we understand the climate well enough for this kind of fine tuning. If one looks at the technician who is adjusting the central heating system, one will notice that part of his skill still consists of trial and error: turn the knob a little and see how the system reacts. For the climate system with its long reaction times and many variables, even the interpretation of the effects of the measures will be difficult.

It will be safer to reduce the emissions of greenhouse gases, by a change in the way of life (which is not popular), by technological tricks or by a combination of both. The following chapters (4 to 7) will discuss energy and transportation of pollutants. There we will discuss some of the technological solutions.

A related solution is to catch the greenhouse gases before they enter the environment and neutralize them. For CO_2 for example, one is discussing the possibilities of *sequestration*, which means to 'retire it into seclusion'. We come back to this in Chapter 4.

3.4.4 *Adapting to change*

When one looks at Figure 3.14, the most realistic view is probably that the best one can achieve is a slowing down of global climate change. In that way, changes will occur more gradually and there is more time to adapt. If sea level rise is unavoidable, the long-term planners of water protection may take this into account. At places where droughts are to be expected, one may study how the local agriculture can best cope with it. If storms and hurricanes will increase in frequency and power, one may install and design early warning systems together with contingency planning to mitigate their effects [41].

EXERCISES

3.1 Look up in an encyclopaedia or a textbook on astronomy, the definition of year. Find at least three astronomical definitions and compare them with the common definition of the year.

3.2 In Figure 3.7 the rule for steady state *energy in* = *energy out* should hold for the earth's surface, for the atmosphere and for deep space separately. Check this for the atmosphere and for deep space (see Example 3.1).

3.3 Identify the terms of Equation (3.1) with Figure 3.7 and find the values quoted in the text.

3.4 Combine equations (3.2) and (3.3) to find an equation in x. Use simple software to calculate temperatures for transmissions $t = 0.05$ and $t = 0.04$. Also calculate the temperatures for $t = 0.061$ and $a = 0.35$.

3.5 Compare Figures 3.7 and 3.8. Correlate the left of Figure 3.8 with the corresponding terms in Figure 3.7. Indicate the numerical value of S_1 on the left of Figure 3.8. Find the correction term Δ.

3.6 Equation (3.2) holds independent of what happens in the atmosphere. Assume that Equation (3.5) is correct with $G_f \approx 1.0\,[\mathrm{K\ W^{-1}\ m^2}]$. For $\Delta I = 4.29\,[\mathrm{W\,m^{-2}}]$ calculate the surface temperature T_s and apply (3.2) to find the atmospheric temperature T_a. Notice that Equation (3.3) is no longer approximately correct.

3.7 Ref. [38] makes the statement that 1 [°C] temperature rise gives 6 percent more water in the atmosphere, implying more rain and infections. Reproduce the number of 6 percent.

3.8 Some scientists want to block incoming sunlight by mirrors in outer space or particles in the upper atmosphere [42] in order to reduce the surface temperature by 2.5 [°C], to counteract global warming. Use equations (3.2) and (3.3) to find the necessary reduction in the solar constant S as about 4 percent. Ref [42] quotes more accurate model calculations, which give a reduction of 1.8 percent. Note that there is no guarantee that this geo-engineering will work.

4 Conventional Energy

It is perhaps characteristic for human beings to delegate hard manual labour to others. It started with the use of other human beings as slaves and the domestication of oxen or horses to pull the ploughs. The mechanical energy a human slave can deliver during a day is not very big. A friend once told the French physicist Coulomb that a group of French marines had reached the peak of Tenerife on foot in $7\frac{3}{4}$ hours [43]. This amounted to an elevation of 2923 metres and, with an average body weight of 70 [kg], the potential energy gained may be calculated as 2 [MJ] (Exercises 4.1 and 4.2). When one takes this number as the usable energy output for a working day (24 hours) the power output is 23 [J s^{-1}] or 23 [W]. This may be compared with the total power consumption per human being in the present-day society, which for several countries is given in Fig. 11.1. It is slowly increasing, but looking at the data for 1995, it appears that the world average was 2000 [W], for the USA much higher at 12000 [W] and for a low-income country like India still some 400 [W]. So even the average Indian has 18 servants available in terms of energy.

> Manpower may be compared with horse power. Continental writers tended to find higher man-to-horse ratios than English writers. It appeared that 5 Englishmen were equal to one horse, while it took 7 Frenchmen or Dutchmen to equal the same horse.
>
> [43], p. 97

This simple numerical example illustrates that modern society is very much dependent on external sources of energy. The complex organisation of energy supply of the industrial state is depicted schematically in Figure 4.1. It is easiest to understand by looking at the *tasks* that the system has to fulfil. That is the application of mechanical power (top right) or the uses of heat for keeping space at a pleasant temperature or as process heat to provide high temperatures to speed up chemical reactions in industry (middle bottom).

To simplify the diagram, we take mechanical and electrical energy as interchangeable. After all, this conversion can have an efficiency as high as 90 percent. Electrical energy can be produced directly by solar cells (top left) or by the intermediary of wind turbines (top left as well). In most power stations, however, electricity still is produced from mechanical energy in coils

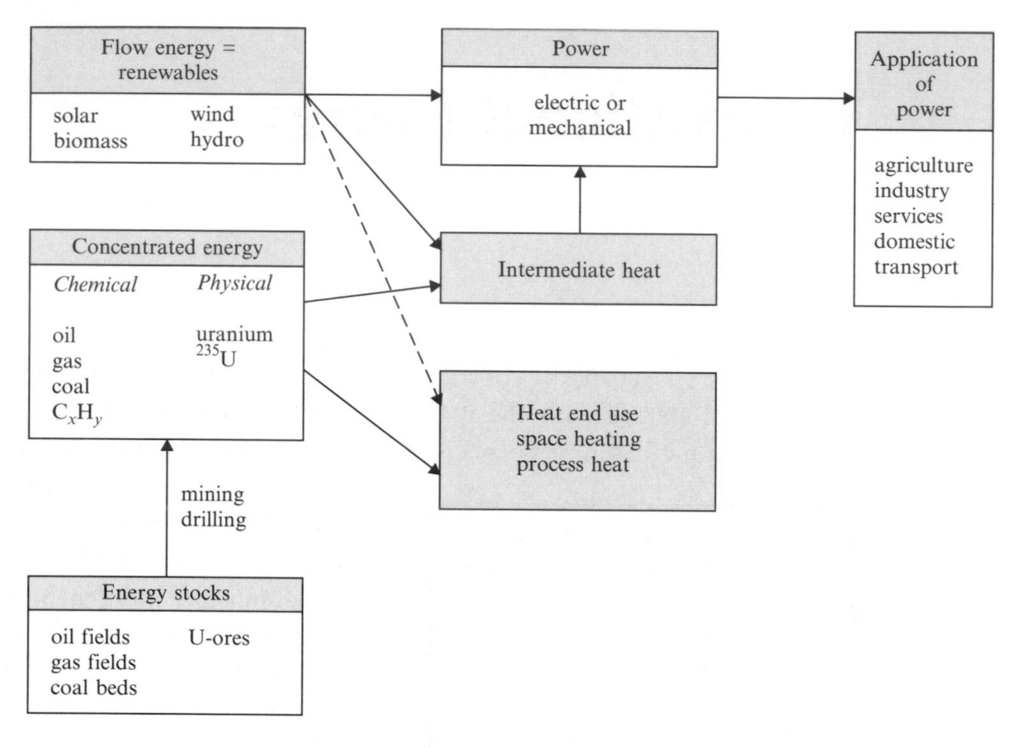

Figure 4.1 Energy supply in the industrial state. On the top right, the application of power in various activities is depicted. In the lower middle one views heat as end use, above that heat as intermediary to the creation of electric or mechanical power. Renewable or flow energies are put at the top on the left and the energy stocks on the bottom left. The arrows indicate activities to go from one box to another, which almost always produces pollution. The dashed arrow refers to solar heat and biomass heat

rotating in a magnetic field. That mechanical energy in its turn is produced by heat, which we call intermediate heat in the diagram.

Sources of heat may be the combustion of biomass (top left) or the combustion of fossil fuels (middle left) or by using the heat from nuclear reactions (also middle left). One may interpret fossil fuels as stocks of chemical energy, summarized as C_xH_y and nuclear fuels as stocks of physical energy, symbolized by ^{235}U. The word 'stock' expresses that these resources are finite as opposed to the renewables which, as the word suggests, are infinite.

In this chapter we devote our attention to heat and its many applications in industrial society. As much of the heat is delivered by fossil fuels, we give this chapter the name 'Conventional Energy'. Fossil fuels produce CO_2 and therefore contribute to the greenhouse effect, discussed in Chapter 3. Chapter 5 will be devoted to options which give little or no net production of carbon in the atmosphere, hence the name Carbonfree Energy. Also here, heat may act as an intermediary, as shown in Figure 4.1. Therefore, many of the concepts of Chapter 4 will be used in Chapter 5 as well.

In Section 4.1 we start with the physics of heat and, in particular, the way in which heat as a form of energy is transported. In Section 4.2 the conversion of heat in mechanical energy is discussed, which was indicated as the top arrow in the middle column of Figure 4.1. This section may be seen as a quick introduction into thermodynamics. Finally, the environmental consequences of the use of fossil fuels and ways to mitigate them are discussed in Section 4.3.

4.1 PHYSICS OF HEAT

> His water mill never quite answered industrialist Boulton's expectations. In the summer, particularly as the mill pond lowered in drought, he was constantly afraid for his source of power. Boulton (1765) decided to install a steam engine to pump water into his mill-pond.
>
> [44], p. 60

In this section we discuss the properties of heat: how it moves, how it can be converted into mechanical power, how to define a concept like efficiency and the less well known concept of exergy. This brings us into contact with the Laws of Thermodynamics. Incidentally, the connection of heat and power drew the attention of industrialists and technicians in the England and Scotland of the Industrial Revolution. It was due to their interaction rather than academic research, that the laws of heat and thermodynamics were discovered [44].

4.1.1 Heat transport

Heat does not stay at a single place but moves from one place to another, either by itself or forced by external causes. Examples from everyday life easily illustrate this. If you have made a trip on a motorcycle and touch the exhaust pipe, it will be very hot. The hot exhaust gases of the engine heat the interior of the pipe and the metal conducts the heat from the inside to the outside, where you will burn your fingers. This is heat transport by *conduction*.

If you put your hand besides the exhaust pipe, just a few [cm] away, you will not burn your fingers, but you will still feel the heat of the pipe. This is heat transport or, better, energy transport by *radiation*. If you put your hand a few [cm] above the exhaust pipe, the radiation is still there, but you will also experience rising hot air. In fact, the density variations in the rising air often make it visible. This is heat transport by *convection*.

The exhaust gases are the result of oxidation of petrol (gasoline), which brings water vapour into the air. As long as the vapour does not condense into small droplets its latent heat will be transported to another location. So, the *transport of latent heat* may be viewed as a fourth way to transport heat. In Chapter 2 we have already encountered the example of latent heat in cloud formation and transport of water vapour appeared to be an effective method for vertical or horizontal heat transport as long as all the vapour has not condensed.

Below, we discuss the physics of the first three methods of heat transport. In each case we apply what we have learned to the solar collector in Examples 4.1, 4.2, 4.3 and 4.4. Please note that in the beginning we will assume that the temperatures of the bodies are kept constant by some external means. Of course, in many practical cases heat transport will result in varying temperatures, but we begin by ignoring that.

Conduction

Imagine a slab or sheet of a homogeneous material with two parallel plane surfaces at a distance d, such as is sketched in Figure 4.2. Let the surfaces have a surface area A, big enough to be able to ignore the effects of the edges. The left surface has a constant temperature T_1 and the right surface has a constant but lower temperature T_2. Then a *heat current* **q** will flow from the higher temperature to the lower, as you may experience by sitting with your back against a hot chimney. The heat current is defined as the amount of energy crossing the wall per second [J s^{-1}]. It is printed in bold letters to indicate that it is a vector with a direction, in this case perpendicular to the surfaces.

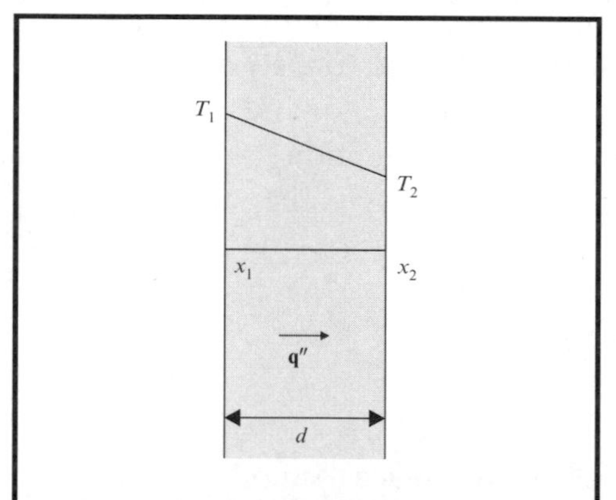

Figure 4.2 Heat flowing from a higher to a lower temperature over a distance $d = x_2 - x_1$

> *About the nature of heat.*
> On 15 February 1776 a paper by John Roebuck, titled 'Experiments on ignited bodies' was read to the Royal Society. This paper describes a series of experiments performed at Soho to test whether a ball of iron weights more when heated than when cold. Roebuck reports no consistent indication of an increase of weight by heating.
>
> [44] p. 168.

The *heat current density* $\mathbf{q}''$ is defined as the heat current through a surface of 1 [m^2], placed perpendicular to the flow with units [J m^{-2} s^{-1}] = [W m^{-2}]. For a problem with only one direction as in Figure 4.2 we may omit the bold print and just discuss the magnitude of the heat current density. This follows a simple relation

$$q'' = k\frac{T_1 - T_2}{d} \tag{4.1}$$

Note that the heat current density $\mathbf{q}''$ is proportional to the temperature difference between the two surfaces and inversely proportional to the thickness d of the sheet. The constant k in Equation (4.1) with dimension [$\mathrm{W\,m^{-1}\,K^{-1}}$] is called the *thermal conductivity* of the material. The name is understandable, as a bigger conductivity k causes a larger heat current to flow. Values of k for some common materials are given in Table 4.1.

The total heat current is easily found from the relation $q = Aq''$ as

$$q = Aq'' = \frac{kA}{d}(T_1 - T_2) = \frac{T_1 - T_2}{R} \tag{4.2}$$

where

$$R = \frac{d}{kA} \tag{4.3}$$

is called the *heat resistance* of the slab of material [$\mathrm{W^{-1}\,K}$]. This name of course triggers the memory of *Ohm's Law*, which we know from electricity theory. Indeed, the parallel becomes clear if we rewrite Equation (4.2) as

$$T_1 - T_2 = qR \tag{4.4}$$

If we read voltage difference $(V_1 - V_2)$ instead of the temperature difference $(T_1 - T_2)$ and electric current instead of heat current and electric resistance instead of heat resistance, then Equation (4.4) becomes the equivalent of Ohm's Law, as we know it.

The parallel even goes a little bit further. Consider two sheets of possibly different materials as indicated by Figure 4.3. In the left sheet the temperature drops from T_1 to T_2 with heat resistance R_1 and in the right sheet it drops

Table 4.1 Typical values of the thermal conductivity k for some materials ([45], Appendix A)

Material	Thermal conductivity k [$\mathrm{W\,m^{-1}K^{-1}}$]
air	0.026
glass fibre	0.043
glass	1.4
wood	0.12–0.19
brick	0.72
concrete	1.4
iron	80.2
steel	52
soil	0.52
sand	0.27
human skin	0.37

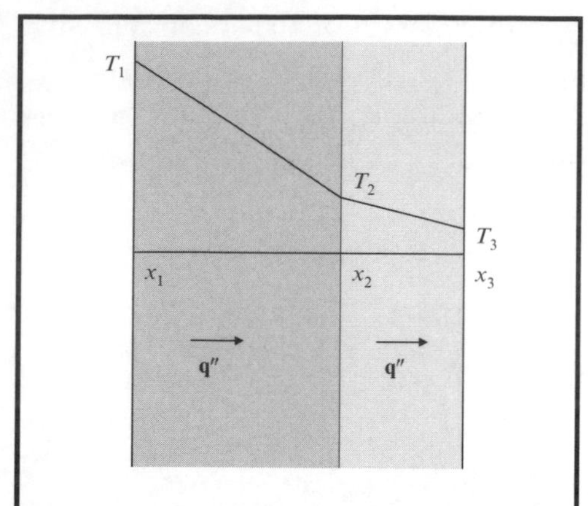

Figure 4.3 Two sheets in series. As the heat current is the same in both sheets, the heat resistances add

from T_2 to T_3 with heat resistance R_2. As temperatures are kept constant and heat cannot pile up or disappear at the interface between the two sheets, the heat current q must be the same in both sheets. From Equation (4.4) we then have the following equations

$$T_1 - T_2 = qR_1 \tag{4.5}$$

$$T_2 - T_3 = qR_2 \tag{4.6}$$

By adding equations (4.5) and (4.6) one obtains

$$T_1 - T_3 = q(R_1 + R_2) = qR \tag{4.7}$$

Therefore, one may replace the two heat resistances R_1 and R_2 by a single $R = R_1 + R_2$. This is precisely the relation that holds for two electric resistances placed in series.

Although the heat current **q** is easy to compare with the electric current through a wire, the more fundamental quantity is the heat current density $\mathbf{q}''$. From Equation (4.1) we may write it in the form of a derivative

$$q'' = -k\frac{T_2 - T_1}{x_2 - x_1} = -k\frac{\mathrm{d}T}{\mathrm{d}x} = -k(\mathrm{grad}\,T)_x \tag{4.8}$$

In Equation (4.8) the x-derivative of the temperature T was written as the x-component of the gradient as in Equation (B.11). In the example of Figure 4.3 the y- and z-components are zero, so we may generalize Equation (4.8) as

$$\mathbf{q}'' = -k \text{ grad } T \tag{4.9}$$

This expresses in the most general way the dependence of the heat current density on the temperature. From the discussion following eq. (B.11) we remember that the gradient is perpendicular to the surfaces with constant temperature T. So, for a any surface with a certain temperature, the heat current density will flow perpendicular to that surface.

The solar collector A solar collector is a device where the incoming solar radiation is converted into heat. In its simplest form it looks like Figure 4.4. A black surface, which is called the absorber, is heated on top by the solar radiation. The heat is transported down by conduction to water tubes, indicated in white. The tubes are embedded in an insulator to avoid losing

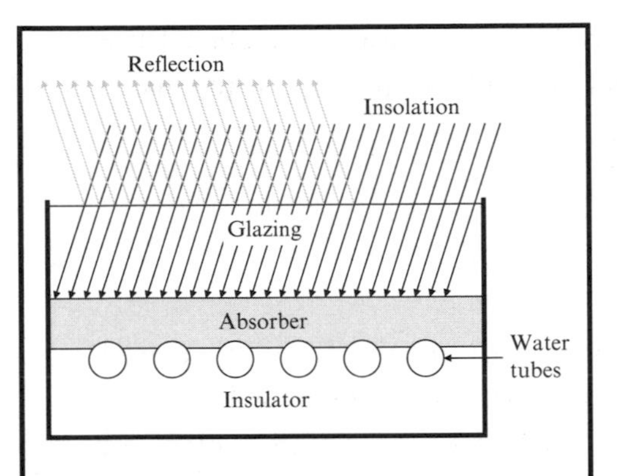

Figure 4.4 A simple solar collector. Incoming solar radiation from top right is absorbed by the black surface and conducted down to the water tubes. A small fraction is reflected from the glazing

the heat again. In Figure 4.4 the absorber is a black, flat plate with a thickness of about 10 [cm], which absorbs almost all solar radiation. In modern designs black water tubes absorb the radiation directly; aluminium mirrors, which reflect the radiation to the tubes, then surround them.*

The incident solar radiation at the top of the atmosphere is defined by the solar constant S. According to Figure 3.7 about 47 percent will reach the surface of the earth. Therefore, under the best conditions, with the collector perpendicular to the solar radiation, $0.47\,S \approx 650[\mathrm{W\,m^{-2}}]$ will be absorbed. The question then is how this is conducted down.

Example 4.1: Conduction in an absorber

Calculate the difference between the temperature T_1 on top of the absorber of Figure 4.4 and the temperature T_2 at the tubes for an iron absorber 10 [cm] lower.

Answer

Above we found $q'' = 650\,[\mathrm{W\,m^{-2}}]$. For iron from Table 4.1 we have $k = 80.2\,[\mathrm{W\,m^{-1}K^{-1}}]$ and it is given that $d = 0.1$ [m]. Equation (4.1) may be rewritten as

$$T_1 - T_2 = \frac{q''d}{k} = \frac{650\,[\mathrm{Wm^{-2}}]\,(0.1)\,[\mathrm{m}]}{80.2\,[\mathrm{Wm^{-1}K^{-1}}]} = 0.8\,[\mathrm{K}]$$

This is very small indeed and we conclude that iron is a good heat conductor. Students who touch the exhaust of a car or motorcycle will know this.

The calculation in Example 4.1 shows that one should have a small thickness d and a high thermal conductivity k in order to get the temperatures at the tubes as high as possible.

In a solar collector one should of course avoid losses of heat. That is the reason why the insulator in which the water tubes are embedded should have

*Data on modern designs are from CPC-2000, Solel, Solar Systems Israel.

a very low thermal conductivity. From Table 4.1 we note that glass fibre could do the job. There are, however, other losses as well. This brings us to the next way of heat transport.

Convection

When one puts a hot pot of tea on the table, the air around it will be heated from the contact; it will become less dense and rise. New cold air will takes it place and pick up some heat as well. During the process the tea of course will lose its heat. This process is called *convection.* It generally happens when an object of a certain material is in contact with a fluid (liquid or gas) of different temperature. The case we just described is an example of *free convection.* It is also possible to blow a fluid along an object. Obviously the process of heat exchange will then proceed more effectively as we know from the experience in cooling a hot cup of tea by blowing on it. This is called *forced convection.* In first approximation *Newton's law of cooling* describes this convection by

$$q'' = h(T_s - T_\infty) \tag{4.10}$$

Here, T_s is the temperature of the surface of the material and T_∞ is the temperature of the fluid. The subscript ∞ indicates that one has to take the temperature of the fluid at some far away point. Finally, h is the convection heat transfer coefficient [W m^{-2}K^{-1}]. Some values are given in Table 4.2. As is expected, the values of h for forced convection are much higher than those for free convection. It is also understandable that liquids, being denser, are able to transfer more heat than gases.

Example 4.2: Loss of heat by convection

Look again at the solar collector of Figure 4.4. Suppose that the absorber was in open connection with the air. Assume, for the sake of the argument, that the absorber has a temperature of 80 [°C] and the surrounding (ambient) air a temperature of 20 [°C]. What would be the heat loss by convection?

Answer

From Table 4.2 we take a value of $h = 10\,[\mathrm{W\,m^{-2}K^{-1}}]$. According to Equation (4.10) we have

$$q'' = h(T_s - T_\infty) = 10\,[\mathrm{Wm^{-2}K^{-1}}](60)[\mathrm{K}] = 600\,[\mathrm{Wm^{-2}}]$$

Table 4.2 Heat transfer coefficients h for convection in [W $m^{-2}K^{-1}$]. Reproduced by permission of John Wiley & Sons from [45], p. 9*

Phase	Free convection	Forced convection
gases	2 – 25	25 – 250
liquids	50 – 1000	50 – 20 000

*If the convection is accompanied by a phase change (boiling or condensation) the heat transfer coefficient may range between 2500 and 100 000.

So, one would lose 600 [W m^{-2}] by convection, which is essentially all of the incoming radiation, which was estimated above as at best 650 [Wm^{-2}]. This implies that a temperature of 80 [°C] for the absorber cannot be reached unless precautions are taken.

A simple solution is the use of the top glazing in Figure 4.4. Some remaining convection between the absorber and the glazing may bring the glass to a temperature of 30 [°C], say. This would reduce $T_s - T_\infty$ in Example 4.2 to 10 [K]. The loss to the outside air then would still be 100 [W m^{-2}]. Although much less, this still represents a formidable loss. Consequently, the interaction between absorber and glass must be reduced, for example by evacuating the space between both. That should effectively lower the temperature of the top cover. But there is still loss of heat by radiation, as we shall see presently.

Radiation

The third and last example of heat exchange is radiation. Every physical body emits a radiation spectrum, depending on its temperature T and its special properties. In a region of wavelengths between λ and $\lambda + d\lambda$ an amount of radiation $I_\lambda d\lambda$ is emitted with units [W m^{-2}]. If λ is measured in [μm] the dimension of the intensity I_λ is [W $m^{-2}\mu m^{-1}$]. As reference for an emission spectrum one takes the black body spectrum, depicted in Figures 1.9 and 1.10 for two different temperatures T.

As discussed in Section 1.3, a *black body*, by definition, absorbs all incoming radiation and reflects or transmits nothing. One may say that the absorption coefficient α_λ, which is the fraction of radiation absorbed at wavelength λ, obeys the relation

$$\alpha_\lambda = 1 \quad (\text{all } \lambda) \tag{4.11}$$

Let $I_{\lambda,\text{black}}$ be the emission spectrum (1.5) for a black body at a certain temperature T, and $I_{\lambda,\text{body}}$ the real emission spectrum for an arbitrary body at the same temperature. The *emissivity* ε_λ of the body then is defined as the fraction of the black body radiation it emits at that wavelength:

$$I_{\lambda,\,\mathrm{body}} = \varepsilon_\lambda I_{\lambda,\,\mathrm{black}} \tag{4.12}$$

Kirchhoff's laws, which can be proven under very general conditions ([45], Chapter 12) express that, for an arbitrary body,

$$\varepsilon_\lambda = \alpha_\lambda \quad (\text{all } \lambda) \tag{4.13}$$

For the special case of a black body $\varepsilon_\lambda = \alpha_\lambda = 1$. Before we consider application to the solar collector, we recall the total emission (1.7) of a black body, which is the integral of the curves of Figures 1.9 or 1.10:

$$I = \int I_\lambda \mathrm{d}\lambda = \sigma T^4 \quad [\mathrm{W\ m^{-2}}] \tag{4.14}$$

In practice, a body with a temperature T will not only emit radiation, but also receive radiation from its surroundings. In calculations, one often represents the surroundings by a single temperature T_{surr}. Then the net emission of the body will be

$$\sigma(T^4 - T^4_{\mathrm{surr}}) \quad [\mathrm{W\ m^{-2}}] \tag{4.15}$$

Finally, we remark that in cases where the absorption coefficient α_λ and the emissivity ε_λ do not depend on wavelength λ they are represented by α or ε without a subscript.

The solar collector Once again, let us have a look at Figure 4.4. As everybody will notice who looks at a solar collector, the surface of the absorber is black. Obviously, that is required to absorb as much as possible of the incoming solar radiation to obtain a high temperature. But this has a difficulty as we calculate in the following example.

Example 4.3: Radiation losses

Assume that the absorber really is a black body. Calculate the emission for a surface temperature of $T = 80\,[^\circ\mathrm{C}]$.

Answer

We should work in kelvin, so $T = 353$ [K]. From Equation (4.14) we have

$$I = \sigma T^4 = 5.671 \times 10^{-8}\,[\mathrm{Wm^{-2}K^{-4}}]\,(353\,[\mathrm{K}])^4 = 881\,[\mathrm{Wm^{-2}}]$$

In this example the emission would be 881 [W m^{-2}]. This is more than the incident radiation estimated at 650 [W m^{-2}], so a solar collector of this type could not give the required high temperatures inside the collector.

From a solar collector advertisement:

- an innovative product
- major efficiency and cost advantages
- sturdy, zero-maintenance design is all-weather safe
- aesthetic appearance
- avoids harmful emissions

In practice, one therefore uses the fact that the absorption coefficient α_λ and the emissivity ε_λ are functions of the wavelength λ. One needs to absorb as much as possible in the peak of the solar spectrum. From Figure 1.9 we deduce the peak region as the wavelength region between 400 and 800 [nm]. So, in that region α_λ and consequently ε_λ should be big and for humans it would look 'black' in the visible. For a so-called black-chrome coating one has $\alpha_\lambda \approx 0.94$. In the same wavelength region we have $\varepsilon_\lambda \approx 0.94$ as well*. But for a temperature of 80 [°C] the black body emission spectrum at that wavelength is very small. This may be verified by looking at Figure 1.10. The temperature of 80 [°C] or 353 [K] gives a spectrum, shifted somewhat to the left of that figure. But as the curve goes down very rapidly for short wavelengths λ, emission and absorption in the region where the solar spectrum of Figure 1.9 peaks is very small. Consequently, a large emissivity at visible wavelengths does not matter.

On the other hand, in the wavelength region where the absorber with a temperature of 80 [°C] emits, around 8 [μm], (see Figure 1.10 and Equation (1.6)), the coating has the property that the absorption coefficient and consequently the emissivity are small. For the black-chrome coating one has $\alpha_\lambda = \varepsilon_\lambda \approx 0.10$. (By special sputtering techniques it may be reduced to 0.07.) This escapes our eyes as it is far outside the visible region of the spectrum. Still, if we replace the black body spectrum by its integral (4.14), the loss through radiation would be roughly $\varepsilon_\lambda \sigma T^4$, which is a fraction $\varepsilon_\lambda \approx 10\%$ of the 881 [W m^{-2}], calculated in Example 4.3, or 88 [W m^{-2}]. To be precise, we now also must take into account the inflow from the surroundings, according to Equation (4.15). However, as there is a glass cover between the absorber and the outside, almost all of the radiation emitted by the absorber is absorbed by the glazing and increases its temperature. So, even if the space between absorber and cover is evacuated, they still experience heat exchange by radiation.

Let us, for the last time, discuss Figure 4.4, starting from the top. We have $S' = 650\,[\text{W m}^{-2}]$ entering the collector. By using special anti-reflective coating, a large fraction $t_c = 0.97$ will pass the cover, so $t_c S'$ is entering the

*For the best coatings these coefficients may be higher than 0.97.

collector. Viewed from the outside, it is just the glass cover, which has interaction with the surroundings as we assume that it absorbs all emissions from the absorbing plate. There are three relevant temperatures: the temperature of the glass cover T_c, the temperature of the outside air T_a, which determines the heat loss by convection and the radiative temperature of the surroundings T_{surr}, which reduces the heat loss by radiation. The net heat flux, available for warming the water in the tubes becomes

$$q'' = t_c S' - \varepsilon\sigma(T_c^4 - T_{surr}^4) - h(T_c - T_{air}) \quad [\mathrm{W\ m^{-2}}] \tag{4.16}$$

where $\varepsilon \approx 0.94$ is the emissivity of the glass cover at wavelengths around 8 [μm].

Example 4.4: A realistic solar collector

(a) Calculate the heat flux q'' into the solar collector for $T_c = 30[^\circ\mathrm{C}]$, $T_{surr} = -10[^\circ\mathrm{C}]$, $T_{air} = 20[^\circ\mathrm{C}]$, $h = 10\,[\mathrm{W\ m^{-2}K^{-1}}]$.
(b) The heat current density q'' is used to heat water in the tubes. A mass m of water is passing per second [kg s^{-1}] and is heated from T_{in} to T_{out}. Calculate the sensible heat taken away with a surface area $A = 3\,[\mathrm{m^2}]$ of the collector and a water circulation of 0.01 [kg s^{-1}].

Answer

(a) Direct substitution in Equation (4.16) gives $q'' = 335\,[\mathrm{Wm^{-2}}]$.

(b) If c_p is the specific heat of water [J kg^{-1}K^{-1}], given in Appendix A, we find that the sensible heat taken away per second [J s^{-1}] = [W] becomes

$$q = m\,c_p(T_{out} - T_{in}) \quad [\mathrm{kg\ s^{-1}}] \times [\mathrm{J\ kg^{-1}K^{-1}}] \times [\mathrm{K}] = [\mathrm{J\ s^{-1}}] \tag{4.17}$$

Note that it is helpful to write out the dimensions of the variables as a check. Using the result of (a) one finds $T_{out} - T_{in} = 24\,[^\circ\mathrm{C}]$.

From Equation (4.16) it is a clear that the output of the solar collector depends rather sensitively on the parameters. It obviously pays to keep the temperature of the glass cover as low as possible (Exercise 4.7).

4.2 *HEAT AND POWER: THERMODYNAMICS*

In ancient days prehistoric humans made fire by rigorously rubbing materials together. We would now interpret this as conversion of mechanical work into

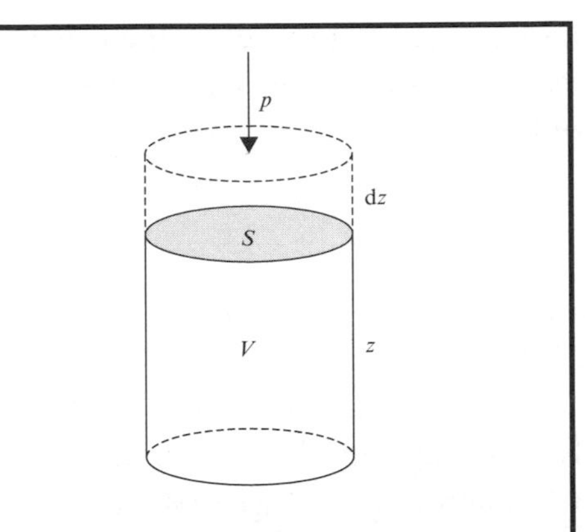

Figure 4.5 Work. A cylinder with surface area S, volume V, has a lid at position z and experiences a pressure p. If the lid moves upward by dz, the gas performs work $\delta W = p\mathrm{d}V$

heat. On the other hand, the steam in a boiling pot of water may push up the lid. This we would interpret as the conversion of heat into mechanical work.

It has taken some hundred years, from the 1750's to the 1850's for the experience of engineers and technicians to be interpreted in the present scientific way [46]. One had to understand how to measure and define heat with the *calorie* as the unit and how to define mechanical energy with the *Joule* as the unit before one could define what was called '*the mechanical equivalent of heat*'. This is the number of Joules corresponding to the conversion of one calorie of heat into mechanical energy [47]. Currently the calorie is forgotten and everything is expressed in Joules (Exercise 4.8).

4.2.1 Mechanical work

> Exactly as in the case of waterpower, wherever there is a difference in level – or temperature – there is the possibility of generating motive power. In other words where you have a hot body and a cold body you can derive useful power by means of a heat engine.
>
> Carnot, 1824 from [46], 130

Let us apply the concept of mechanical work to an amount of gas with volume V, which is enclosed in a cylinder as shown in Figure 4.5. The lid on top may move. The surrounding air exerts a pressure p on the lid, which has a surface area S. Assume that the lid goes upward from position z to $z + \mathrm{d}z$. The work the gas inside the cylinder has to do is called δW*, where the symbol δ indicates it is a small amount of work, just as $\mathrm{d}z$ is a small change in position.

There is a difference, though. The mathematicians call $\mathrm{d}z$ a *differential*, which means that $\int_1^2 \mathrm{d}z = z_2 - z_1$. In other words, the sum of all the small changes, which is represented by the integral is just the value at the end minus the value in the beginning. As we shall see below, the sum of all the small amounts of work cannot be represented by a work function W at the end minus that function at the beginning. It depends on the intermediate states. Therefore we use the symbol δW for the small change of W. Such small changes of δW or $\mathrm{d}z$ are also called *infinitesimal*, as they are infinitely small; they appear under an integral sign or in limits $\rightarrow 0$. The finite amount or finite change of any quantity we will occasionally indicate by the symbol Δ, such as ΔW, ΔU etc. Also, ΔQ will be the addition of a finite amount of heat.

*Note that in many thermodynamics texts, such as [47], the symbol W indicates the work done on the system, which therefore has the opposite sign.

Work is defined traditionally as *force times distance*. In Figure 4.5 the force equals pS, as the pressure p is the force per unit area. For the work δW performed by the gas against force pS we obtain

$$\delta W = p\,S\,\mathrm{d}z = p\,\mathrm{d}(Sz) = p\,\mathrm{d}V \tag{4.18}$$

Let the volume of Figure 4.5 expand from a volume V_1 to a volume V_2. In the meantime the pressure p on the volume may also change from p_1 to p_2. We assume that the change is done in a *reversible way*. This means that the change proceeds in small steps, in fact continuously, between well defined situations (values of p, V), which may be shown in a pV diagram as in Figure 4.6 (left)† The total work done by the gas in the volume may be written as

$$W_{1\to 2} = \int_1^2 p\,\mathrm{d}V \tag{4.19}$$

We immediately see that this is the grey area below the curve, which therefore depends on the intermediate positions between the initial situation, indicated by 1 and the final situation, indicated by 2. Figure 4.6 (middle) shows the work (indicated in grey), which is performed by the gas if it returns to the initial position by a different path.

$$W_{2\to 1} = \int_2^1 p\,\mathrm{d}V \tag{4.20}$$

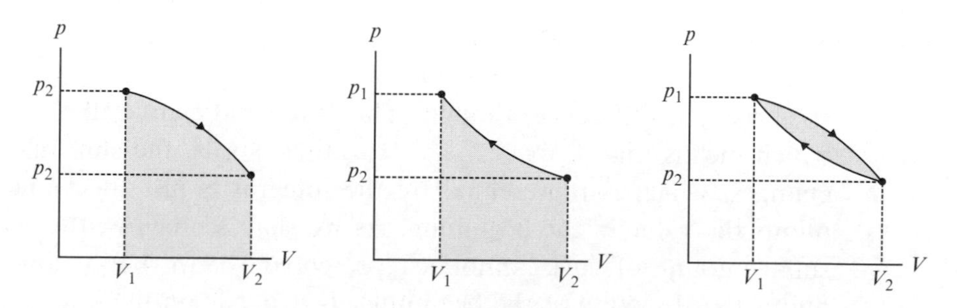

Figure 4.6 Work depends on the path. On the left the volume expands and delivers work against the outside with a magnitude indicated in grey. In the middle the volume is acted on and returns to the first position by a different path and it receives work, indicated in grey. On the right, the cycle is shown with the work performed by the gas as the enclosed grey area

†A more precise definition of a *reversible change* is given in [47], Chapter 7. It is a change of the system and its surroundings from situation 1 to situation 2, such that it is possible to bring back the system and its surroundings to situation 1. In this situation (1) the system, surroundings, and the rest of the universe are in exactly the same state (1) as before ([47], Chapter 7).

This is the grey area of Figure 4.6 (middle), taken from right to left. That is the area underneath the curve with an extra minus sign. The positive area itself can be interpreted as the work done on the gas by the outside world. For the complete cycle the net work performed by the gas is the grey area between the two curves of Figure 4.6 (right), which therefore is called the 'integral for a closed path' indicated by a circle across the integral sign

$$W_{1\to 2\to 1} = \oint p\,\mathrm{d}V \tag{4.21}$$

In Figure 4.6 the path is run clockwise and gives a positive result. When it would have run the other way round the result would have been negative. From the context it should be clear which way is meant.

4.2.2 *Heat*

> In terms of energy, 1 [kWh] of electricity and the sensible heat contained in 43 [kg] of 20 [°C] water are equal (i.e. 3.6 [MJ]). At ambient conditions it is obvious that 1 [kWh] electricity has a much larger potential to do work (to turn a shaft or to produce light) than the water.
>
> Ref. [37], p. 81 footnote

Heat flows from a reservoir with a higher temperature to one with a lower temperature. To be precise, in Equation (4.1) we wrote down the heat *flow*, the amount of Joules passing an interface between two bodies of constant temperature per second per square metre [J $s^{-1}m^{-2}$]. The cylinder, filled with gas in Figure 4.5 will have a certain temperature T. When it is in contact with a reservoir of a higher temperature, heat will flow in, thereby increasing its temperature. As the temperature is changing, Equation (4.1) has to be adapted, but there will be a heat flow q [J s^{-1}] anyway.

Assume that the contact with a hot reservoir takes the gas from a situation 1 to a situation 2, and then the contact finishes. The total amount of heat added to the system will be called Q, which can be measured or calculated. This is the addition of sensible heat to the system. If the volume of the gas is kept constant there is no work done, and the added energy must appear as an increase in the *internal energy* U of the system. In other words

$$Q = U_2 - U_1 \tag{4.22}$$

The energy function U is a so-called *state function*, which can be described as a function of the variables, which uniquely define the system. For a homogeneous gas (one kind of molecule only) with one phase (no liquid or solid around) these parameters just are p, V and T, which are related by the equation of state. For an ideal gas this was given in Equation (2.2). The internal energy of the system may be a large number, as in Figure 3.7, but not all of it can be used to produce mechanical energy.

If there is no work done, Equation (4.22) shows that the amount of heat added is just dependent on the initial and final situation and apparently not on the path. If part of the heat is used to perform work, one may show that the added heat depends on all the intermediary situations between initial and final states. This means that Q cannot be defined as a state function, which would only be dependent on the variables of the system. 'The heat' of a state cannot be defined, only the added heat can. We therefore indicate a small, infinitesimal addition of heat by δQ, just like δW, which was defined in Equation (4.18).

4.2.3 *First law–conservation of energy*

Equation (4.22) is an expression of the law of conservation of energy for the case without work done. If work is done as well, the added heat $Q_{1\to 2}$ has to cover both the increase in internal energy and the performance of work. Conservation of energy then implies

$$Q_{1\to 2} = U_2 - U_1 + W_{1\to 2} \tag{4.23}$$

which is the *First Law of Thermodynamics*. For small, infinitesimal changes this may be written as

$$\delta Q = \mathrm{d}U + \delta W \tag{4.24}$$

Note that δW indicates all sorts of work, not only the expansion work $p\,\mathrm{d}V$ of a gas. We may write this explicitly as

> One of the most serious disadvantages of steam is that it is not possible to employ it at high temperatures without using containers of extraordinary strength. The use of atmospheric air for generating the motive power of heat raises very serious difficulties in practice, but these are perhaps not insurmountable.
>
> Carnot (1824), quoted in [46], 137

$$\delta Q = \mathrm{d}U + p\mathrm{d}V + \delta W_\mathrm{e} \tag{4.25}$$

where δW_e summarizes all other work done by the sample under consideration.

The first law (4.23) implies that work can only be done when heat is supplied or at the expense of a reduction in the internal energy U, or both. Mechanical energy cannot be created out of nothing (conservation of energy), or, as the Americans say, there is not such a thing as a 'free lunch'. In thermodynamics this is expressed by saying that there does not exist a perpetual motion machine of the first kind ([47], Section 6–5).

4.2.4 *Entropy*

In the first law (4.24) we notice the differential $\mathrm{d}U$ and the infinitesimal changes δQ and δW. The last two expressions are not differentials, as the value of the integrals

$$\int_1^2 \delta Q \quad \text{or} \quad \int_1^2 \delta W \text{ depend on the path.}$$

One may prove ([47], Ch. 8) that for a *reversible* change, indicated by the subscript rev, $\delta Q_{\mathrm{rev}}/T$ again is a differential, which we denote by $\mathrm{d}S$, so

$$\mathrm{d}S = \frac{\delta Q_{\mathrm{rev}}}{T} \tag{4.26}$$

The integral $\int \mathrm{d}S$ between two points 1 and 2 is therefore independent of the path. It may be written (if δQ is added *reversibly*) as

$$\int_1^2 \frac{\delta Q_{\mathrm{rev}}}{T} = \int_1^2 \mathrm{d}S = S_2 - S_1 \tag{4.27}$$

where S is a state function, which is sharply defined for any set of variables that define a state completely. This is called the *entropy* S of the state. We note in passing, that just as with the internal energy U, one may add an overall constant to the entropy S without changing its defining Equation (4.27).

Reversible changes are a good instrument for theoretical analysis. In practice, changes are usually *irreversible*. If the same δQ as in Equation (4.26) is added irreversibly we get the *Clausius inequality*, which states that the entropy change for an irreversible process is always larger than for a reversible change

$$\mathrm{d}S > \frac{\delta Q}{T} \tag{4.28}$$

In an *adiabatic change* of a system no heat is exchanged with its surroundings, so $\delta Q = 0$. For a reversible change Equation (4.26) implies $\mathrm{d}S = 0$, so the entropy remains constant. For an irreversible change Equation (4.28) says that $\mathrm{d}S > 0$, an entropy increase. Both results together may be expressed as

$$dS \geq 0 \tag{4.29}$$

for any adiabatic change, where the = sign refers to a reversible, adiabatic change.

Spontaneous changes usually correspond to an entropy increase $dS > 0$. This may be made plausible by the following example

Example 4.5: Heat transfer with increase of entropy

A system, consisting of two reservoirs of heat, one at a high temperature T_H and another at a lower temperature T_C are brought into thermal contact. Heat δQ flows spontaneously from the hot reservoir to the cold one. Then the contact ends. Calculate the total entropy change of the system.

Answer

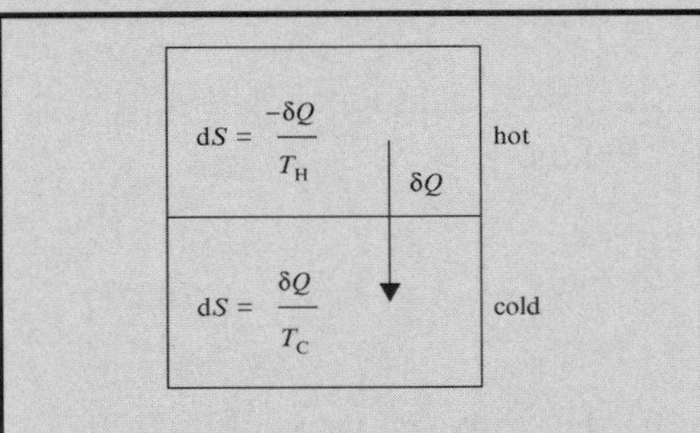

Figure 4.7 Spontaneous flow of heat δQ from a hot reservoir to a cold, increasing entropy

In thermodynamics a reservoir is a big mass whose temperature does not change by adding or subtracting quantities of heat. When one adds δQ to a reservoir and then takes out the same quantity δQ from the same reservoir, it is assumed that the reservoir returns to the same thermodynamic state ([47], Section 8–5). In other words, these changes are reversible. The entropy change of the reservoirs therefore is given by (4.26).

The two reservoirs are brought in thermal contact as in Figure 4.7. Heat $(-\delta Q)$ is added to the hot reservoir and heat $(+\delta Q)$ added to the cold one. The total change of entropy in the system, assuming $T_H > T_C$, becomes

$$(dS)_{total} = \frac{-\delta Q}{T_H} + \frac{\delta Q}{T_C} > 0 \tag{4.30}$$

Following (4.29) this is interpreted as an irreversible change of the total system.

When one looks at the total universe, any change of that total system happens adiabatically, by definition, and in general the changes are irreversible. So, the total entropy of the universe is increasing.

4.2.5 *The second law – no 'perpetual motion'*

From ancient time man has hoped to find inexhaustible sources of energy, better than slaves or horses. Could such a source exist? Consider an aeroplane flying from America to Europe at a height of 10 [km]. From Figure 3.7 one knows that there is a lot of internal energy available in the atmosphere. Why not use some of that energy, cooling the atmosphere a little and use the mechanical work to power the engines? This would be allowed by the first law (4.23) as energy could be conserved in this process.

In this example heat (or energy) would be withdrawn from a single reservoir, the atmosphere, and converted into work. A machine that achieves this is called a perpetual motion machine of the second kind – or, in Latin a *perpetuum mobile*. The *second law of thermodynamics* sums up common experience in making the statement that such a machine does not exist. In other words, one will always need two reservoirs with two different temperatures to get a heat flow, resulting in work, and flying from America to Europe on a single reservoir of atmospheric heat is impossible.

4.2.6 *Heat engines*

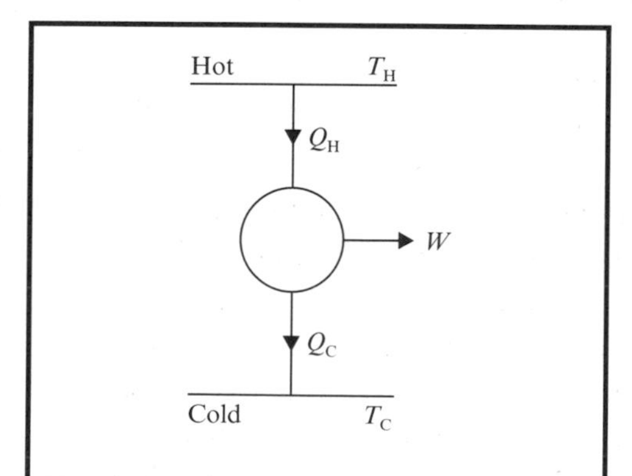

Figure 4.8 Heat engine. Heat flows from a high-temperature reservoir to one with low temperature performing work $W = Q_H - Q_C$

Heat engines are devices that operate between two or more reservoirs and thereby produce mechanical work. Heat engines work in cycles, after which the engine again is in its initial state. Let us consider one such cycle and the simplest possible scheme, illustrated in Figure 4.8. An amount of heat Q_H, taken from a hot reservoir with temperature T_H, enters a device which converts heat into work W. That device produces waste heat Q_C at temperature T_C, which leaves the device.

We apply the first law (4.23) to a complete cycle of the engine. As the initial and final states are the same, the internal energy does not change. So the added heat $Q_H - Q_C$ equals the work W:

$$Q_H - Q_C = W \tag{4.31}$$

The system as a whole, including the reservoirs, does not exchange heat with the outside world, so its changes are adiabatic and obey Equation (4.29). The entropy changes take place in the two reservoirs, as the work producing device has run a complete cycle and has not changed its entropy. So the increase ΔS of the total system (reservoirs and work producing device) becomes

$$\Delta S = \frac{-Q_H}{T_H} + \frac{Q_C}{T_C} = -\frac{Q_H}{T_H} + \frac{Q_H - W}{T_C} \geq 0 \qquad (4.32)$$

where Equation (4.31) was used. One may rewrite the last inequality in (4.32) as

$$-\frac{Q_H}{T_H} + \frac{Q_H}{T_C} \geq \frac{W}{T_C} \qquad (4.33)$$

or

$$W \leq Q_H\left(1 - \frac{T_C}{T_H}\right) \qquad (4.34)$$

The *thermal efficiency* η of the device is defined as the useful output divided by the required input

$$\eta = \frac{\text{useful output}}{\text{required input}} = \frac{W}{Q_H} \leq \left(1 - \frac{T_C}{T_H}\right) = \eta_{\text{Carnot}} \qquad (4.35)$$

At the right of Equation (4.35) the *Carnot efficiency* is defined as

$$\eta_{\text{Carnot}} = \left(1 - \frac{T_C}{T_H}\right) \qquad (4.36)$$

This clearly is the maximum efficiency possible, as the maximum η will correspond to the equality sign. This equality sign in Equation (4.35) holds when one would have started with an equality in Equation (4.32), which means $\Delta S = 0$ and a reversibly operating engine. So when the engine works with Carnot efficiency there is no net increase of entropy, and vice versa: when there is no net increase in entropy the machine will operate with the Carnot efficiency (4.36). Consequently, if the efficiency is smaller than the Carnot efficiency there has been entropy production somewhere and irreversible changes must have taken place, because the inequality sign (<) in Equation (4.35) originates from the inequality sign (>) in Equation (4.28), which implies irreversibility. Of course, in practice all changes are irreversible.

4.2.7 *Refrigerator*

One may put the heat engine of Figure 4.8 into reverse, putting in work W and moving heat Q_C from a cold reservoir to a hot reservoir, as indicated in

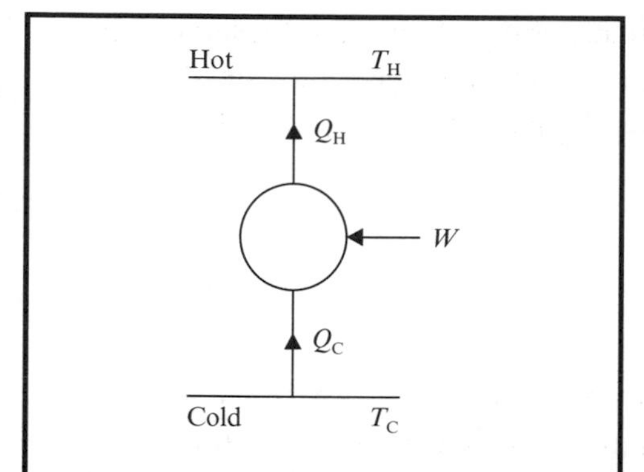

Figure 4.9 The refrigerator. Heat is taken out of a low-temperature reservoir, work is put in and heat is added to the high-temperature reservoir

Figure 4.9. The first law (4.23) implies that $Q_H = Q_C + W$ and consequently more heat goes to the hot reservoir than is leaving the cold. This is a simple representation of the *refrigerator*. In the household, Q_C represents the heat that is leaking into the refrigerator or brought in from the warmer supermarket articles, and has to be taken out. A small electric pump usually performs the work W there. A fluid evaporates on the inside of the refrigerator (the low-temperature reservoir) under low pressure, picking up heat. The pump brings it under high pressure, where it condenses on contact with the outside air (the high-temperature reservoir), thus releasing the heat. If you put your hand above the outlet of the refrigerator when the pump is running, you should feel the heat leaving it.

The performance of a refrigerator is measured by a so-called *coefficient of performance*, COP. It is defined as

$$\mathrm{COP} = \frac{\text{useful output}}{\text{required input}} = \frac{Q_C}{W} = \frac{Q_C}{Q_H - Q_C} = \left(\frac{Q_H}{Q_C} - 1\right)^{-1} \tag{4.37}$$

It is more convenient to write this in terms of temperatures. Then one again considers a cycle in which the total entropy increase $\Delta S \geq 0$. Looking at Figure 4.9 and Equation (4.26) one finds

$$-\frac{Q_C}{T_C} + \frac{Q_H}{T_H} \geq 0 \tag{4.38}$$

or

$$\frac{Q_H}{Q_C} \geq \frac{T_H}{T_C} \tag{4.39}$$

So one has, with Equation (4.37),

$$\mathrm{COP} \leq \left(\frac{T_H}{T_C} - 1\right)^{-1} \tag{4.40}$$

The equality sign holds here, when it holds in Equation (4.29), i.e, when there is no net increase in entropy. For the household refrigerator one may substitute $T_H = 300$ [K] and $T_C = 270$ [K] and find COP ≤ 9.

One may use the refrigerator of Figure 4.9 in a slightly different set-up as a *heat pump*. In that case the objective is to get heat inside a home for house warming. The cold reservoir may be the outside cool air, or a cold reservoir in the ground. The useful output in that case is the heat Q_H and one should adapt (4.37) and the following equations a slightly to find the COP in that case (Exercise 4.9).

4.2.8 *Examples of heat engines*

We give a few examples of heat engines by specifying what happens in the device, indicated by a simple circle in Figure 4.8. These devices contain a working fluid (liquid or gas) which may change phase during the cycle, but the initial and final states are the same and the changes occur reversibly. This cannot be true in practice, so we describe 'ideal' devices. We explain the device by its cycle, starting with the simplest example: the *Carnot cycle*. There we will illustrate that a visual representation of the cycle gives the most comprehensive information when one displays it both in a pV-diagram and a TS-diagram.

Carnot cycle

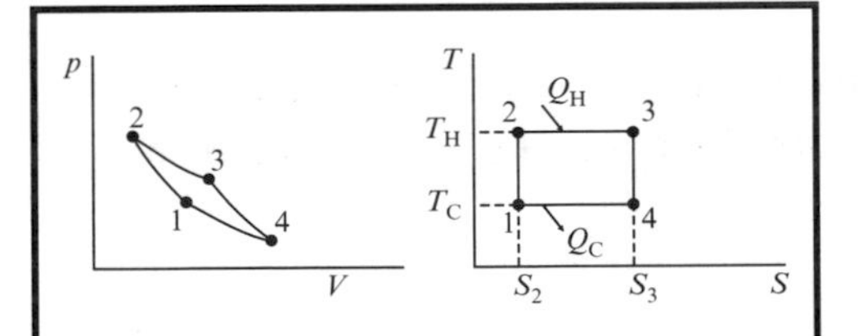

Figure 4.10 The Carnot cycle. On the left it is shown in a pV-diagram and on the right in a TS-diagram

Take a gas with pressure p_1 and volume V_1, indicated as point 1 in the pV-diagram of Figure 4.10 (left). The gas is compressed adiabatically, going to point 2 in the diagram. The term 'adiabatically' implies that there is no exchange of heat with the surroundings. Assuming that the gas is behaving as an ideal gas, this means that the curve in the pV-diagram follows the Poisson relation

$$pV^{\kappa} = \text{constant} \tag{4.41}$$

where κ is the ratio of the specific heats of the gas with constant pressure c_p and with constant volume c_v. For an ideal gas like air* one has $\kappa = c_p/c_v \approx 1.4$. It is clear that this value must be bigger than 1, as the gas under constant pressure will perform work, whereas the gas with constant volume will not.

From point 2 the gas is expanded *isothermally*, which means with constant temperature. From the ideal gas law (2.2) one knows that

$$pV = \text{constant} \tag{4.42}$$

*This value holds for gases with molecules, which consist of two atoms ([47], p. 110)

From point 3 to point 4 the gas expands adiabatically and from 4 to 1 it is compressed along an isotherm.

> The thermodynamics of a hurricane can be modelled as an idealized heat engine, running between a warm reservoir – the sea around 300 [K] – and a cold reservoir, 15–18 [km] up in the tropical troposphere, at about 200 [K]. Turbulent heat exchange at the surface drives the hurricane.
>
> *Nature* **401**, 14 October 1999, 649–50

From the diagram on the left in Figure 4.10 one cannot deduce the adiabatic or isothermal expansion or compression unless the drawing is accurate (see Example 4.6, below). This is easier, when one represents the same cycle in a *TS*-diagram, as shown on the right in Figure 4.10. It is immediately obvious that from 2 to 3 and from 4 to 1, the temperature T remains constant. It is also obvious from Figure 4.10 (right) that between 1 and 2 or between 3 and 4 the entropy S remains constant. From definition (4.26) it follows for a small entropy change that $\mathrm{d}S = \delta Q_{\mathrm{rev}}/T$. As $\mathrm{d}S = 0$ along each little piece of the path, it follows that $\delta Q_{\mathrm{rev}} = 0$, or in other words, there is no heat exchange with the surroundings and the change is adiabatic. There is heat exchange, however, during the isothermal expansion and compression, and its value follows easily from Figure 4.10 and Equation (4.27). As during isothermal changes the temperature T remains constant, one has for path $2 \to 3$

$$T_{\mathrm{H}}(S_3 - S_2) = \int_2^3 \delta Q_{\mathrm{rev}} = Q_{\mathrm{H}} \tag{4.43}$$

Here Q_{H} is the heat added to the system at the high temperature T_{H}. Similarly

$$T_{\mathrm{C}}(S_1 - S_4) = \int_4^1 \delta Q_{\mathrm{rev}} = -Q_{\mathrm{C}} \tag{4.44}$$

Symbolically, the meaning is exactly the same as in Equation (4.43), so $(-Q_{\mathrm{C}})$ is the heat added when the system goes from 4 to 1. Obviously this is negative as $S_1 - S_4$ is negative. Therefore Q_{C} itself is the energy withdrawn from the system at the low temperature T_{C}. The arrows in Figure 4.10 already indicated this. From the first law (4.23) the net head added during a cycle equals the work W, so

$$W = Q_{\mathrm{H}} - Q_{\mathrm{C}} = T_{\mathrm{H}}(S_3 - S_2) + T_{\mathrm{C}}(S_1 - S_4) = (T_{\mathrm{H}} - T_{\mathrm{C}})(S_3 - S_2) \tag{4.45}$$

where Figure 4.10 was used to see immediately that $S_1 - S_4 = -(S_3 - S_2)$. The thermal efficiency again is defined as output/input (see (4.35)), which gives in this case

$$\eta = \frac{W}{Q_H} = \frac{(T_H - T_C)(S_3 - S_2)}{T_H(S_3 - S_2)} = \left(1 - \frac{T_C}{T_H}\right) = \eta_{Carnot} \tag{4.46}$$

This indeed is the Carnot efficiency of Equation (4.36). This is the answer any reversible heat engine should give, which is operating between the same two reservoirs without creation of entropy.

Example 4.6: pV plot of the Carnot cycle

Use a simple software package to plot the pV-diagram for the Carnot cycle of Figure 4.10. Take units in which $p_1 = 1$[atm], $V_1 = 1$[L], $V_2 = 0.3$[L], $V_3 = 3$[L]. Find the corresponding pressures. Find point 4 as an intersection between an adiabat and an isotherm.

Answer

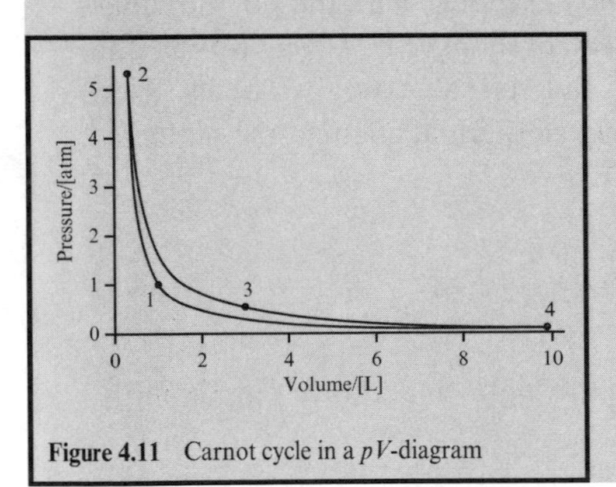

Figure 4.11 Carnot cycle in a pV-diagram

As we need a package like Mathematica anyway, we use it to find our points. With Equation (4.41) we find from $p_1, V_1 : p_2 = 5.39$ [atm]. From p_2, V_2 we find with (4.42) that $p_3 = 0.539$ [atm]. For point 4 one has two equations $p_3 V_3^\kappa = p_4 V_4^\kappa$ and $p_1 V_1 = p_4 V_4$. By dividing the two equations one obtains an equation in V_4 only. One finds $V_4 = 10$ [L] and $p_4 = 0.1$ [atm]. The resulting plot is shown in Figure 4.11.

As is known from Equation (4.21) the work may be calculated in the pV-diagram, giving

$$W = \oint p \, dV \tag{4.47}$$

This is the surface area of a cycle in the pV diagram and numerically equal to the value of (4.45). The curves in Figure 4.11 are close, which originates in the fact that the ratio $\kappa \approx 1.4$ is only slightly bigger than 1. Therefore, the work for each cycle is quite small and the Carnot cycle is only used for theoretical considerations. Since in practice any machine will work irreversibly with a lower efficiency, any losses will soon reduce the work output to zero. Therefore one needs other cycles to produce practical heat engines.

Stirling cycle

In 1816, before the science of thermodynamics had even begun, a minister of the Church of Scotland named Robert Stirling designed and patented a hot-air engine that could convert some of the energy liberated by fuel into work. The modern Stirling engine has low exhaust emissions and high engine efficiencies for varying loads, but high manufacturing costs.

[47], p. 137

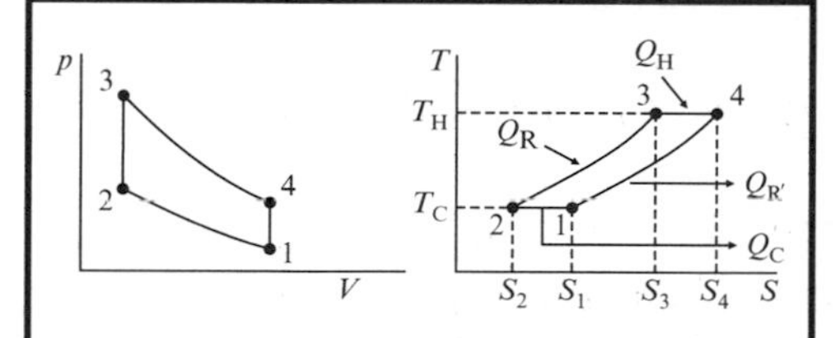

Figure 4.12 Stirling cycle. The heat Q'_R leaving the cycle between 4 and 1 is added again between 2 and 3 as Q_R. This results for the 'ideal' Stirling engine in the Carnot efficiency

The *Stirling engine* is defined by its cycle in the pV- and TS-diagrams, as shown in Figure 4.12. From point 1 to 2 the process is isothermal and heat is taken out, as is clear from the TS-diagram. From point 2 to 3 the volume remains constant (see the pV-diagram), but the pressure rises, so heat is added and the entropy increases (see the TS-diagram). From point 3 to 4 heat is added isothermally (see the TS-diagram) and from 4 to 1 the pressure falls with constant volume, so heat is taken out (see both diagrams).

One may not simply apply the rule that all heat engines have the same efficiency, that is the Carnot efficiency. For one no longer has two reservoirs at two distinct temperatures. In the case of the Stirling engine one has a continuous set of temperatures between points 2 and 3 and similarly between points 4 and 1. So, for the Stirling engine one needs to take a fresh look.

Fig. 4.12 suggests that the curves $2 \rightarrow 3$ and $4 \rightarrow 1$ in the TS-diagram have the same shape and that happens to be real. One can even prove that the heat taken out between 4 and 1 is approximately equal to the heat added between 2 and 3. The trick then is to 'park' the heat Q'_R leaving the cycle between 4 and 1 for a moment. If the cycle has reached point 2 one adds that same heat again as Q_R. In that way one effectively just has two reservoirs at a high temperature T_H and a low temperature T_C with the Carnot efficiency η_{Carnot} of Equation (4.36).

Stirling engines are still in use and are being improved all the time. They operate between reservoirs of heat independent of the origin of the heat and may be used in reverse as a heat pump or refrigerator.

Steam engine

Soon shall thy arm, unconquered steam, afar. Drag the slow barge, or drive the rapid car.

Charles Darwin, 1791, [44], p. 31

As last example of a heat engine we discuss the steam engine in the form of the Rankine cycle. In this case we only give the TS-diagram on the right of Figure 4.13 accompanied by a sketch of the set-up on the left. In the figure one notices a dashed line, which separates the phases. At the extreme left

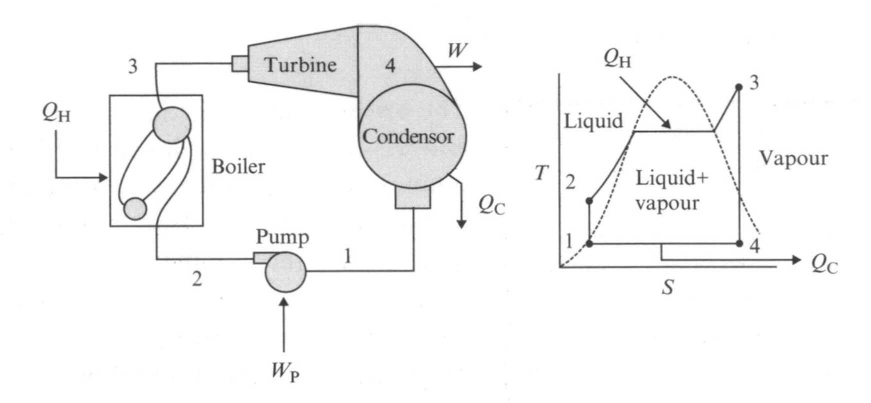

Figure 4.13 Steam engine according to the Rankine cycle. The numbered points on the left and right correspond. Notice the pump, the boiler, the turbine where the useful work is produced and the separate condensor

one has the liquid phase, in the middle two phases together with an interface and on the extreme right, vapour only.

The cycle starts at point 1 with the liquid phase. That is put under pressure adiabatically by pumping, going to point 2. Then heat Q_H is added: the temperature increases as well as the entropy. Notice in Figure 4.13 that one first approaches the mixed phase. When that is reached, the temperature remains constant when all liquid evaporates, while finally the vapour is heated until point 3. Then the high-pressure vapour drives a turbine (adiabatically) against a low pressure and reaches point 4. The vapour condenses in a separate condensor where heat Q_C is taken out of the system and we return to point 1.

Using the first law (4.23) one may write down the thermal efficiency of the steam engine following definition (4.35) as

$$\eta = \frac{W}{Q_H} = \frac{Q_H - Q_C}{Q_H} = 1 - \frac{Q_C}{Q_H} \tag{4.48}$$

As the heat Q_H is added at many different temperatures one does not obtain the Carnot efficiency (4.36). Indeed, technicians make use of steam tables to calculate the maximum efficiency. Steam engines working with pistons were the cradle of the Industrial Revolution, but they are hardly in use any longer. The modern equivalents, the steam turbines, are still widely in use in power stations, as we shall see later.

> Watt's interest in steam engines revived in 1764. Then a model of Newcomen's machine, which had been sent to London for repair, returned to the University of Glasgow. The repaired model still failed to work effectively. Watt worked on it and eventually discovered the principle of the separate condensor.
>
> [44], p. 63

4.2.9 *Internal combustion*

With an *internal combustion engine* one needs fuel inside the device itself. The fuel (gasoline, petrol, diesel oil) is oxidized with air, giving a pulse of heat and high pressure that drives a piston. In this way chemical energy is converted into mechanical work. So there are at least two differences with heat engines: in the internal combustion engine the source of heat is combustion and that combustion takes place inside the device. No external reservoirs of heat are present.

The set-up for a simple engine is sketched at the top of Figure 4.14. A single cylinder contains a piston, which performs a horizontal motion between two extreme positions. At position 1 the piston is on the far right. The inertia of a rotating flywheel pushes it to the left, a fuel/air mixture gets in, but once the piston is half way, that supply is cut off. The mixture is compressed a little more and at position 2 an electric spark ignites the mixture. While the volume remains constant for a short time, the pressure rises as well as the temperature. Then the volume starts expanding from point 3 to point 4, delivering work. At the end, the exhaust gases are drawn out, the pressure falls with constant volume and one arrives in point 1 again.

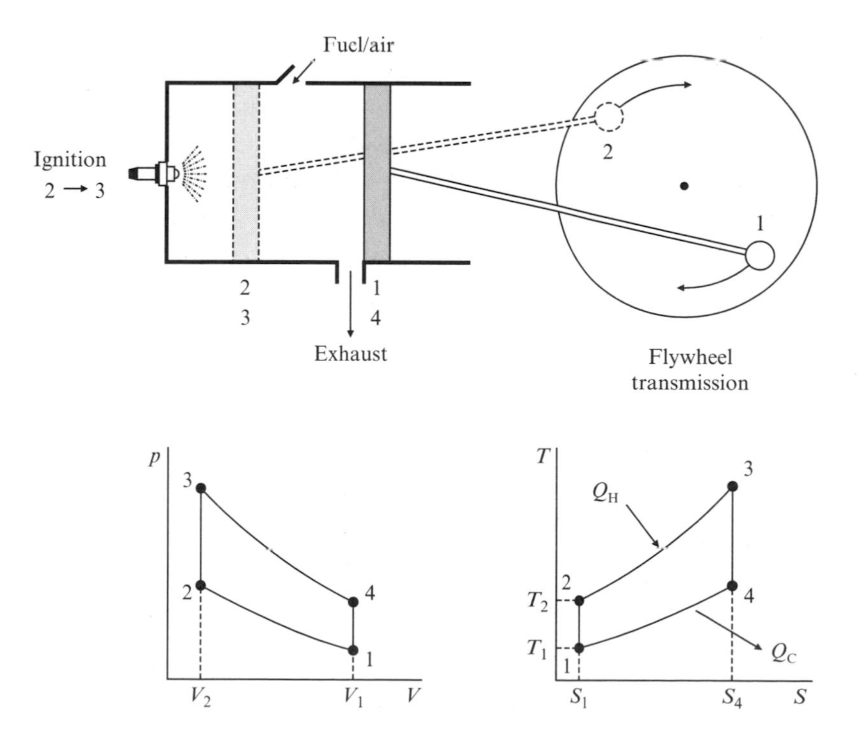

Figure 4.14 Internal combustion engine (Otto cycle). At the top one notices the two extreme positions of the piston. The linear motion is converted into circular by the flywheel. At the bottom the pV-and TS-diagrams are shown, imitating the cycle of a working fluid

The description just given is represented in the lower half of Figure 4.14 in a pV- and TS-diagram. Although the fuel/air mixture entering the cylinder is different from the exhaust leaving, one just proceeds as if one has a single fluid performing a cycle. Recognize the pressure increase between positions 2 and 3 and the decrease between 4 and 1. Also notice that between 1 and 2 or between 3 and 4 no heat is exchanged with the surroundings. The combustion between 2 and 3 is represented as addition of heat and the emission of exhaust gases between 4 and 1 as extraction of heat.

As to the efficiency η, one does not expect the Carnot efficiency (4.36), as even in the approximate TS-diagram there are no two reservoirs with a definite temperature. From the diagram one may derive ([1], p. 91–92)

$$\eta = \frac{W}{Q_H} = 1 - \frac{1}{r^{\kappa - 1}} \tag{4.49}$$

Here $r = V_1/V_2$ is the ratio of the two extreme volumes of the cylinder which is called the *compression ratio*. In Equation (4.49) a high ratio r implies a large efficiency. It is plausible that the compression ratio r enters into the equation in this way, as a high compression would result in a large surface area in the pV-diagram.

Compression ratios cannot be increased arbitrarily as when they are larger than about $r = 10$, the pressure of the fuel air mixture becomes so high that it will ignite by itself, causing knocking of the engine. By using additives in the fuel one can postpone this effect somewhat, but environmental regulations are outlawing the addition of additives like lead.

One may turn the problem of self-ignition into an advantage by using it for ignition instead of a spark. This is done in a *diesel engine*. Compression ratios of 18 to 25 may be reached making this engine, in principle, more efficient than a spark-ignition engine ([1], p. 93–94).

4.2.10 *Exergy and energy*

> Since 1996 China has shut down 60,000 smoky, inefficient industrial boilers across the country and cut coal use by 30 percent to around a billion tonnes. In Bejing some 28,000 boilers and stoves were converted to run on natural gas. In 1999 it had fewer than 100 days of dangerously filthy air.
>
> *New Scientist*, 22 January 2000, 16–17

The discussion of heat engines and internal combustion engines indicates that not all of the energy in a system can be converted into mechanical work. One also presumes from Equation (4.29) that there is a relation with the increase of entropy S. One can make these notions more precise by defining the *exergy* of the system, which will become the *energy* one can *ex*-tract from the system in the form of mechanical work.

Let the system be defined by the values of its internal energy U, volume V and entropy S. The system has an interaction with the atmosphere, which has a pressure p_0 and a temperature T_0. The atmosphere is big and the system is small, therefore the properties of the atmosphere do not change by the interaction. At the 'end of the day' the system reaches equilibrium with the atmosphere, with final values U_f, V_f, S_f for its internal energy, volume and entropy. The *exergy* B of the system then is defined by

$$B = (U - U_f) + p_0(V - V_f) - T_0(S - S_f) \tag{4.50}$$

All variables in Equation (4.50) are uniquely defined properties of the system. So the exergy is a property of the system as well, it is a state function. There may be the slight difficulty that, in practice, the atmospheric properties p_0 and T_0 vary somewhat. But one may take some standard, average values for them.

By carefully looking at the exchange of heat between the system and the atmosphere, one may derive a relation for the mechanical work W that can be done by the system as ([48]; [1], p. 85)

$$W = B - T_0(\Delta S)_{s+a} \tag{4.51}$$

where $(\Delta S)_{s+a}$ is the finite entropy change of the system and the atmosphere together. If all interactions are done reversibly by passing many intermediate states, which are all in equilibrium with the surroundings, then no entropy is generated. It follows that $(\Delta S)_{s+a} = 0$ and the work equals the exergy. This is the reason why the exergy is also called the *available work*. In practice, some or all of the interactions will happen irreversibly, so then the work the system performs is smaller than the exergy.

Example 4.7: Exergy of heat

Calculate the exergy of an amount of heat Q at temperature T_H.

Answer

The heat is contained in a system that is brought into equilibrium with the atmosphere, using all the heat but without changing its volume, as we want a simple calculation. So, $V = V_f$ and the exergy is found from (4.50) as

$$B = (U - U_f) - T_0(S - S_f) \tag{4.52}$$

From the first law (4.23) $-Q = U_f - U$, as Q leaves the system and therefore $(-Q)$ is the added heat. The heat Q is taken from the system at constant temperature T_H, so from (4.26) for a finite change one has

$$T_H(S_f - S) = (-Q) \tag{4.53}$$

Substituting this in Equation (4.52) it follows that

$$B = Q - Q\frac{T_0}{T_H} = Q\left(1 - \frac{T_0}{T_H}\right) \tag{4.54}$$

which again expresses (4.36), stating that the Carnot efficiency is the maximum efficiency for converting heat into work.

Loss of exergy in combustion

> When it comes to renewable energy we have a lot of choices. A variety of chemicals as well as electricity can be made from biomass, depending on how it is processed. For maximum impact of renewable energy, we must match the renewable technology with regional resources and local and national needs.
>
> *Science* **285**, 20 August 1999, 1209

Consider some fuel/air mixture such as in an internal combustion engine. There is an amount of chemical energy available, which produces heat by combustion. From Figure 4.14 one knows that, after this, sudden production of heat expansion starts and work is produced.

The sudden combustion not only produces heat but also entropy [48]. Therefore a fraction L of the exergy is lost. This may be calculated as ([1], p. 86–87)

$$L = \frac{T_0 \ln(T_c/T_0)}{T_c - T_0} = \frac{\ln(T_c/T_0)}{(T_c/T_0) - 1} \tag{4.55}$$

Here T_c is the temperature of the mixture after combustion and T_0 is the atmospheric temperature. When one substitutes values $T_0 = 300\,\mathrm{K}$ and $T_c = 2240\,\mathrm{K}$ ([48], p. 43) one finds a loss of about 30 percent (Exercise 4.11).

So combustion is not the best way to use the exergy of the fuel/air mixture. The reason is that the combustion process is irreversible. The challenge is to find better ways to use the exergy properly.

Second law efficiency

If one wants to judge the performance of some device, the efficiency or COP, defined as (useful output)/(required input) as in Equation (4.35) and (4.37) is not very appropriate. For a heat engine, that value is always lower than 100 percent as the Carnot efficiency (4.36) is the maximum possible efficiency, which from the definition is always smaller than 100 percent. In that case it is the fraction of the Carnot efficiency (4.36) that is a significant indicator of performance.

For a refrigerator or a heat pump it followed from Equation (4.40) that the COP value may be as high as 9. In this case it is the fraction of the theoretical maximum that is interesting. These considerations have resulted in the definition of a *second law efficiency* ε as [48]

$$\varepsilon = \frac{\text{useful output(work or heat)}}{\text{maximum possible useful output for any system with the same input}} \tag{4.56}$$

This value will always be smaller than one or equal to one, as the useful output is never bigger than its maximum. The denominator can be calculated for simple cases. For a heat engine it will be the Carnot efficiency (4.36) and for a refrigerator or heat pump, the maximum COP. For home heating it can be shown that heating by combustion of gas in a furnace yields a small second-law efficiency ([1], p. 83–84), as the detour gas → work → heat pump makes better use of the contained exergy.

If one applies definition (4.56) one needs a value judgement. What does the word 'useful' mean? If one wants to feel comfortable in a castle or a palace, is it then necessary to heat the whole building? Discussing a careful use of exergy is more than a numerical exercise. The tasks we put ourselves to in society should be questioned as well ([48], Chapter 2). In Chapter 6 we come back to this point using the example of comfort at home.

4.2.11 *Useful thermodynamic functions*

In the preceding pages, we have discussed a simple gas as a thermodynamic system. Its state could be defined by three variables p, V and T, of which two are independent. We have introduced a state function as a function, which is determined by the variables of the state of the system. We have noted three examples, the internal energy U, the exergy B and the entropy S. They turned out to be useful to describe physical processes. There are two more functions which it is essential to know: the enthalpy H and the Gibbs free energy G.

Enthalpy H

As discussed before, added heat cannot be found as the difference between the values of a state function in the two states. The reason is that, in general, the added heat depends on the path, just as work does. There is, however, a state function, for which the difference between two states in certain cases equals the added heat. That is the *enthalpy* H of the system, defined as

$$H = U + pV \tag{4.57}$$

All functions on the right, U, p and V, are uniquely defined in any state, so is H. For a small (infinitesimal) change in the system one has

$$\mathrm{d}H = \mathrm{d}U + p\,\mathrm{d}V + V\,\mathrm{d}p \tag{4.58}$$

and for a finite change

$$H_2 - H_1 = U_2 - U_1 + \int_1^2 p\,\mathrm{d}V + \int_1^2 V\,\mathrm{d}p \tag{4.59}$$

In practice, many processes and experiments are *isobaric*, that is at constant pressure, usually the atmospheric pressure. This means that $\mathrm{d}p = 0$ for all infinitesimal steps and the last terms in Equation (4.58) and (4.59) are zero. When all work is expansion work, one finds from Equation (4.19) and the first law (4.23)

$$H_2 - H_1 = U_2 - U_1 + W_{1\to2} = Q_{1\to2} \tag{4.60}$$

So for isobaric processes, the increase in enthalpy is equal to the heat added.

When hydrogen and oxygen react to form water according to the reaction

$$2H_2 + O_2 \rightarrow 2H_2O \tag{4.61}$$

heat is leaving the system. Usually this process is isobaric and the heat leaving is minus the heat added, so one has a negative enthalpy change of the system. In textbooks of (physical) chemistry *reaction enthalpies* are tabulated ([49], Appendix C, Table 2.12). For reaction (4.61) one finds two values* $-285.83\,[\mathrm{kJ\ mol^{-1}}]$, when the result is liquid water and $-241.82\,[\mathrm{kJ\ mol^{-1}}]$, when the result is water vapour. The difference is due to the evaporation enthalpy of water (Exercise 4.12).

Gibbs free energy G

As noted before, from the point of view of thermodynamics, the best way to use chemical energy is to convert it directly into mechanical work, without

*Reaction enthalpies depend on temperature. The data given are for 298 [K] or 25 [°C].

Josiah Willard Gibbs spent most of his working life at Yale and may justly be regarded as the originator of chemical thermodynamics. He reflected for years before publishing his conclusions, and then did so in precisely expressed papers in an obscure journal. He needed interpreters before the power of his work was recognized and it could be applied to industrial processes.

[49], p. 242

the detour of combustion, with the resulting loss of exergy (4.55). We will meet one such example, the fuel cell, in Section 4.3.

Other ways may be found in the future, so it makes sense to define a state function, which is connected with this possibility, the *Gibbs free energy G*, often just called free energy†. The Gibbs free energy is defined by

$$G = H - TS = U + pV - TS \quad (4.62)$$

This obviously is a state function as all its defining quantities are defined uniquely. The meaning of G becomes clear when one considers an infinitesimal change dG

$$\mathrm{d}G = \mathrm{d}U + p\,\mathrm{d}V + V\,\mathrm{d}p - T\mathrm{d}S - S\,\mathrm{d}T \quad (4.63)$$

For a reversible change it follows from (4.26) that $T\mathrm{d}S = (\delta Q)_{\mathrm{rev}}$ and from (4.25) one knows that $(\delta Q)_{\mathrm{rev}} = \mathrm{d}U + p\mathrm{d}V + \delta W_{\mathrm{e}}$, where δW_{e} summarized all non-expansion work done by the system under consideration. Substituting these two relations in Equation (4.63) one finds

$$\mathrm{d}G = V\mathrm{d}p - S\,\mathrm{d}T - \delta W_{\mathrm{e}} \quad (4.64)$$

In practice, many changes in the system will be isobaric (constant p) and isothermal (constant T). So, the non-expansion work done by the system δW_{e} equals the decrease in Gibbs free energy

$$\delta W_{\mathrm{e}} = -\mathrm{d}G \quad (4.65)$$

Of course, for finite changes, it follows from (4.65) that the total non-expansion work W_{e} equals the finite decrease in Gibbs free energy G. This property explains the adjective 'free' in the name Gibbs free energy.

In Equation (4.62) the chemical composition of a system is hidden in the internal energy U. One can make this dependence on composition of U and therefore of G, explicit. Suppose that one has n_1 moles of chemical 1, n_2 moles of chemical 2, and in general n_i moles of chemical i. The Gibbs free energy G then may expressed as a function of two independent variables, say p and T, and of the chemical composition, so

†There is possible confusion here, as the term free energy is also used for the state function $F = U - TS$, which is not discussed in the present text.

$$G = G(p, T, n_1, n_2, \ldots n_i \ldots) \tag{4.66}$$

Therefore, an infinitesimal change in G can be written in terms of dp, dT, dn_1, dn_2 etc., or in general

$$\mathrm{d}G = V\mathrm{d}p - S\mathrm{d}T + \sum_i \mu_i \mathrm{d}n_i \tag{4.67}$$

This means that when for constant p, T one adds dn_i moles of substance i the Gibbs free energy increases by $\mu_i\,\mathrm{d}n_i$. More generally, if the chemical composition changes in a known way, one may calculate the change in Gibbs free energy and the non-expansion work δW_e from Equation (4.65). Of course, one needs to know the *chemical potentials* μ_i, which depend on p and T, and these may be found or deduced from tables.

Note that the exergy B, expresses the available mechanical work when one brings the system into equilibrium with the atmosphere. Although it is a state function, it depends on the atmospheric properties. The Gibbs free energy G gives the 'available' non-expansion work for constant p and T, which may or may not have the atmospheric values.

In the same way as the entropy S for a system left alone tends to a maximum, as suggested by example 4.5 and Clausius' inequality (4.29), the Gibbs free energy G tends to a minimum. For chemical equilibrium, this gives the equilibrium constant as shown in the following example.

Example 4.8: Chemical equilibrium

The chemical potential μ_i of substance i depends on temperature T and concentration $\{i\}$ (expressed in $[M] = [M \times 10^{-3}\,\mathrm{kg\,L^{-1}}]$) in the following way

$$\mu_i = \mu_i^0 + RT\ln\{i\} \tag{4.68}$$

Here, R is the universal gas constant and μ_i^0 is the value of the chemical potential under standard conditions*. Take Equation (4.68) as given and derive an expression for the equilibrium constant K for the reaction

$$n_A A + n_B B \leftrightarrow n_C C + n_D D \tag{4.69}$$

Answer

As in equilibrium the Gibbs free energy G has its minimal value, a small change of reaction (4.69) to the right by an amount of dξ should not change

*Standard conditions are usually taken as $T = 25$ [°C] = 298.15 [K] and pressure 10^5 Pa ([49], p. 30). The latter is about the atmospheric pressure 1 [atm] = 1.01325×10^5 Pa.

G. Therefore, we calculate the change $\mathrm{d}G$ with the following conditions (A more general derivation may be found in [49], p. 276)

$$\mathrm{d}n_\mathrm{A} = -n_\mathrm{A}\mathrm{d}\xi, \mathrm{d}n_\mathrm{B} = -n_\mathrm{B}\mathrm{d}\xi, \mathrm{d}n_\mathrm{C} = n_\mathrm{C}\mathrm{d}\xi, \mathrm{d}n_\mathrm{D} = n_\mathrm{D}\mathrm{d}\xi \tag{4.70}$$

From (4.70) and (4.68) it follows that

$$\begin{aligned}\mathrm{d}G &= \begin{pmatrix} -n_\mathrm{A}\mu_\mathrm{A}^0 - RTn_\mathrm{A}\ln\{A\} - n_\mathrm{B}\mu_\mathrm{B}^0 - RTn_\mathrm{B}\ln\{B\} \\ +n_\mathrm{C}\mu_\mathrm{C}^0 + RTn_\mathrm{C}\ln\{C\} + n_\mathrm{D}\mu_\mathrm{D}^0 - RTn_\mathrm{D}\ln\{D\} \end{pmatrix}\mathrm{d}\xi \\ &= \left((-n_\mathrm{A}\mu_\mathrm{A}^0 - n_\mathrm{B}\mu_\mathrm{B}^0 + n_\mathrm{C}\mu_\mathrm{C}^0 + n_\mathrm{D}\mu_\mathrm{D}^0) + RT\ln\frac{\{C\}^{n_\mathrm{C}}\{D\}^{n_\mathrm{D}}}{\{A\}^{n_\mathrm{A}}\{B\}^{n_\mathrm{B}}}\right)\mathrm{d}\xi\end{aligned} \tag{4.71}$$

As $\mathrm{d}G = 0$ for small $\mathrm{d}\xi$, the expression between brackets should be zero. For constant temperature, all variables are constant, except the concentrations. Consequently the expression with the concentrations behind the log should be constant as well, from which it follows that

$$\frac{\{C\}^{n_\mathrm{C}}\{D\}^{n_\mathrm{D}}}{\{A\}^{n_\mathrm{A}}\{B\}^{n_\mathrm{B}}} = K(T) \tag{4.72}$$

which is the usual relationship between concentrations for chemical equilibrium. Note that the concentrations themselves may change; for example, when adding some chemical, but that the expression (4.72) as such should remain constant.

4.3 *CONVERTING HEAT INTO POWER: PROBLEMS AND 'SOLUTIONS'*

In this section we list the major sources of pollution and then discuss how to deal with them in the powering of vehicles and in electricity production. Table 4.3 shows the sources of air pollution in the European Union as an example of an industrialized region. Although the figures are dated (1990) and apply to the EU of the time, the general picture will hold for all industrialized regions. Most of the pollution clearly originates from the combustion of fossil fuels in transport or in the production of electric power.

4.3.1 *Use of fossil fuels*

The combustion process of fossil fuels is summarized in Figure 4.15. Pollution may arise from the mining and processing of the fuels, from the insufficient

Table 4.3 Air pollutants, their origins and dangers. From [50], p. 3[†]

Pollutant	Principal Origin (1990)	Dangers
CO	road transport (65%)	– affects tissue oxygenation
NO_x	road transport (51%)	– respiratory effects – inhibits plant growth – acid rain
SO_2	burning of fossil fuels (66%)	– exacerbates respiratory pathology – acid rain
Fine particles and heavy metals	industrial emissions and exhaust gases	– irritation and damage of respiratory functions – occasional carcinogenic
Volatile organic compounds (VOC)	– road transport (36%) – solvent use (25%)	– olfactory complaints – mutagenic and carcinogenic (benzene)

[†]The EU source also mentions ozone and dioxins.

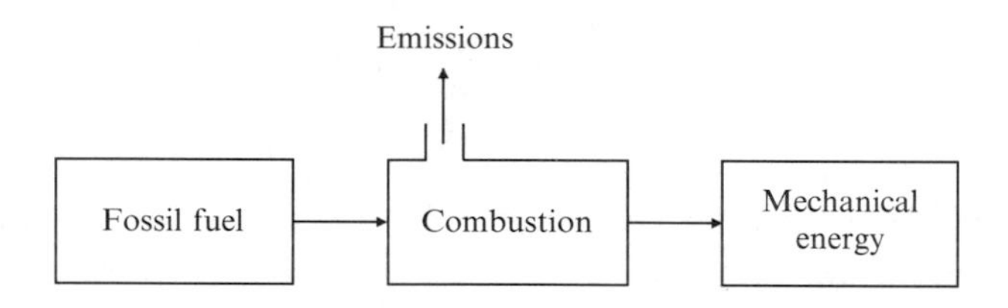

Figure 4.15 Producing mechanical power from fossil fuel, pollutants are generated

control of the emissions originating from the combustion and from the leftovers of the process, for example coal waste. All steps contribute to the final pollution and for most steps the pollution can be reduced by technical measures. One should not conclude that technology is the answer to pollution and that the use of fossil fuels has no limitations. It is not only necessary to reduce pollution, but also to reduce the consumption of raw materials. Even natural gas, which is a rather clean fossil fuel will eventually become scarce. Consequently, it pays to look for a reduction in fuel use.

> In Delhi, the levels of particulates are extremely high in the city's ambient air. Diesel is the biggest source of extremely tiny particles which are, from a public health point of view, the most dangerous. Moreover, diesel produces sulphate particles which are the most dangerous of all particles.
>
> CSE 4 October 1999

Returning to Table 4.3 one will notice that much of the air pollution originates from road transport with internal combustion engines. Also burning of fossil fuels for electricity production appears in the table (in particular for SO_2) and, finally, emissions from all kinds of industrial processes.

Before discussing the physical background of the polluting processes and ways to reduce them, it should be realized that an internal combustion engine is a remarkable machine. It has to operate under quite different conditions: it will have a cold start, then heat up, produce power during acceleration but much less during idling and deceleration. It has to perform its cycle of ignition, flame, energy conversion and emission of exhaust gases hundreds of times each minute. Finally, everything happens in a few compact cylinders with a size of a few hundred [cm^3] each, which form the heart of the engine. Under all these conditions the engine has to obey official emission standards. It is a technological *tour de force* that this indeed is possible. In fact, it is not the single car that poses an environmental problem, but their vast numbers.

Controlling the emissions of new cars to be built is not the only issue. During use, engines will get dirtier and have to be maintained. So, regular checks on the cars should be compulsory and have to be enforced. Furthermore, particularly in low income countries there will be many old cars on the roads. Adjusting the input fuels for old cars may help.

> A whole series of Directives has been aimed at motor vehicle emissions, as a result of which our cars, trucks and buses are already significantly cleaner than they used to be. Despite this progress, Europe's roads remain a steadily worsening source of pollution as traffic volumes continue to rise.
>
> EU, Air Quality-Fact Sheet, p. 5

Industrial processes and electricity production usually take place in large units where fossil fuel combustion is kept at a constant rate. Consequently emission control is easier to achieve and to regulate than in many scattered units, albeit at a cost. In homes and buildings, fossil fuels are mainly used for heating. Below we will consider some examples of pollution and control measures, while we devote a special subsection to power stations. We now follow the pollutants of table 4.3, although it is impossible to give a comprehensive account. See [51].

CO and CO_2

Fossil fuels mainly consist of carbon compounds, which on combustion with oxygen produce CO and CO_2. After a short period of time an equilibrium will be established

$$CO_2 \leftrightarrow CO + \frac{1}{2}O_2 - \text{heat} \tag{4.73}$$

In tables of molar enthalpies one will find that for the reaction from the left to the right the enthalpy increases (Exercise 4.13). So heat has to be taken from the environment to convert CO_2 into CO. Hence the words (– heat) were added in Equation (4.73). Such a reaction is called *endothermic* and shifts to the right when the temperature is raised. This is an example of *Le Châtelier's principle*, which is known from physical chemistry: a system in equilibrium that is subjected to a stress will react in a way that counteracts the stress.

The CO_2, which is produced by combustion, increases the greenhouse effect, which is a long-term problem by itself. The CO, which is formed, is immediately dangerous as it poisons the blood* and is lethal in high concentrations. One cannot smell it, so the human senses do not warn against this hazard. Indeed, fatal accidents frequently occur when somebody is running a car engine in a closed garage, or is burning a stove with a malfunctioning chimney.

How can one reduce the amount of CO in the exhaust gases from combustion? From Le Châtelier's principle it follows qualitatively that addition of oxygen will shift the equilibrium to the left. Therefore, one may reduce the amount of CO produced by increasing the air/fuel ratio in the combustion process. This is being done in the internal combustion engine, but a smaller percentage of fuel results in a reduction of power and performance of the engine.

NO_x

> Chinese vehicles are among the most polluting in the world. Pollution has a price in failed crops, ruined land, illnesses and lost work time. The cost has risen as high as 14 percent of China's gross national product in recent years.
>
> *Bulletin Atomic Scientists*, Sept/Oct 1999, 58–66

The symbol NO_x comprises the two nitrogen oxides NO and NO_2, which are formed during any combustion process as we shall see below. The oxides themselves are not very dangerous: NO does not pose a health hazard at common concentrations and NO_2 only during long exposures [52]. Nevertheless, their concentrations are limited by regulations, particularly as they participate in the photochemical reactions that produce *photochemical smog* which affects the respiratory functions (see Chapter 9).

NO_x may be formed in two ways [52]. The first way starts with the nitrogen compounds, present in the fossil fuel itself. As these fuels originate from plants

*The CO strongly binds to the haemoglobin (Hb) in the blood preventing the binding of oxygen to Hb. Consequently the Hb can no longer transport oxygen, which results in suffocation. When a victim is discovered in time, fresh air and extra oxygen may save him or her ([51], pp 119–120).

containing nitrogen, the fuel itself may contain remaining nitrogen compounds. On combustion these compounds are oxidized to NO_x. Coal, oil and gas may contain up to 1.5 percent of nitrogen ([53], Appendices G and H). So, the choice of fossil fuel will influence the formation of NO_x along this route.

A second method of formation of NO_x is *thermal* by endothermic reactions: when N_2 and O_2 are brought together at high temperatures (1000 [K] or so) they combine to nitrogen oxides, almost all of them in the form of NO. As virtually all combustion processes use air (and not pure oxygen) their high temperatures will always result in NO_x formation.

Thermal NO_x formation increases with temperature. So a way to reduce its formation is to avoid hot spots in the flames and let the fuel burn quietly. Obviously, in an internal combustion engine with many short flames in a minute, this will be difficult. So, having a constant flame, which produces heat as input for a heat engine may reduce NO_x emissions.

SO_2

The main source of SO_2 pollution is the use of fossil fuels in combustion. Plants not only contain the nitrogen mentioned above, but also sulphur. This remains in the fuel to the order of a few % by weight. There are big differences between coals from different origins, while gas is usually free from it ([53], Appendices E, G and H). Petrol is fairly free from sulphur ($< 0.05\%$) and diesel fuel as well ($< 0.25\%$)[†] so coal fired power stations are the main source of sulphur oxides in the air.

The SO_2 gas is very soluble in water and therefore easily absorbed in the moist tissues of mouth, nose and lungs. In high concentrations or with long exposure times, the resulting acid can do considerable damage to man [52].

Less acute but as bad is the transportation of SO_2 by the winds in the form of gas, or absorbed in water particles, which precipitate far from their origin, causing *acid rain*.

Severe government regulations have reduced the emission of SO_2 by power stations and industries in industrial countries. The solutions combine measures in all parts of Figure 4.15: use fuel with low sulphur content, eliminate sulphur out of the coal fuel, apply combustion methods in the presence of chalk

> Winter smog is created when pollutants are trapped by the mass of cold air above our cities, causing them to build up. The air quality guidelines of the World Health Organization, WHO, are broken at least once a year in three quarters of Europe's big cities. The primary culprit is the steady growth of car traffic.
>
> EU, Air Quality-Fact Sheet, p. 2

[†]Ref. [51], p. 1045

(CaO) giving $CaSO_4$, and finally clean the exhaust gases by bringing them in contact with water (wet scrubbers).

Particles

In the field of air pollution the terms particles, particulate matter and *aerosols* are used interchangeably. These words comprise all compositions, masses, sizes and shapes. The chemical composition and the mass of a particle are uniquely defined, size and shape are not. So the latter two are taken together in the definition of an *equivalent diameter, d*, which is the diameter of a water sphere which under the influence of gravity will settle to the ground with the same speed as the particle under consideration ([51], p. 116).

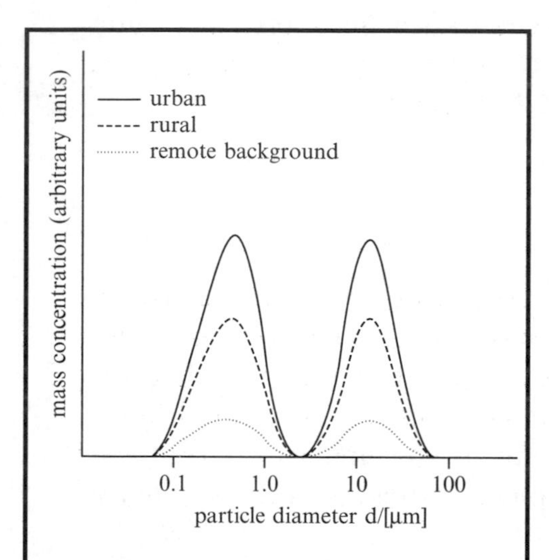

Figure 4.16 Mass distribution of polluting particles in the atmosphere. Note the horizontal logarithmic scale. Reproduced by permission of John Wiley from [51], p. 32

The masses, which occur at a given (equivalent) diameter d can be measured and displayed. This is done in Figure 4.16 in a qualitative way with a logarithmic scale for the d-axis. Notice the two peaks, one at a diameter of somewhat less than 1 [μm] and one at a diameter of around 10 [μm]. The smaller particles pose the highest health risk, as they penetrate deepest in the lungs. When they are chemically or biologically active, for example when they contain sulphuric acid, they may do much local damage ([51], p. 116). As the smallest particles are most dangerous, regulations define a quantity PM_{10}. This is the particulate matter [μg m^{-3}] with diameter $d < 10$ [μm], ([51], p. 32)

> Death rates in the 90 largest US cities rise on average 0.5 percent with each 10 [$\mu g/m^3$] increase in particles less than 10 [μm] in diameter, known as PM_{10}. The tinier the particle, the more likely it is to lodge in the lungs. The EPA, the Environmental Protection Agency, has to determine what sources of particles it should regulate and needs to know which components of PM – sulfates, metals, acids, or ultrafine particles, inflict harm.
>
> *Science* **289**, 7 July 2000, 22, 23

In Figure 4.16 one notices that particles are even present in the 'remote' natural background. That probably implies that, during evolution, the human body will have developed some defence mechanisms against the particles and can remove or neutralize them. Again it is the high concentrations which pose the health hazard and those parts of the population which are physically weakest are most vulnerable.

At first glance, the burning of coal and wood would seem the 'bad guy' in the production of small particles. This indeed is true for open fires or malfunctioning stoves. Industrial installations such as modern power stations, however, use all kinds of technologies to remove particles from their exhaust or smoke gases. An old method lets the gases pass through a chamber where the particles settle down by gravity. More modern is the use of a cyclonic device by which the particles are deposited on the wall and subsequently removed. Finally, we mention electrostatic precipitators, where the particles are charged and removed ([15], p. 88).

In countries which can afford clean technologies and have the power to enforce them, a remaining source of particles is urban traffic. Incomplete combustion of diesel will produce carbon particles. The wear and tear of the engine will produce small particles, consisting of carbon and metals, which may be emitted into the ambient air. Lubricating oil may add to this problem. In these respects, modern diesel still performs worse than modern petrol driven cars.

Volatile organic compounds

The class of pollutants called volatile organic compounds, abbreviated as VOC, comprises the many organic compounds which easily evaporate from the liquid phase. Many of them have the chemical form of hydrocarbons, written as C_xH_y with various combinations (x, y). Sometimes they are just written as CH. Their volatility is apparent from the persistent smell near gas filling stations due to spent petrol (often by careless drivers). Besides the ill effects shown in Table 4.3, the VOC's combine with NO_x under the influence of sunlight to form ozone (see Chapter 9) and smog. This is the main reason to regulate the emissions of VOC's.

The measures comprise all stages of the combustion process, shown in Figure 4.15. One may use fuels which are less volatile themselves or which produce fewer volatile compounds. Alternatively, one may use clever designs of the combustion process or clean the exhaust gases. One of several methods of cleaning is catalytic oxidation of the VOC's to CO_2 and water ([51], p. 738). This method is extensively used in cars, so we describe it in a separate subsection.

> In Delhi, India, benzene levels are extremely high, a problem that comes from petrol vehicles, especially two-stroke vehicles. Though catalytic converters reduce benzene substantially, all Indian cars and scooters are unlikely to have these devices fitted for a very long time. The only option, in the interim, is to reduce the benzene content of petrol.
>
> CSE, 4 October 1999

The three-way catalytic converter (TWC)

In the discussion of CO (Equation (4.73)) and of NO_x we mentioned that they are formed in endothermic reactions. This would imply that at temperatures lower than the high ones occurring during combustion, reaction (4.73) would run to the left and similarly NO_x would return to N_2 and O_2. Indeed, at 600[K] the equilibrium (4.73) would be almost completely shifted to the left.

Obviously, one does not have equilibrium situations. The exhaust gases are quickly removed by a tailpipe and there is no time for a new equilibrium to establish at the lower temperatures of the outlet. Therefore, one puts a *catalyst* in the tailpipe between the combustion chambers and the outlet, which would speed up the return reactions with extra air added. The catalyst usually consists of precious metals (Pt, Pa, Rh) with a large surface area to optimize the contact of the exhaust gases with the catalyst. In principle, the catalyst does not change itself by the reaction process, but it may become dirty and has to be maintained.

In cars, one uses a TWC, *a three way converter*, as it fulfils three tasks at the same time: CO and C_xH_y are oxidized to CO_2 and H_2O, while the nitrogen oxides NO_x are reduced to N_2 and O_2 ([51], pp 223, 224, 718). Of course, electronic devices regulate the flow of the exhaust gases, the air/fuel ratio at the inlet and the additional air into the converter.

Technology has done a lot to reduce the emissions of individual cars. It must be emphasized again that good maintenance and regular check-ups are essential to keep the emissions low during the lifetime of the vehicle. This, of course, has to be enforced as not everybody will do it voluntarily ([51], p. 226).

Automobile fuels

The most common fossil fuel for cars is still petrol and for trucks it is still diesel fuel. Other fuels are increasingly used, however. We will discuss them briefly, leaving electric traction to the section on storage, below.

Alcohol fuels are in use and being improved, particularly methanol and ethanol. Chemically they may be understood as methane or ethane in which one hydrogen atom is replaced by an –OH bond, as shown in Figure 4.17. Ethanol is the alcohol, contained in beverages like beer, wine and spirits.

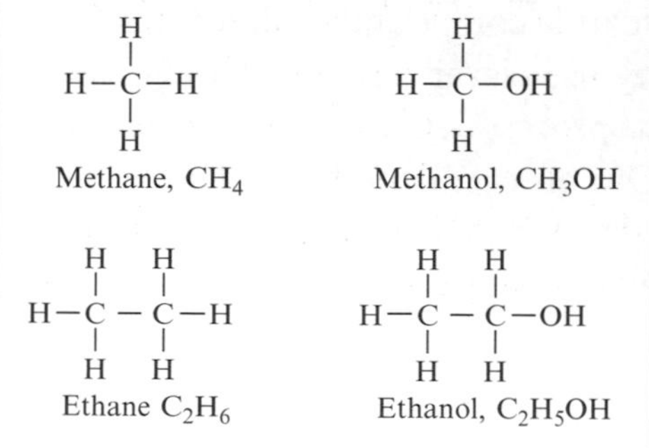

```
     H                    H
     |                    |
   H-C-H                H-C-OH
     |                    |
     H                    H
 Methane, CH4       Methanol, CH3OH

   H   H               H   H
   |   |               |   |
 H-C - C-H           H-C - C-OH
   |   |               |   |
   H   H               H   H
 Ethane C2H6        Ethanol, C2H5OH
```

Figure 4.17 Structure of methane, ethane, methanol and ethanol

Methanol does not contain sulphur, therefore SO_2 emissions are absent ([51], 1045). There are small emissions of particles, not from the fuel, but from the lubricating oils. Its flame temperature is rather low, resulting in small NO_x emissions. Emissions of CO cannot be avoided, but have to be limited by a catalytic converter. The principle

Asthma is a chronic inflammation of the airways in the lungs, marked by attacks of wheezing and shortness of breath. In 1980 about 3.1 percent of the US population suffered from it; in 1994 it had risen to 5.4%, and more so in poor urban areas with lots of traffic, bus stations etc.

Research indicates that vehicle exhaust can indeed acerbate asthma's symptoms, even if it is not the underlying cause of the disease.

Scientific American, November 1999, 13–14

pollution problem is the formation of the poisonous formaldehyde H_2CO, in which the –OH bond in methanol is replaced by a double =O bond. As source of methanol coal or natural gas are used, so it does deplete fossil fuel reserves.

Ethanol has similar properties as methanol, e.g. no sulphur content. It has the additional advantage that it can easily be produced from biomass, so the resulting CO_2 is again bound in biomass for the next round of use in transportation. In this way there is no net increase in the concentration of this greenhouse gas in the atmosphere and the use of ethanol does not deplete fossil fuel reserves except what is needed for its industrial production. At present it is mainly used in Brazil (96 percent ethanol, 4 percent water, [51], p. 1435) or as admixture to petrol or gasoline with the name gasohol = gasoline × alcohol.

Natural gas consists mainly (> 90%) of methane gas. It may be liquefied by compression in the form of LNG. Similarly, liquefied petroleum gas LPG is used as fuel for cars. Both LPG and LNG have no particle emissions, except from the lubricating oils and they have moderate emissions of hydrocarbons. The NO_x emissions depend on the flame temperature, which may be somewhat lower than for petrol ([51], p. 164).

4.3.2 *Electricity production*

Electricity is a widely used form of power and very convenient. You just have to plug an appliance into its socket and it plays without any visible environmental consequences ('*plug and play*'). Its use is clean, but problems may occur during its production in power stations and during transportation from station to user. Table 4.4 shows the relative importance of electricity by fossil fuels compared with other sources of electricity and the total production of fossil fuels. The subscript 'e' refers to electric Joules and the subscript 'th' to thermal Joules (heat). Their relation is discussed below. The data are from 1998 ([54], [55], [56]) and increase in subsequent years, but the relative contribution of the different sources only changes slowly with time.

In the table, one notices the total production of fossil fuels expressed in their heat value [J_{th}]. It follows that $92/344 = 27\%$ of all fossil fuels are used for electricity production, which is a good reason to discuss it separately.

Table 4.4 Electricity generation from several sources in 1998 (from [54], [55], [56])†

	Electricity 10^{12} [kWh]	Electricity 10^{18} [J_e]	Electricity 10^{18} [J_{th}]
Thermal (fossil)	8.54	30.7	92.2
Hydropower	2.56	9.2	28.1*
Nuclear power	2.32	8.4	25.8*
Geothermal + biomass	0.18	0.6	1.9
Wind	0.022	0.08	0.24
Photovoltaic	0.001	0.004	0.01
Solar thermal	0.001	0.004	0.01
Total electric	13.62	49.0	148*
Total use of fossil fuel, including electricity production			344×10^{18}[$J_{th}yr^{-1}$]

*The wind data are based on [55], assuming 2500 productive wind hours per year on maximal capacity. The data for photovoltaic and solar thermal electricity are estimated, assuming 1000 productive hours at peak capacity [56]. The other data are from [54]; for the data with an *we used the thermal Joules of Ref. [54], for the others 1 [J_e] = 3[J_{th}].

In order to compare the electricity production by fossil fuels and by hydropower stations, one could look at the electric Joules produced (the third column of Table 4.4) or at the number of [kWh] produced (the second column). Fossil fuels clearly are the major source of electric power on a world scale, with considerable contributions from hydro and nuclear power.

From an environmental point of view it is relevant how many fossil Joules are saved by producing one electric Joule [J_e] by hydropower or the other renewable sources of energy that we will discuss in Chapter 5. We use the fact that on a world scale the thermal efficiency (4.35) of the fossil fuel power stations is about 33 percent. So one electric [J_e] corresponds with 3 thermal [J_{th}]. This number is used to produce the last column of Table 4.4, except where indicated by an asterisk (*). One immediately sees that one has to go a long way before all or most of the 344×10^{18}[J_{th}] of fossil fuel is replaced by renewables. This, of course, is no reason to become desperate, but just to work harder and to mitigate as much as possible the adverse effects of the fossil fuel use.

> As China's economy develops, its demand for energy will be met by the country's abundant coal reserves. Coal fired power plants and industrial boilers will spew out large amounts of SO_2, one of the main sources of acid rain. Scrubbers, energy efficiency and switching to less polluting fuels does not help much, unless 2 percent of China's BNP is spent on these measures.
>
> *New Scientist*, 13 February 1999, 25

In Chapter 5 we shall see that much of the renewable energy is produced in the form of electricity. Therefore we will now discuss two aspects which are of general interest: the concept of grid load and the way to transport electric power. This leads to the question of storage of (electric) energy and the connected problem of the electric car. Then we return to a few points which are specific for fossil fuel production of electricity: the emissions causing acid rain, the CO_2 emissions and the solution of CO_2 sequestration. Finally, we mention the way to get more out of the exergy contained in the fuel: combined production of heat and power.

Grid load

One expects from an electricity supplier, a constant voltage all the time at the lowest possible cost. For the grid manager this is a problem, as illustrated in Figure 4.18, which shows the grid load of electricity for a midsummer and a midwinter day. The grid load measures the electric power transported by the grid. It is somewhat lower than the total production as small industrial installations, which only service the own plant, do not charge the grid. For a typical midsummer day one observes a peak around noon and a minimum during the night. In midwinter there are fluctuations as well, but they are less extreme. This particular graph is inspired by the data for Germany in 1990/1991.* For other countries they are different, depending on climate, on the relative contributions of electricity for heating and cooling compared with other ways, etc. But in all cases there are fluctuations; there are even fluctuations on a timescale of minutes, not shown on the graph.

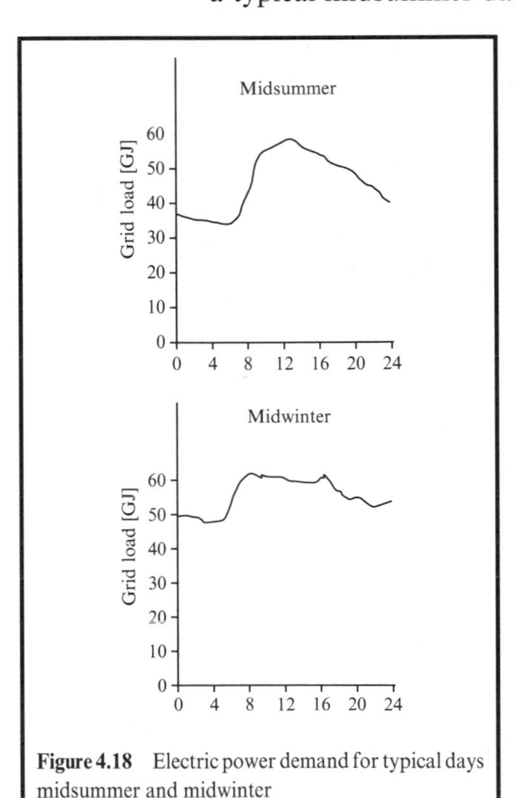

Figure 4.18 Electric power demand for typical days midsummer and midwinter

Utilities tackle the problem of supplying the electricity request by defining a base load, which is constant during summer and winter. In Figure 4.18 this would be 35 [GJ]. Power stations, which are expensive to build, but are cheap in fuel are allowed to run constantly, in order to distribute the high building costs over a great many [J_e] produced. One often gives nuclear power stations this role. An additional reason is that they are not easily switched on and off, so that it pays to let them run permanently. Indeed, on a global scale they operate for 75 percent of the time, the rest being used for maintenance [57]. Also big coal-fired power plants

*19 June 1991 (left) and 19 December 1990 (right). See [1], p. 96.

with their cheap fuel but costly facilities to clean the exhaust gases are used for base loads [53].

At the other end of the scale one has power stations, which are easily switched on and off and may use costly fuels as they only operate for short times. Gas turbines are an example of this type of plant.

In between one will have the intermediate-load power systems. Combined cycle systems, to be discussed later, are used for that service.

It follows that power stations based on renewables like sun or wind, which may change quickly in time pose an extra, albeit soluble, problem for the grid manager. This extra power has to be adapted in the total system – and will result in switching on and off other plants. This of course needs a reliable weather prediction to make it workable.

Transmission of electricity (power lines)

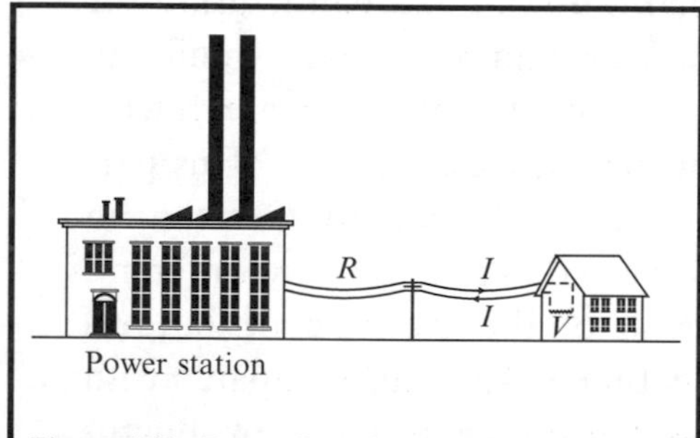

Figure 4.19 Power lines from a plant to an end user. In practice the high voltage V is transformed down to 220 V or lower before entering the private homes

From the power stations, electricity has to be transported to the individual user. This is done by high voltage power lines, which cross the countryside. The high voltage of $V = 380\,000$ [V], or even more, is meant to reduce the losses during transmission in the lines. This can be understood by considering a simple power line with direct current I, as sketched in Figure 4.19. Power lines with resistance R run to a user who takes off the transmitted power

$$P_{\mathrm{T}} = IV \tag{4.74}$$

The losses P_{L} in the lines are mainly due to heat production in the resistance R of the lines due to

$$P_{\mathrm{L}} = I^2 R \tag{4.75}$$

Note that the resistance R applies to the transmission line and not to the load of the end user. The relative transmission losses are the ratio

$$\frac{P_{\mathrm{L}}}{P_{\mathrm{T}}} = \frac{I^2 R}{IV} = \frac{IR}{V} = IV\frac{R}{V^2} = P_{\mathrm{T}}\frac{R}{V^2} \tag{4.76}$$

The power that is to be transmitted is determined by the end user. The utility therefore should make the ratio R/V^2 as small as possible, which in this case implies high voltage V.

In practice an alternating current is used, as it is easy to transform. In the countryside one will observe sets of three heavy wires with often a fourth, thinner ground wire. In the heavy wires the peaks of the current are reached at different times; this is called a three-phase current. If the load between each of the wires and the central ground wire at the end are equal, the return current in the ground wire is zero. The ground wires may be made thin or even omitted, which reduces the costs.

Health effects People worry about the influence of the transmission lines on their health living close to the lines. The alternating currents in the wires may induce currents in the human body. They indeed do, but in all but exceptional cases they are much smaller than the currents and voltages already present in the human body to keep the brain and the heart going.

This fact may not be conclusive, though. Many studies have therefore been made on the influence of electromagnetic radiation on health. Not only for high voltage lines, but also for magnetrons and mobile phones. In the latter two cases the heat production in human tissue may do some damage. During normal operation of these devices no adverse effects have been shown conclusively, so far.

Energy storage

From the discussion of Figure 4.18 it may be clear that it makes sense to find a way to store electric energy. For then one could produce the power during the night, and use it during the day, thus diminishing fluctuations in the grid and reducing the need for expensive 'stand-by' power stations. Another reason to study storage is to make the electric car a feasible option. The pollution of the car is then shifted to the power plant, where it may be easier to control.

An idea of the possibilities of energy storage may be obtained by studying Table 4.5. There one finds the energy storage of several options [kJ kg^{-1}]. Look at the table, starting from the bottom. In a capacitor, positive and negative charges are separated. The potential energy [kJ kg^{-1}] is very low. In batteries electrical energy is converted into chemical energy, which give somewhat higher energy storage possibilities. 'Falling water' refers to the conversion of the potential energy of 1 [kg] of water at a height of 100 [m]. We will look more closely at this possibility when discussing hydropower. Petrol (or gasoline), methane and hydrogen have an increasing storage capability in this order, but one should realize that 1 [kg] of gases like methane or hydrogen has a higher volume and is more difficult to handle than a fluid like petrol.

Table 4.5 Energy storage of several materials and constructions/[kJ kg^{-1}]*. Reproduced by permission of McGraw-Hill from [53], Table 9.1, p. 482

Hydrogen	12×10^4
Methane	5.0×10^4
Petrol (gasoline)	4.4×10^4
Falling water over 100 [m]	980
AgO-Zn battery	437
Pb-acid battery	119
Capacitor	0.016

*The energy content of hydrogen, methane and gasoline is realized by combustion with oxygen. The data given refer to the case when the latent heat of the resulting water vapour is not used productively, but disappears into the air.

To compare storage options, one may look at the electric energy produced by a power station of 500 [MW_e] during 6 hours. That is 500×10^6 [$J_e s^{-1}$]$\times$ 6×3600 [s] $= 1.08 \times 10^{13}$[J_e]. This would need roughly 10^5 [kg] of hydrogen to store, assuming that one chemical Joule of hydrogen could be converted into one electrical Joule. This method of energy storage is not impossible, but the reverse reaction is still expensive, as we will see below.

The obvious method then is to pump up 10^7 [kg] of water (or 10^4 [m^3]) over 100 [m]. The hydropower technology to get the energy back, has been in long use and is cheap. In fact, this method is used during the night in the coupled grids of Europe.

Let us now look at the table from the point of view of an automobile. We notice that it is not amazing that liquid petrol is used as an energy source to drive a conventional car. It has a rather high content of energy and is easily used in the internal combustion engine.

The most obvious way to power electric cars is to use hydrogen with its high-energy content (top row of Table 4.5). One may produce the hydrogen from water by the reaction

$$2H_2O \rightarrow 2H_2 + O_2 \tag{4.77}$$

which needs an input of electricity. At a later stage the hydrogen could be combined with oxygen from the air to run the reaction in the reverse.

The electricity needed for reaction (4.77) may be produced in power stations of any kind, even in arrays of photovoltaic cells, which need not be coupled to the grid. One could put the photocells in a sunny desert and transport the hydrogen to places where it is used.

The technology for a controlled reverse reaction is the *fuel cell*, to be discussed below. For the required hydrogen, its source is irrelevant.

Iceland has got a lot of hydropower to produce electricity. It is now conducting a countrywide experiment to use hydrogen to power its buses, next its fishing fleet, then its trucks and cars. The chief problems holding back the technology, are costs, the difficulty of storing liquid hydrogen safely and packing enough on board to give the vehicle adequate range and power.

New Scientist, 1 May 1999, 20–21

So, there are developments where the hydrogen is produced in the car itself from methanol in a little chemical unit. Or it could be produced from nafta, a liquid similar to petrol, also in the car itself [58]. The drawback of producing hydrogen in the car is that the emissions are not negligible. In both cases, methanol or nafta, the greenhouse gas CO_2 is produced, which is difficult to catch in a car exhaust. Besides, one has to consider pollutants contained in the methanol or the nafta fuels.

Fuel cell

The basic reaction of the hydrogen fuel cell is the reverse of (4.77) that is

$$2H_2 + O_2 \rightarrow 2H_2O \tag{4.78}$$

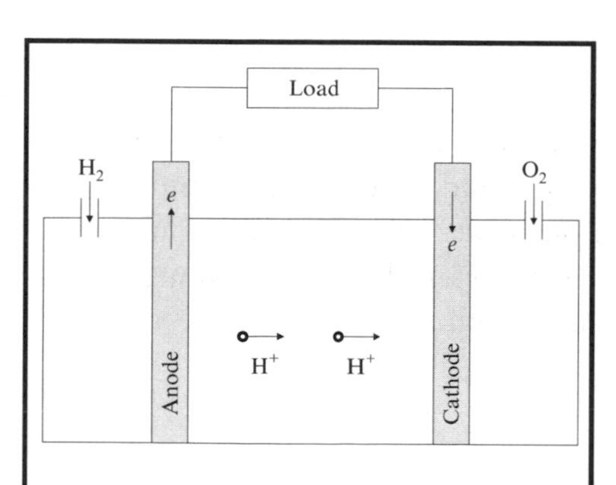

Figure 4.20 Fuel cell. Hydrogen enters at the anode and produces electrons and ions. They combine with oxygen at the cathode to form water

One uses the two gases hydrogen and oxygen as input at different locations in the cell, sketched in Figure 4.20. The hydrogen enters an *electrolyte*, which is a liquid in which ions can move. There the hydrogen loses an electron. The hydrogen H^+ ions move in the electrolyte from anode to cathode, while the electrons produce an outside current. (With an alkaline electrolyte such as NaOH it will be the OH^- ions which move in the opposite direction. These fuel cells cause less corrosion than the above described acidic ones.)

At the cathode, the electrons, hydrogen ions and oxygen combine to form water, indicated on the right of Equation (4.78). As in a battery, chemical energy is converted into electrical energy, but in a battery the stored chemical energy is exhausted during its lifetime. In a fuel cell, the chemical energy is supplied from the outside in the form of hydrogen.

The efficiency of a fuel cell will be defined as the output of electrical energy divided by the input of chemical energy. The output is found from (4.65). The cell will work with constant temperature and pressure, so the output is the non-expansion work ($-\Delta G$). Remember that the symbol Δ indicates the difference of Gibbs free energy G between final state and initial state. The chemical energy which is converted is the decrease of enthalpy ($-\Delta H$) as discussed below Equation (4.61). So the efficiency can be written as

$$\eta = -\Delta G / -\Delta H \tag{4.79}$$

For the hydrogen fuel cell one finds from the tables an efficiency of $\eta = 83\%$ (Exercise 4.14). This means that 1 [J] of chemical energy delivers 0.83 [J_e]. This is better than the Carnot efficiencies for all practical cases, as the step of producing heat from chemical energy is bypassed.

Until recently, the fuel cell was not regarded as a realistic commercial option, for an expensive catalyst such as Au or Pt was needed to speed up the reactions. Also, the cell would only operate at a high temperature. Currently, new electrolytes from polymer are developed with less need for catalysts and working at lower temperatures of around 200 [°C] or even less [59]. Still, in 1998, the construction cost for a fuel cell per [kW] of power was in the order of 5000 $, to be compared with only 50 $ per [kW] for an internal combustion engine ([58], [59])

Acid rain

Acid rain from power stations is caused by the sulphur and nitrogen in the fossil fuels, and by the reactions between nitrogen and oxygen in the combustion flames. Technological improvements of the power stations, backed up by international agreements are reducing the emissions in developed countries. The chemical and biological recovery of heavily polluted sites is proceeding very slowly, however [60].

CO_2 *sequestration/reduction*

> Statoil, the Norwegian state owned petroleum company has been pumping 1 million tonnes of liquefied CO_2 annually into an aquifer below the sea floor since 1996.
>
> *New Scientist* 15 May 1999, 14

The greenhouse gas carbon dioxide is produced in any combustion of fossil fuels. Equation (1.1) gave an example for combustion of complete carbohydrates. From Figure 1.15 we get an impression of the flux concerned (about 5.5×10^{12}[kg yr^{-1}]). Figure 3.6 showed the measured, steady increase of CO_2 since the beginning of the industrial revolution. As a consequence, the induced global greenhouse effect will increase as well. In Table 3.1 we showed the present effect, and it is clear from that table that CO_2 is one of the major man-made greenhouse gases.

A more efficient use of fossil fuels in all processes, combustion and others, will clearly help. One can also try to catch the produced CO_2 and store it somewhere. This process is called *sequestration* [61].

A simple way to catch CO_2 on a global scale is to plant more trees, or in general to expand the forests. For some time this will reduce the accumulation of CO_2 in the atmosphere, but eventually (after tens of years) the forests will achieve equilibrium with their surrounding ecosystems. This solution will help postpone more drastic measures or offer opportunities for more research into other options, but it offers no final solution to the problem of sequestration.

Below about 2700 metres, liquid CO_2 is denser than seawater. And at low temperatures and high pressures, it reacts with water to form a solid, ice-like hydrate. Deep-sea disposal could form very large 'icebergs' at the ocean bottom that, while not permanent, would endure for a very long time.

New Scientist 15 May 1999, 14

There exists some industrial experience with removal of CO_2 gas from a gas mixture. For example, one may let the mixture react with a chemical absorber. In a later stage the absorber is heated and the CO_2 is removed and can be stored. This method costs energy and some absorber material at each step. Another way is to put the CO_2 gas into physical contact with an absorbing fluid, without a chemical reaction. A small reduction in pressure will then lead to expulsion of CO_2 which can be done at a suitable location.

The concentrated CO_2 may be liquefied and placed deep down in an ocean. This option is under investigation. One is looking at the costs, the effectiveness and the environmental consequences. Storing large quantities of CO_2 in the ocean may change its chemical composition or physical and chemical properties and have consequences for the food chain of the oceanic fish.

In another option the CO_2 gas is pumped into empty mines or exhausted fields of natural gas. In all cases the CO_2 should not find a way to enter the atmosphere again.

Using waste heat

In Section 4.2 we discussed the concept of exergy as the available work in a sample of material. Equation (4.55) showed that combustion already spoils one-third of the exergy and one loses more as waste heat Q_C in the next step of running a heat engine between a hot reservoir and a cold one. So, direct conversion of exergy into mechanical work or into electricity sounds very attractive. However, a direct conversion of the chemical energy of a fossil fuel is only possible in a few cases and still expensive. So, fossil fuel power stations will be around for a long time and the challenge is to use them in the cleverest way. We discuss two options.

Combined cycle systems The Carnot efficiency (4.36) reads

$$\eta_{\mathrm{Carnot}} = 1 - \frac{T_{\mathrm{C}}}{T_{\mathrm{H}}} \tag{4.80}$$

This thermal efficiency should be made as large as possible. The temperature T_{C} of the cold reservoir cannot be made smaller than ambient air or water, around $T_{\mathrm{C}} = 300$ [K]. High flame temperatures can be reached in a gas turbine, in which combustion of natural gas may give temperatures as high as $T_{\mathrm{H}} = 1530$ [K]. At first sight an efficiency of 80 percent would seem possible. Unfortunately, this is not the case as the exhaust temperatures of the gas turbine are in the order of $T_{\mathrm{C}} = 850$ [K]. This would give a Carnot efficiency of 44 percent. In practice, an efficiency of $\eta_A = 40\%$ can be reached.

It would be a shame to let the exhaust gases with all their exergy escape into the chimney. In such a case one uses the output heat of the gas turbine as input heat for a steam turbine. The output temperature T_{C} of that turbine will still be higher than the 300 [K] mentioned above, as the heat is dumped in the low-temperature reservoir where it cannot be removed immediately. Therefore, in the best case, the steam turbine would operate between $T_{\mathrm{H}} = 850$ [K] and $T_{\mathrm{C}} = 400$ [K] with a Carnot efficiency of 53 percent. In practice this value is around $\eta_{\mathrm{B}} = 35\%$.

The combination of the two turbines with thermal efficiencies η_A and η_B is called a *combined cycle*. Its thermal efficiency is found from Figure 4.21. There one has heat q_{A} as input and work $\eta_{\mathrm{A}} q_{\mathrm{A}}$ as output. The 'waste' heat $(1 - \eta_{\mathrm{A}})q_{\mathrm{A}}$ is not wasted but used as input to the second loop. There it produces a second amount of work $\eta_{\mathrm{B}}(1 - \eta_{\mathrm{A}})q_{\mathrm{A}}$. The total thermal efficiency is given by

$$\eta = \frac{\eta_{\mathrm{A}} q_{\mathrm{A}} + \eta_{\mathrm{B}}(1 - \eta_{\mathrm{A}})q_{\mathrm{A}}}{q_{\mathrm{A}}} = \eta_{\mathrm{A}} + \eta_{\mathrm{B}} - \eta_{\mathrm{A}}\eta_{\mathrm{B}} \tag{4.81}$$

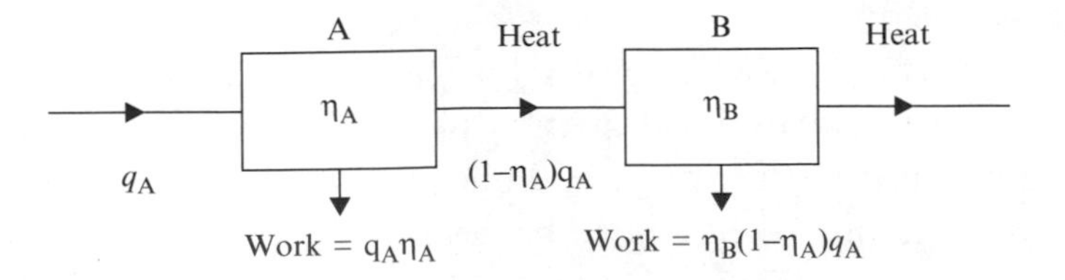

Figure 4.21 Combined cycle. The waste heat of gas turbine A is used as input for steam engine B

With the data $\eta_A = 40\%$ and $\eta_B = 35\%$ given above, one finds $\eta = 61\%$. This value of the thermal efficiency is approached by the best possible combined cycles on the market.

Cogeneration of heat and power The final loop of the combined cycle still produces exhaust gases that are considerably higher in temperature than the ambient temperature of 300 [K], so there is still some exergy left. This can be used for heating the offices and halls of the power stations, or it can be transported to nearby residential areas for heating of homes and tap water.

Hot water will lose heat by conduction through the walls of the connecting pipes and by convection and radiation at the outside of the pipes. Therefore, one has to insulate the pipes with materials of low thermal conductivity (see Table 4.1). It is then a matter of economics whether transportation of the available heat is cheaper than household combustion of gas or oil.

One may also look at the problem from another angle. In a complex of houses, an apartment building, or a university campus, one needs both electricity and heat. So one could produce both at the same time locally with power stations of a moderate size. Increasingly, this *cogeneration* is applied. The use of a fuel cell system to power a building and at the same time provide heat for heating space and tap water [59] is of current interest.

The energy plant at the authors' university, for example, produces 5.7 [MW_e] of electricity and 9.1 [MW_{th}] of heat. The plant is optimized so that, on average, it comes close to the needs of the campus. In any such system, it is economic to have a connection to the central grid in which one may dump surplus electricity, or from which one may tap electricity when the supply is not sufficient. The final dimensioning of a cogeneration plant will depend on the price the utility is willing to pay for surplus electricity. Indeed, that decides whether the concept is feasible from a financial point of view.

EXERCISES

4.1 Marines of bodyweight 70 [kg] reach an altitude of 2923 [m] in 7.75 hours. Calculate the total potential energy gained. What is the resulting power? What is more sensible, to calculate the power based on 7.75 hours, or based on 24 hours?

4.2 In his fitness centre, one of the authors (mass m = 75 [kg]) walks 4 [km] during an hour on a conveyor belt with a slope of 10 percent. According to the fitness centre he looses 2.4×10^6 [J] of energy in body fats. (a) Estimate the potential energy gained when walking a mountain in the same way. (b) Compare with Exercise 4.1. (c) What is the efficiency with which body energy is converted into potential energy? (d) What happens to the rest of the energy?

4.3 Find some more examples from daily life of heat transport: conduction, convection, radiation and transport of latent heat.

4.4 From Equation (4.1) check the dimension (units) of the thermal conductivity k. Check the dimension of the heat resistance from Equation (4.3). Compare with the result that one obtains from Equation (4.4).

4.5 Can you write down the equivalent of Equation (4.9) for electricity? If so, derive Ohm's law from that equation.

4.6 A thermopane glass window (with an area $A = 3$ [m^2]) consists of two layers of glass with a thickness of 2 [mm] each with 4 [mm] of air in between. (a) Calculate its heat resistance. (b) Calculate how thick a single glass window would be with the same heat resistance. (c) Calculate the heat loss by conduction if the inside temperature is kept constant at 20 [°C] and the outside temperature is 5 [°C]. (d) Calculate the heat loss by conduction for an ordinary glass window of 4 [mm] thick.

4.7 Calculate the heat output of Equation (4.16) for a lower glazing temperature $T_c = 25$ [°C], instead of $T_c = 30$ [°C], while keeping all other variables the same. What would happen on a windy day when the convection heat transfer coefficient goes to $h = 25$ [W m^{-2} K^{-1}]?

4.8 The calorie is defined as the amount of heat required to heat one gram of water by one degree. Use the specific heat of water (Appendix A) to calculate the corresponding number of joules. This is 'the mechanical equivalent of heat'.

4.9 Use the set-up of Figure 4.9 as a heat pump. Derive the COP for this case.

4.10 Comment on the following statement (used as an appetiser). In terms of energy, 1 [kWh] of electricity and the sensible heat contained in 43 [kg] of 20 [°C] water are equal (i.e. 3.6 [MJ]). At ambient conditions 1 [kWh] of electricity has a much larger potential to do work (to turn a shaft or to produce light) than the water ([37], p. 81 footnote).

4.11 Plot the loss fraction L of Equation (4.55) against the ratio T_c/T_0. Note that the loss fraction goes down with increasing combustion temperature, so it pays to keep T_c high.

4.12 From the data given below Equation (4.61) for the reaction enthalpies of water, find the heat of evaporation in [kJ mol^{-1}] and [J g^{-1}]. Comment on the sign of the result.

4.13 Consult tables for enthalpies of a mole of substance (for example in [49]) and find that for reaction (4.73) the enthalpy increases by 283 [kJ mol^{-1}].

4.14 Check the maximum efficiency for the hydrogen fuel cell, using Equation (4.78) and (4.79). Do the same for a methane fuel cell, assuming all CH_4 is oxidized into CO_2 and H_2O (liquid).

5 Carbonfree Energies

This chapter discusses ways of producing heat and work without the emission of greenhouse gases, or at least with a very low emission. Hence the name 'carbonfree'. In Figure 4.1 they were called renewables, to be discussed in Section 5.1 and Uranium. The latter points to nuclear power to be dealt with in Section 5.2.

The fact that renewable energies or nuclear power do not produce greenhouse gases, does not automatically imply that they are socially acceptable. That is a political decision, depending on the consequences of their use, to be discussed in the final chapter.

5.1 *RENEWABLE ENERGIES*

By definition, *renewable energies* are those of which the supply is inexhaustible. Application of renewable energies, therefore, is necessary for a *sustainable* way of life. The application of renewable energies is not sufficient for sustainability since they may be polluting or produce waste (e.g. heavy metals from solar cells). Only when the pollution of renewables is dealt with, may one have sustainability. As solar radiation will only change over billions of years, one may for all intents and purposes regard it as renewable. So we will consider it here. This energy from the sun can be captured directly by solar collectors, solar cells or photosynthetic organisms. The sun is also responsible for winds that may be tapped as wind energy. The sun evaporates water causing rain which produces hydropower high up in the mountains and the sun supplies the energy for the growth of biomass. For simplicity, we will devote separate subsections to these renewable energies. Their present contribution to the electricity generation may be derived from Table 4.4, which shows that promising developments such as wind power and solar electricity still have a very small market share.

> A chimney 1500 [m] high could perhaps be built in North Cape, South Africa. At the bottom, a glass disk of 7 [km] across would collect solar heat. The hot air drawn towards the central chimney and drives some turbines on its way up.
>
> *New Scientist* 6 March 1999, 30–35

Other renewable energies exist that are not discussed here, as only small-scale applications are expected. We mention the energy from the waves (deriving from the winds and therefore ultimately from the sun) and from the tides (deriving from the rotation of the earth and the gravity pull from sun and moon). Geothermal energy, where heat from the interior of the earth is tapped, will not be discussed. In this case the contribution to satisfying the world's energy needs may be significant (Table 4.4), but the technology is quite straightforward (see Example 1.1).

5.1.1 *Solar energy*

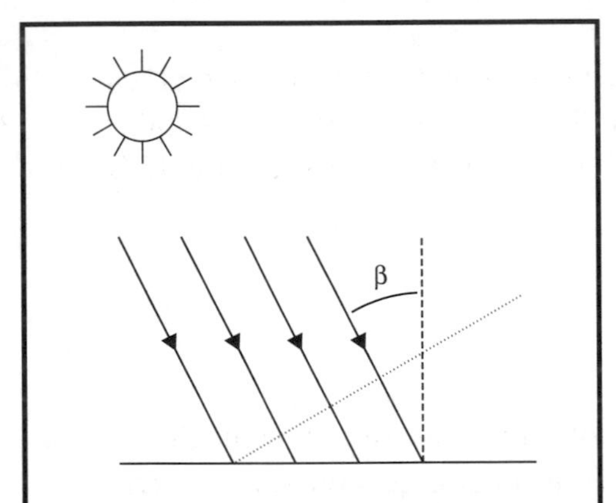

Figure 5.1 Incoming solar radiation will make an angle β with the vertical on a horizontal surface

A solar lantern kit costs £70. It is a purpose-built sealed unit containing its own rechargeable battery and has a microchip charge regulator. It constantly monitors the battery to ensure it remains charged. At night it will switch off the lantern rather than allow the battery to go flat.

New Scientist, 22 July 2000, 22

The incoming solar radiation will make an angle β with the vertical as shown in Figure 5.1. If the solar irradiation per [m^2] perpendicular to the solar rays is S' [J m^{-2}s^{-1}] then the irradiation I_{hor} on a horizontal surface is given by

$$I_{hor} = S' \cos\beta \tag{5.1}$$

The rising sun starts at the horizon with $\beta = 90°$, reaches its highest position in the skies at midday, and sets again at the horizon with $\beta = 90°$. The value of β at its highest position is a function of the geographical latitude λ and the time of the year. The latter can be indicated by the number of days since the beginning of spring in the Northern Hemisphere or equivalently by the *declination* δ of the sun. This declination may be regarded as the latitude at which the sun is in the zenith position at noon that day. During a year the declination changes from $\delta = 23.5°$ South, the Tropic of Capricorn (around 22 December) to $\delta = 23.5°$ North, the Tropic of Cancer (around 22 June). This angle of 23.5° equals the tilt axis of the earth.

Example 5.1

Figure 5.2 shows the equatorial plane of the earth (MVAB) and the ecliptic of the sun (MVSunC). Both are viewed from the fixed centre of the earth and projected on to the same sphere. Show from Figure 3.5 that V determines the beginning of spring or autumn. Show that the tilt angle ε defines the extreme declinations δ of the sun. Take the latitude where you live and draw a curve, indicating your daily motion, as projected on to the sphere.

Answer

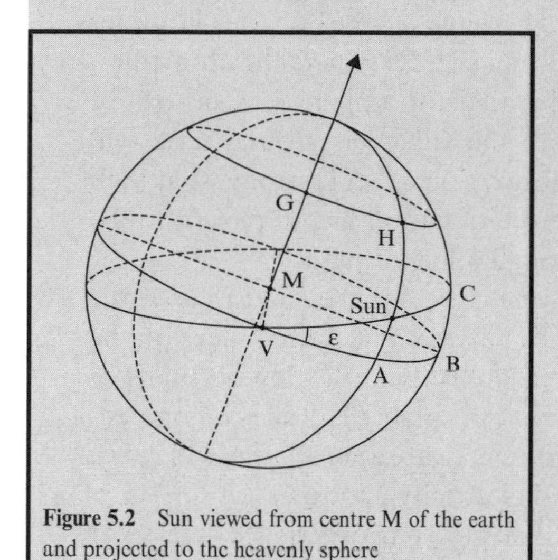

Figure 5.2 Sun viewed from centre M of the earth and projected to the heavenly sphere

In Figure 3.5 the beginning of spring and autumn were defined as the dates when day and night are equally long everywhere on earth. That is where the earth equator plane cuts the ecliptic at the Sun's location. From Figure 5.2 one sees that V is one of these points. When V denotes spring, it is called the *vernal point* with an Arab symbol, resembling a capital V.

In a year the sun makes one round along the ecliptic. For a certain day one may assume the sun fixed. We drew G as the location where the authors live. The daily motion then is a small circle around the axis. It is clear that the angle β = G–M–Sun is smallest when G equals H on the meridian of the sun. This is midday or noon. In the plane CMB, the plane of the axis, the angle CMB equals the angle between the normal to the ecliptic and the normal on the equatorial plane. This is the tilt ε, which we also have indicated near V. From Figure 5.2 it is clear that in C the sun has its highest declination ε, which is the tilt angle.

To determine the solar irradiation at the earth's surface, the relevant quantity according to Equation (5.1) is cos β. One may show ([1], p. 118) that, outside the Arctic and Antarctic regions,

$$\cos\beta = \cos\lambda\cos\delta\,(\cos\omega t - \cos\omega T) \tag{5.2}$$

In Equation (5.2) $t = 0$ defines noon at latitude λ with declination δ. For $t = T$ or $t = -T$, one obtains $\cos\beta = 0$. Therefore $t = -T$ corresponds with sunrise and $t = T$ with sunset. The time T may be found from the relation*

$$\cos\omega T = -\tan\lambda\tan\delta \tag{5.3}$$

*Students may use this relation to calculate the length of the day at their homes. In reality, the daylight lasts somewhat longer because of the dispersion of light at sunrise and sunset. Note that because of the time zones, noon ($t = 0$) will only coincide with 12 o'clock at certain places.

For practical purposes, e.g. to determine the influx to solar collectors, the total solar irradiation during a day on a horizontal surface is a relevant measure. It is found by integrating eq. (5.2) from sunrise to sunset. The result will depend on the geographical latitude λ and the declination δ. The declination in turn depends on the number of days since the beginning of the spring in the Northern Hemisphere (N.H.) when $\delta = 0$. For four latitudes the result is shown in Fig. 5.3, where the absorption of solar light by the atmosphere is ignored. In practice therefore the values at ground level will be lower. This absorption effect is strongest at high latitudes, where the radiation has to penetrate the atmosphere through a longer path.

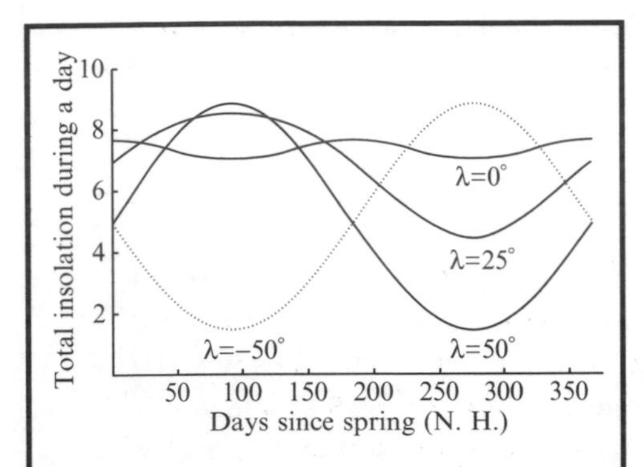

Figure 5.3 Insolation of a horizontal surface for four latitudes λ. The curve for the Southern Hemisphere is dotted. The vertical scale is the number of hours the sun should shine perpendicularly, to give the same total insolation. The unit equals 4.9×10^6[J m^{-2}]

In Figure 5.3 one notices that at midsummer the total irradiation is not very much dependent on the latitude. The reason is that the lower angle of the sun at the sky at the higher latitudes is compensated by a longer day. At midwinter the differences are appreciable, which is a pity because then the demand for heat and electricity are highest.

As one sees from Equation (5.2) and knows from experience, the solar irradiation depends on the time of the year and the time of the day. One may, of course, try to improve the solar input in a device by orienting it perpendicular to the solar rays using some servomotor mechanism. Then one enforces $\beta = 0$ or $\cos \beta = 1$. For devices coupled to a single house or building this is rather complicated and expensive. One therefore puts them at some angle α with the horizontal surface and orients the devices to the South, where the sun is at its highest point anyway. The angle α may be optimized for use in winter or in summer, depending on the application. And, of course, one has to be careful that nearby trees or buildings do not cast their shadows over the solar devices.

Solar heat

In Section 4.1 and Figure 4.4 the solar collector as a device to heat water, was discussed. The big differences between summer and winter at moderate latitudes, illustrated in Figure 5.3, imply that hot water will not always be available when required. Therefore, as sketched in Figure 5.4, a back-up system is needed. There, the hot water from the collector exchanges its heat with a boiler. The boiler has its own supply of cold water, which enters at the bottom and leaves at the top, heated by the boiler. If the water is not hot

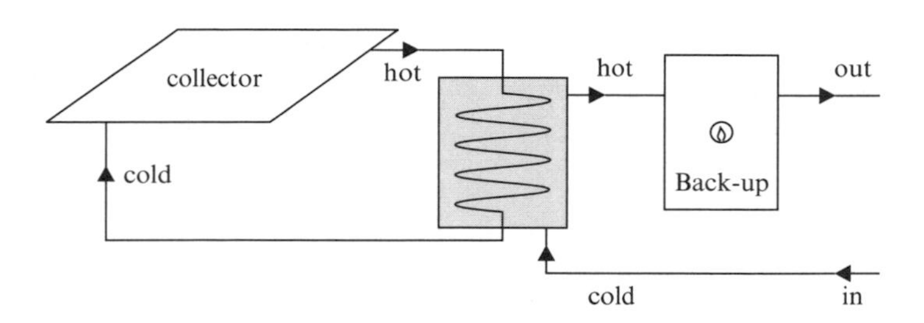

Figure 5.4 Solar collector system. In a boiler the heat from the collector is exchanged with cold water to be used for the tap. If necessary, the tap water receives extra heating in a back-up system, e.g. the central heating furnace. A simple collector was sketched in Figure 4.4

enough, an afterburner will provide additional heating until the required temperature is reached.

At moderate latitudes the solar heating in winter time is not sufficient for modern comfort, so one will need a central heating furnace anyway. This furnace may then heat the water from the boiler, if necessary. With some government subsidies (or tax facilities) the boiler may be competitive with hot water from a gas furnace (Exercise 5.2).

> Solar thermal power or CSP (= concentrating solar power) provides the lowest cost solar energy in the world. In California 354 [MW_e] has been connected to the grid since the 1980's in plants sized from 14 [MW] to 80 [MW]. It is expected that new plants could produce electric power for 10 to 12 [$cts/kWh].
>
> solarpaces@dlr.de, January 1999

In another application, mirrors concentrate solar radiation in a focal point, where the heat creates a high temperature reservoir. Water or molten salt may transport the heat to a nearby place, where it is used to run a heat engine. Both steam turbines and Stirling engines are in use for this purpose. The mechanical energy produced by these engines is then converted into electricity.

A nice feature of this device is that the hot water or molten salt can be stored for a few hours in an insulated container so that, to some extent, the electricity can be produced when the demand is at its highest.

Direct solar electricity: PV

The method of producing solar electricity by generating heat, then converting the heat to mechanical energy and finally to electric energy is indirect, with losses at each step. Therefore a major research effort is focussed on the direct conversion of solar radiation to electric power by so-called *solar cells* or *photovoltaic cells* [62]. The latter refers to the mechanism of converting photon energy into voltaic energy, hence the abbreviation PV for photo-voltaic.

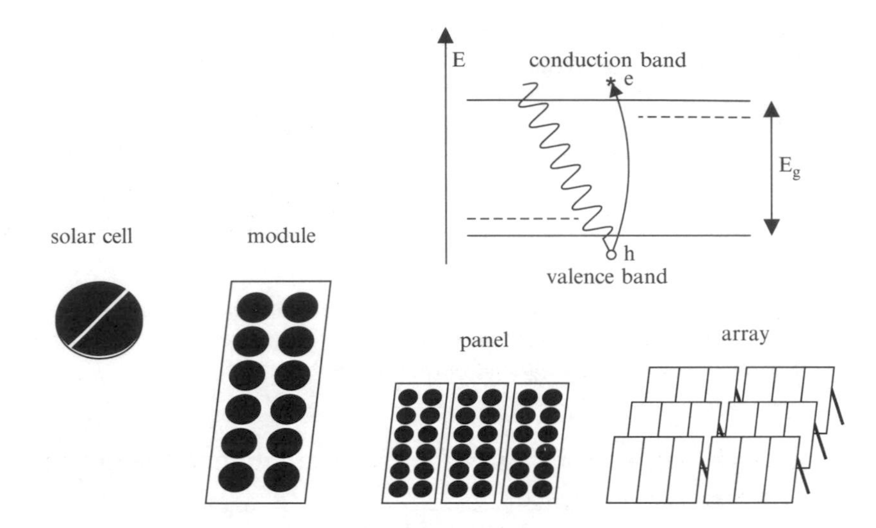

Figure 5.5 The photovoltaic hierarchy. In the upper right corner the principle of the solar cell is given. In part reproduced with permission of John Wiley from [62], Figure 4.2, p. 79

A solar cell is the smallest unit. In practice, one combines solar cells into modules, modules into panels and panels into an array. This hierarchy is illustrated in Figure 5.5.

In Figure 5.5 the principle of a solar cell is shown in the upper right corner. The material of the cell is a semiconductor, which means that at temperature $T = 0$, all electrons are in the lower, *valence band*. In that case the higher, *conduction band* is empty and the conductivity of the material is zero. For higher temperatures, the kinetic energy of the electrons makes it possible that some cross the *energy gap* between both bands. This results from competition between the energy of the thermal motion, expressed in units kT, where k is the Boltzmann constant, and the energy gap E_g. At room temperature $kT = 0.025$ [eV], which is much lower than the energy gap E_g of around one [eV]. Therefore, under ordinary circumstances there are very few electrons in the conduction band giving a very low electric conductivity.

In a solar cell a photon with an energy $E = h\nu$, which exceeds the energy gap E_g is absorbed by an electron in the valence band, which is transferred to the higher conduction band. This leaves a positive hole h in the valence band, which can move around as freely as the electron in the conduction band. We note that the energy absorbed by an electron with charge e, if it crosses a potential V equals eV. Therefore in discussing PV one uses the electronvolt [eV] as a unit, which is much smaller than the [J] (cf. Appendix A).

Before discussing the photovoltaic effect, we consider a single crystal of Si, which is one of the major materials used for present day solar cells. The low conductivity of the crystal at room temperature is increased by 'doping' the Si with a low concentration of another element, which replaces a single Si atom

> Solar electricity is particularly appropriate for Nepal since it is convenient and cost-effective to install on the scale of a single household. Over 2500 systems have already been deployed across Nepal and they have proved to be extremely popular with villagers.
>
> *Renewable Energy World*, May 1999, 91–95

in the crystal. If one uses an atom with five outer electrons instead of the four of Si, then one has one electron too many. This electron occupies the upper dashed line of the level scheme in Figure 5.5 and it can move freely around. This is called *n-type* Si.

Alternatively, one can dope with an element with only three outer electrons. This gives an extra possibility for the electrons in the valence band to move around, indicated by the lower dashed line in the level scheme. One interprets this as an extra, positive hole in the valence band: *p-type* Si. The conduction in the doped Si is still quite small, but essentially the conduction is due to the extra electrons or the extra holes, produced by the doping.

On the left of Figure 5.6, a solar cell is sketched, which consists of a thin round waver, made up of a layer of n-type Si on top ($\approx 1\,[\mu m]$) and p-type Si on the bottom ($\approx 100\,[\mu m]$). Look, still in the absence of light, at the left of Figure 5.6 and assume the external voltage is absent. Assume that the two layers are produced separately and then are brought together.

As soon as the two thin layers of the waver touch at their junction, various things will happen. The surplus of electrons at the n-side relative to the p-side, will result in a *diffusion current* of electrons from n to p. Similarly, a diffusion of holes will occur from p to n. In terms of an electric current, both flows contribute to a positive* current I_0 from p to n. By this diffusion current a charge density will build up, shown in the middle of Figure 5.6 and indicated by + and − signs. This charge density gives an electric field from n to p. Under the influence of the electric field, some of the small number of holes at the n-side will move to the p-side. Similarly, from the few electrons on the p-side some will move to the n-side. This results in a so-called *field current*. In the steady state, which is reached very quickly, both currents will be equal in magnitude but with different signs. This is indicated on the right in Figure 5.6.

> A solar battery charger may be very flexible, polymer laminated, with powers ranging from 5 [W] for an area of 0.14 [m^2] to 32 [W] for an area of [0.6 m^2]. unisolarinfo @ ovonic. com, 1999

Let us now connect the electrodes as on the left in Figure 5.6. The positive electrode on the p-side will support a positive current from the p-side to the

*In the following, the sign of the current is occasionally needed. This is most easily found by looking at the direction in which the holes are moving.

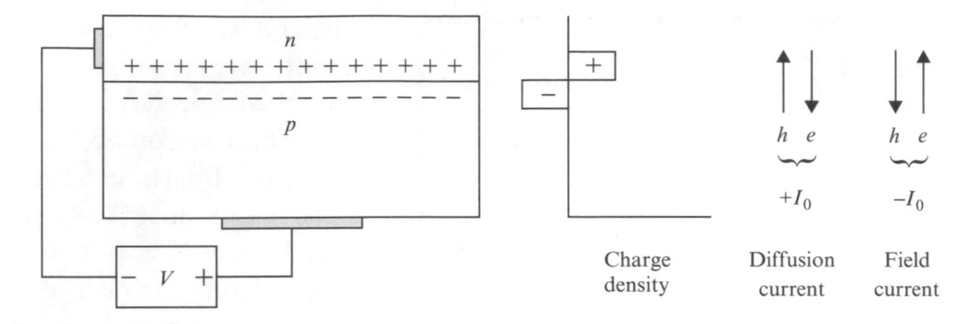

Figure 5.6 Semiconductor junction in the absence of light. When the junction is formed with the electrodes unconnected, a diffusion current will cause a charge density (middle). This results in a field current, which in equilibrium is as large as the diffusion current but with opposite sign

n-side, which supports the diffusion current. The field current in the opposite direction is not influenced as long as the potential V is not very high. This is the case, as the local field from the charge density (Figure 5.6 middle) remains strong enough to support the field current. The resulting current I_D therefore can be written as

$$I_{\mathrm{D}} = I_0\,\mathrm{e}^{eV/kT} - I_0 = I_0(\mathrm{e}^{eV/kT} - 1) \tag{5.4}$$

This positive current goes from p to n; the first term is the reinforced diffusion current and the second one the unchanged field current. The fact that one finds an exponential in the diffusion current is connected with the fact that the energy eV of the electrons crossing the potential V has to be expressed in relevant energy units. As the temperature T enters the game the energy unit will be kT. When an energy eV has to compete with a thermal energy kT, it frequently gives rise to an exponential as in Equation (5.4). In fact, this also happened in Equation (2.8) where, in the exponential, the expression z/H was found. This can be written as Mgz/kT where M is the mass of a single molecule [kg] (Exercise 5.3). In that case, the potential energy Mgz of a single molecule competes with the thermal energy kT. In general, in many microscopic phenomena, competition between an energy E, which tends to order, and thermal motion kT, which tends to random disorder, leads to a *Boltzmann distribution*

> 1 [MW] of PV power is integrated into the roof and facade of the Academy Mont-Cenis Herne, near Essen, Germany. Here 3184 solar modules are connected to 569 string inverters. The modules consist of two layers of glass (6 [mm] and 8 [mm]) with in between 2 [mm] resin filling with the solar cells. The solar cells used have efficiencies of 12.5 percent and 16 percent. The PV elements have different levels of translucency to create a pattern resembling a cloud-patterned sky.
>
> *Renewable Energy World*, May 1999, 30–36

$\exp(-E/kT)$. By this argument Equation (5.4) may have gained some respectability.

When a photon enters the crystal, it will produce an electron-hole pair near the junction, say on the n-side. The electron will disappear in the crowd of electrons around, but the hole will feel the field of the charge density and join the field current. The current I_s from the solar photons therefore will be added to I_D with a negative sign, giving

$$I = I_D - I_s = I_0 (e^{eV/kT} - 1) - I_s \tag{5.5}$$

In this equation kT refers to room temperature $kT = 0.025$ [eV]. This has to be emphasized as the solar induced current I_s is calculated from the incident solar spectrum. This is approximated by a *black body spectrum* (1.4) for a temperature $T = 5800$ [K] or $kT = 0.50$ [eV], corresponding to the surface temperature of the sun. For photons one may write $E = h\nu$ and all other factors in Equation (1.4) may be comprised in a factor a. Equation (1.4) may then be rewritten as

$$N(E) = \frac{aE^3}{e^{E/kT} - 1} \tag{5.6}$$

The spectrum is defined such that an amount of incident energy $N(E)\mathrm{d}E$ is received between E and $E + \mathrm{d}E$. If one expresses the energies in [eV], one may normalize the coefficient a in (5.6) such that the total incident energy is precisely 1000 [W m^{-2}]. This gives (Exercise 5.4)

$$a = 1.538 \times 10^{22} \ [[\mathrm{eV}]^{-3}\mathrm{m}^{-2}\mathrm{s}^{-1}] \tag{5.7}$$

Note from equations (5.6) and (5.7) that $N(E)$ has the dimensions [m^{-2}s^{-1}]. However, $N(E)\mathrm{d}E$ has the dimensions [[eV] m^{-2}s^{-1}], which is the energy [eV] entering per second in a square metre, as it should. In Figure 5.7 the energy spectrum (5.6) is displayed with some relevant parameters, such as the visible range and the energy gap E_g of the Si crystal.

> A solar module should be elevated towards the sun in order to receive the maximum irradiation. In areas with a high albedo one can also use the scattering from the ground to the back of the module. Modules are for sale with 60 [W] on front, 37 [W] on the rear with an isolation of 1000 [W m^{-2}] and an area of 0.594 [m^{-2}].
>
> www.pvsolmecs.com, 1999

From the energy picture in Figure 5.5 it will be clear that only photons with energies $E > E_g$ can give a photovoltaic effect. Therefore, only the part of the spectrum to the right of $E_g = 1.12$ [eV] will contribute and create an electron – hole pair. The electrons of the pair,

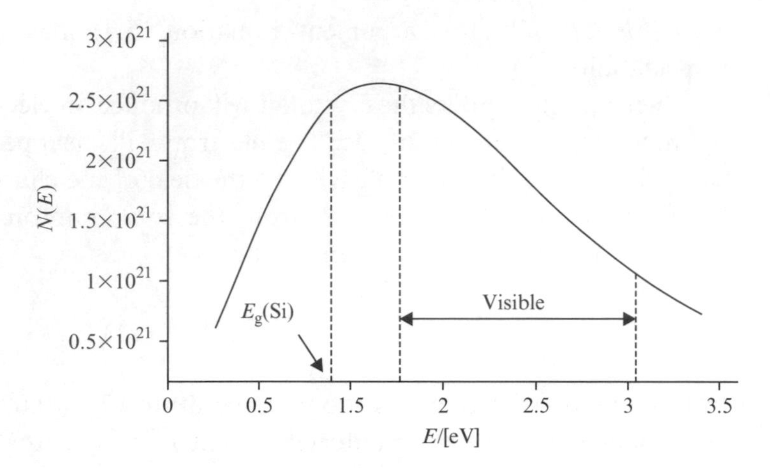

Figure 5.7 Black-body energy spectrum for $T = 5800$ [K] (the sun), normalized on 1000 [W m^{-2}] incident energy. Note the visible region and the Si energy gap

excited in the conduction band, quickly relax to the lowest level of that band, releasing some heat, before they enter the external load to return to the top of the valence band. So for each absorbed photon with energy $E > E_g$ only the energy E_g is useful for the PV effect. For the black body spectrum of Figure 5.7 one may calculate a solar current $I_s = 391$ [A m^{-2}], which should be substituted in Equation (5.5). For the value I_0 in that equation, the literature gives a typical value as $I_0 = 5.9 \times 10^{-8}$[A m^{-2}], which demonstrates the high relative magnitude of the solar-induced current.

With the values given above one can plot an IV-diagram for the solar cell, based on Equation (5.5). This is done in Figure 5.8, where $kT = 0.025$ [eV] was used, as the cell is operating at room temperature. Note that for a voltage $V \rightarrow 0$ the current $I \rightarrow -I_s$, which agrees with Equation (5.5). Also note that the curve is nothing like Ohm's law: $V = IR$, which underlines that one has a complicated system, totally different from a conducting electric wire.

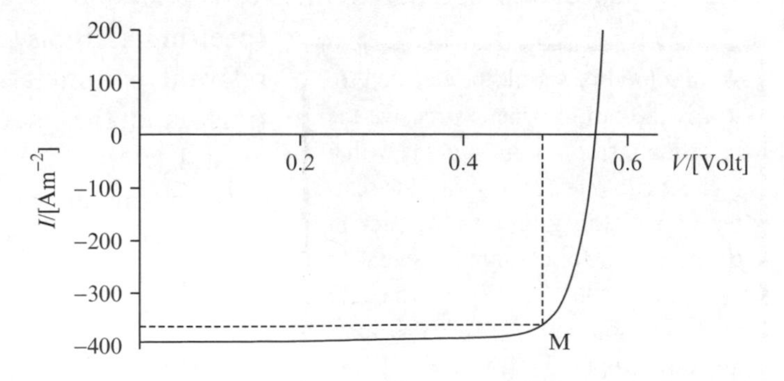

Figure 5.8 IV diagram for the solar cell with parameters given in the text. Point M gives the highest power output IV

The power P of a device such as the one represented by Figure 5.8 is given as

$$P = IV \text{ [W m}^{-2}\text{]} \tag{5.8}$$

Dutch windmill used for polder drainage.

Its value depends on the external voltage V, which is imposed on the system of Figure 5.6. The maximum output is found from the derivative $\partial P/\partial V = 0$, which in the present case results in $V = 0.49$ [V] (indicated as point M in Figure 5.8) and $P = 182.4$ [W m^{-2}]. As the insolation was taken as 1000 [W m^{-2}] it follows that the cell has an efficiency of 18 percent. In reality, as is shown in Figure 1.8, the solar spectrum arriving at sea level shows much more structure than the black body spectrum of Figure 5.7. Therefore, the 1000 [W m^{-2}] is distributed in another way over energies E, resulting in a higher solar-induced current and a somewhat higher maximum efficiency.

Finally, it should be mentioned that besides crystalline Si many other materials for solar cells are in use or in development ([62], p. 173). Here, the efficiency plays a role, but it is mainly the production costs that are important. One of the major challenges for the coming years is to make solar electricity competitive with electricity from fossil fuels.

5.1.2 *Wind energy*

Since olden times the wind has been used to power sailing ships. Also, windmills of many kinds were used to grind grain or to pump water up to a higher level. Since the 1970's, however, special devices have been constructed with the main task of converting wind power into electricity [63]. Although they have blades, just like windmills, they are usually called *wind turbines*, as the word 'mill' suggests too much the handiwork of the miller. Below we will discuss the power [Js^{-1}m^{-2}] contained in the wind, next the *Betz limit*, which is the maximum that can be converted into mechanical power and we finish with a few notes about aerodynamics.

Power in the flow

Consider 1 [m^3] of air with velocity u. The mass of that air equals its density ρ. Its kinetic energy becomes $\frac{1}{2}\rho u^2$ [J m^{-3}]. Next, put an imaginary surface with

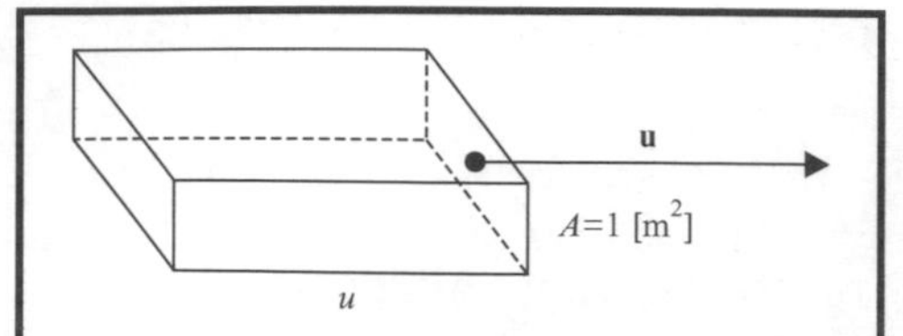

Figure 5.9 All particles in a bag with length u parallel to the wind velocity will pass the surface A in 1 second

an area of $A = 1[\text{m}^2]$ perpendicular to the airflow. Against the direction of the flow an imaginary bag is placed with length u and sides parallel to the flow. This set-up is sketched in Figure 5.9. From the graph it appears that all air particles within the bag will pass the surface area A in 1 second. Consequently, the kinetic flow energy P passing 1 [m²] per second in units $[\text{J s}^{-1}\ \text{m}^{-2}] = [\text{W m}^{-2}]$ equals the kinetic energy contained in the bag which gives (Exercise 5.5)

$$P = \tfrac{1}{2}\rho u^3 \tag{5.9}$$

Thus, the power P contained in a moving volume of air, increases with the third power of the wind velocity u. It is therefore advantageous to find locations with high wind velocities. The most obvious possibility is to put the turbine on a high tower and profit from Equation (2.22), which shows a logarithmic increase of air velocity with altitude. But Equation (2.22) also implies that the velocity at the lowest and the highest point of the blades are quite different (Example 5.2) and technological tricks are required to guarantee the smooth operation of the turbine.

In Germany the so-called electricity feed-in law requires the utilities to buy power in the form of renewables against a reasonable price. From 2000 onwards it will be 8.9 Euroocts per [kWh] for a period of 5 years. After that the compensation will be reduced.

New Energy 2/2000, p. 15.

Example 5.2

The axis of a turbine with two blades, each with a length of 26.3 [m] is constructed at an altitude of 55 [m]. The tower is put on farmland with trees and villages for which z_0 in Equation (2.22) may be taken as $z_0 = 0.3$ [m] ([63], p. 8). At $z = 10$ [m] a velocity $u = 3\,[\text{m}]\text{s}^{-1}$ is measured. Calculate the velocity at the lowest point of the blade and at its highest position.

Answer

From $z = 10$ [m], $z_0 = 0.3$ [m] and $u = 3\,[\text{m s}^{-1}]$ it follows with (2.22) that $u*/k = 0.855\,[\text{m s}^{-1}]$. The lowest position of the blade is at $55 - 26.3\,[\text{m}] = 28.7\,[\text{m}]$ with (from (2.22)) a velocity of $3.90\,[\text{m s}^{-1}.]$ The highest position of the blade equals $55 + 26.3\,[\text{m}] = 81.3$ m with a velocity of $4.79\,[\text{m s}^{-1}]$. The increase of velocity with z may not seem much, but for energy retrieval one should take the third power. The ratio between highest and lowest position then becomes $(4.79/3.90)^3 = 1.85$.

Besides the complication of varying wind velocity with altitude, the velocity usually varies in magnitude and direction as well. So in order to calculate the output of a wind turbine before actually building it, one should know the probability $f(u)$ that a certain wind velocity will occur. This can be done by measurements at the proposed locations. It is obvious that at the seaside in Denmark or high up in the mountains in California, the probability of high velocities will be larger than inland, or in between mountain ranges.

The Betz limit

Let us, for simplicity, assume a homogeneous wind flow with velocity $u = u_{in}$ everywhere, ignoring changes with altitude or in time. The question then is how much of the power (5.9) contained in the flow can be converted into the rotational energy of the turbine. Intuitively one feels that the wind velocity downstream from the turbine cannot become zero, as then the air would pile up at that position. Consequently, not all of the power (5.9) can be converted. Also, clearly some reduction in velocity is required as otherwise no power is tapped from the flow. In our considerations we therefore use the schematic set-up, sketched in Figure 5.10.

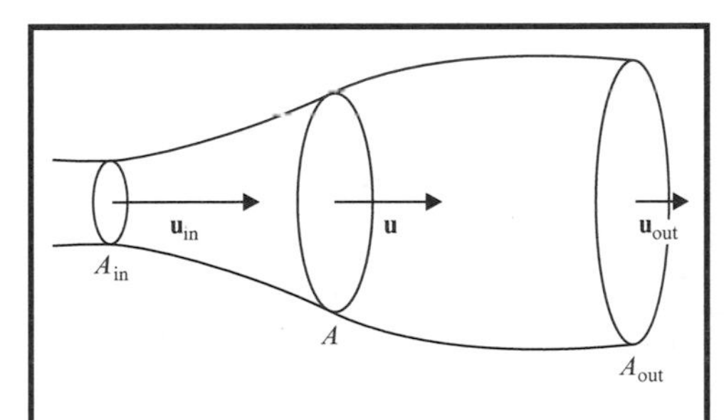

Figure 5.10 Flow through a wind turbine. A is the area of the turbine itself. The curves indicate outer streamlines

The Wind Powering America programme calls for a dramatic increased reliance on wind energy. The American Wind Energy Association recognizes that wind is the biggest success story in the renewable energy field and the closest to being commercially competitive, with 5 [\$ct/kWh].

Nature **400**, 15 July 1999, 201

At the far left in Figure 5.10 one finds the incoming flow with velocity u_{in}, not disturbed by the turbine. Closer to the turbine the streamlines start diverging, resulting in a velocity u at the turbine and an even lower velocity u_{out} at some point behind the turbine. At the three locations shown, the cross-sectional areas are A_{in}, A and A_{out} respectively. The same mass m[kg s^{-1}] passes at the three locations in a coherent flow.

Apparently the interaction between all molecules participating in the flow transmits upstream the knowledge that there is a turbine to pass. Therefore, the streamlines in Figure 5.10 start diverging before the air actually hits the turbine. In this way the 'incoming' energy for the turbine should be taken from the turbine cross-section A with the velocity u_{in} of the furthest flow. The power P, which the turbine actually can take out of the flow, therefore, should be expressed as a number C_p times that energy which is furthest away.

$$P = C_p \tfrac{1}{2} \rho A u_{\text{in}}^3 \,[\text{W}] \tag{5.10}$$

The *coefficient of performance* C_p must be smaller than 1, as not all energy can be taken out of the flow. It has been shown ([1], pp 133, 134) that the maximum possible value amounts to $C_p = 16/27 \approx 0.59$, which is called the *Betz limit* after the German engineer who derived this expression. The modern wind turbines which one finds in the countryside, may come close to the Betz limit with $C_p \approx 0.4$ ([63], p. 21). It may be surprising that turbines with two or three blades are able to capture so much energy as their rotation at any moment covers only part of their cross-section. The physical reason is that the turbine interacts with the field as a whole and not just with individual air molecules.

Aerodynamics

As a turbine interacts with the airflow as represented by the streamlines, it reminds of aeroplanes. Therefore, it is no coincidence that the methods of *aerodynamics* used to calculate the optimal shape for the wings of aeroplanes are also applied to find the best shape for the blades of a wind turbine. The cross-section of such a blade is shown in Figure 5.11 and it indeed resembles the wing of an aeroplane.

On the left of Figure 5.11 the wind with velocity u is coming horizontally from the left. Assume that the blade is in upward motion with a constant velocity u_{up}, indicated in the figure. The interaction between blade and flow is easiest described in a coordinate frame, in which the blade is at rest. In order

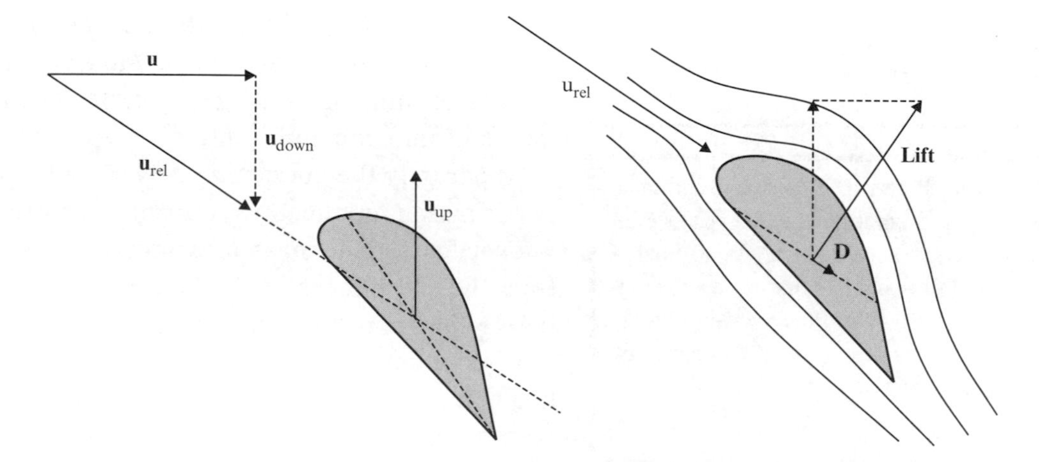

Figure 5.11 Turbine blade. The upward motion of the blade results in a relative velocity of the flow. The **Lift**, shown on the right, is perpendicular to this relative velocity. Its vertical component drives the turbine

to get the flow with respect to the blade one then has to add a downward velocity u_{down} to u, with the same magnitude as u_{up} but with the opposite direction. The two velocities $\mathbf{u}$ and $\mathbf{u}_{\mathrm{down}}$ together give the relative velocity $\mathbf{u}_{\mathrm{rel}}$, also indicated in the left part of Figure 5.11.

The streamlines of the resulting flow (with respect to the blade) are shown on the right in Figure 5.11. From aerodynamics it follows that there is a lift force* **Lift** perpendicular to the main flow and a drag force **D**, in the direction of the flow. The blade is designed such that the lift force is about 100 times bigger than the drag force, and may be ignored. The vertical component of the **Lift** produces the upward motion of the blade and the horizontal component exerts a thrust on the tower, which of course should be strong enough to counter it.

The fact that we talked about 'upward' motion and the 'vertical' is no restriction. For the original flow is assumed to be homogeneous in the horizontal direction and we did not refer to the vertical gravitational field in any way. Consequently, the horizontal direction may be regarded as an axis of symmetry. Figure 5.11 therefore describes an arbitrary situation, where 'upwards' just means the direction of the convex side of the blade.

> It has been said that the best conceivable wind turbine blade aerodynamically would be only 10 percent more efficient than a plank of wood. However a well chosen aerofoil will have a number of characteristics which will improve the machine greatly.
>
> [63], p. 41/43

Figure 5.11 gives rise to three other observations. In the first place, the upward velocity u_{up} will be proportional to the distance to the axis. Near the axis it will be almost zero and at the tips it will be very large. For an optimal performance the blades are twisted such that at any place the maximal *Lift* is obtained. In the second place, the streamline picture is also influenced by the magnitude of the incoming velocity u. In practice, one has a range of velocities and the shape of the blade will be optimized to the dominant value. Finally, the incoming velocity has to be perpendicular to the motion of the blades. To this end the blades are mounted on a nacelle that, by means of a servomechanism, is always oriented perpendicular to the winds.

The rotation of the turbine is coupled to a generator which creates an alternating electric current (AC). In order to couple it to the grid its frequency should be constant. This is done in various ways, for example, by the conversion from AC with variable frequency to direct current (DC) and next by conversion to AC with the right constant frequency ([63], p. 40).

*Note that the word 'lift' betrays its origin in the study of aeroplanes.

5.1.3 *Hydropower*

In China thousands of old dams must be repaired urgently to prevent a disaster like the collapse of a dam in 1975 in which 20 000 people drowned. A big dam on the Yangtze river will move 3 million people from their homes. Small-scale hydropower projects that don't need large reservoirs are now taking off. The 200 [kW] turbine to be installed in a weir on the river Thames could be the first of many in Britain.

New Scientist, 1 May 1999, 55

The physics of hydropower is trivial. We restrict ourselves to the basic equations for the two cases: power from dams and power from flowing rivers.

Power from dams

A hydropower station uses a dam, from which water will go down a height h and pass a turbine. The *potential energy* of the water is converted into kinetic energy of the turbine, which is coupled to an electric generator. A mass m with height h will have a potential energy mgh. If $Q\,[\mathrm{m^3 s^{-1}}]$ passes the turbine its mass will be $\rho Q\,[\mathrm{kg\ s^{-1}}]$, where ρ is the density of water. Consequently, the mechanical power the dam produces will be

$$P = \rho Q g h [\mathrm{J\ s^{-1}}] \approx 10\, h\, Q\ [\mathrm{kW}] \tag{5.11}$$

Power from flowing rivers

The *kinetic energy* of a flowing river at some places is still used to drive waterwheels. The rotational energy is then used directly for powering saws in the woodcutting industries, or looms as in the early British textile industries. For a flow with velocity u and mass m, the kinetic energy equals $\frac{1}{2}mu^2$. If $Q\,[\mathrm{m^3 s^{-1}}]$ passes the waterwheel or turbine its mass again will be $\rho Q\,[\mathrm{kg\ s^{-1}}]$, giving a mechanical power

$$P = \tfrac{1}{2}\rho Q u^2 \approx \tfrac{1}{2} Q u^2\ [\mathrm{kW}] \tag{5.12}$$

Comparison

The height h in Equation (5.11) could easily be 50 [m], while the velocity u in Equation (5.18) is often not much higher than 1 [m s^{-1}]. This implies that, for the same amount Q of water passing, dams are 1000 times as effective as flows. Besides, in a dam one may use all the water in a stream, while in a waterwheel usually only part of it is forced to pass the wheel. All together this means that the kinetic energy of flows nowadays will only be useful in specialized small-scale applications.

5.1.4 *Bio-energy*

When the human species started to make fires from wood, gathered in the forests, they used the energy stored by photosynthesis in the biomass. The basic equation was given in Equation (1.8) as

$$6\,CO_2 + 6\,H_2O + 4.66 \times 10^{-18}\,\text{Joule} \rightarrow C_6H_{12}O_6 + 6\,O_2 \tag{5.13}$$

> Wood processing industry and the paper-and pulp industries will utilize their biogenous wastes if this is economically favourable or if they are forced by legislation.
>
> *Renewable Energy World*, May 1999, 107–15

This chemical equation is very much simplified, as in reality all kinds of complex carbohydrates are formed. For a physical scientist Equation (5.13) is already complicated enough and much of the study of photosynthesis will be in the realm of chemistry and biology, while the contribution of physics lies in the detailed unravelling of the microscopic mechanisms. That is too complicated to discuss here. Still, we will say something about the efficiency, the thermodynamics and the applications of biomass.

Efficiency

On a microscopic scale process (5.13) requires a multitude of electrons to be excited by incoming photons. It seems unbelievable but 60 photons are needed for one turnover of the process (5.13). This number may become plausible, when one divides reaction (5.13) on the left and on the right by a factor of six. One, furthermore, has to realize that the reduction of CO_2 to carbohydrates and the oxidation of water to form O_2 are two different chemical events. (A substance is oxidized if it gives up electrons. It is reduced when it takes up electrons.) This effectively splits Equation (5.13) into two separate processes:

$$2\,H_2O \rightarrow O_2 + 4\,H^+ + 4\,e^- \tag{5.14}$$

$$CO_2 + 4\,H^+ + 4\,e^- \rightarrow (CH_2O) + H_2O \tag{5.15}$$

In the first reaction, four photons are used to extract four electrons from two molecules of water to generate one O_2 molecule (that escapes into the atmosphere) and four protons. In the second reaction four photons generate four electrons that are used to reduce CO_2 and to take up four protons (H^+), while one water molecule is formed. The net effect is one water molecule on the left, but this algebra obscures the mechanism of which we shall say more in the following three paragraphs, to appreciate its complexity.

> Agriculture does more than just produce food–it has a profound impact on many other aspects of local, national and global economies and ecosystems, and these impacts can be positive or negative. Farming can erode soils, reduce biodiversity and poison rivers, as well as adding to global warming.
>
> *New Scientist*, 18 December 1999, 10

Reactions (5.14) and (5.15) occur across a membrane, in which the water oxidation (5.14) occurs on one side of the membrane, while the CO_2 reduction (5.15) occurs on the other side. The turning over of (5.14) and (5.15) will generate a proton gradient across the membrane. This proton gradient is used to produce a compound called ATP from another compound called ADP and inorganic phosphate, by an enzyme called ATP-ase.

ATP is the universal energy carrier of a biological cell and is in fact required to produce the carbohydrates on the right-hand side of Equation (5.15). However, if the reactions really would proceed with the number of molecules as indicated by equations (5.14) and (5.15) too few ATP molecules would be generated. Therefore, some of the photons absorbed by the photosynthetic system must be used to transfer protons across the membrane and to generate the required amount of ATP.

Incidentally, this explains why solely water and CO_2 plus light energy and chlorophyll are not enough to drive photosynthesis. More is needed. First, special proteins are required to transfer those electrons and protons. Second, photosynthesis needs a membrane to generate a stable proton gradient, which is required for the formation of ATP. And third, one needs special enzymes to generate ATP and also to use the ATP for the production of the carbohydrates shown in Equation (5.15).

Counting all together, 10 photons are needed for reactions (5.14) and (5.15) and consequently 60 for the complete process (5.13). The photons required have energies corresponding to the visible region of Figure 5.7 (actually a little to the high-energy side) which is in the neighbourhood of the peak in the solar spectrum. When one takes an average value of 2 [eV] per photon, one needs 60×2 [eV] = 120 [eV] = 19.20×10^{-18} [J]. When one compares this with the 4.66×10^{-18} [J] stored according to Equation (5.13), one arrives at an efficiency of 24 percent.

For a more accurate calculation one needs to take into account that photons outside the blue-green part of the photon spectrum do not contribute (Exercise 5.7). Furthermore, there is some reflection of the incoming sunlight and also the respiration of the plant, which corresponds with process (5.13) running to the left. This finally gives a theoretical maximum efficiency of 5.5 percent. A farmer will be glad if he gets 1 percent of the solar energy converted into bio-energy, not counting the energy content of fertilizers and agricultural machines.

Thermodynamics of storage

From the point of view of thermodynamics one may describe photosynthesis as the light-driven transition between different states of the photosynthetic system, which are indicated in Figure 5.12. The lowest horizontal line indicates the lowest energy level of Chlorophyll associated to a substrate, hence the abbreviation ChlS. Sunlight is absorbed by the system, inducing excitations to a higher energy state of the chlorophyll, indicated as Chl*S. The energy of this state with respect to the ChlS state is indicated as E^*. We use a notation, where {ChlS} indicates the concentration of chlorophyll in thc ground statc [mol L^{-1} = M]. Then the product k_i {ChlS} is the number of transitions per second. The numerical value of the rate k_i depends on the intensity of the light and is shown in Table 5.1 for two cases: bright and cloudy.

Once the system is in the excited state Chl*S there are two possibilities: it can return to the ground state with a loss rate k_l, or it can convert into a product P with a rate k_{st}. The state P can only return to the initial ground state by the detour over Chl*S, with a 'back' rate k_b. So, product P stores the energy of the photosynthetic process.

One may show ([1], p. 142) that at temperature $T = 0$ the energy differences put vertically in Figure 5.12 are equal to the difference in chemical potential μ_i of the states shown. By Equation (4.67) the chemical potentials determine the Gibbs free energy. By Equation (4.65) the Gibbs free energy determines the non-expansion work at constant p, T and in that way the exergy of the system. Therefore, the photo energy stored in P is the difference in chemical

Table 5.1 Typical rates in photosynthesis / [s^{-1}molecule^{-1}]†

k_l	k_{st}	k_i (bright sunlight)	k_i (cloudy day)
10^9	20×10^9	10	0.1

†Note that the rates are given per molecule. For the numbers per mol one has to multiply by Avogadro's number N_A.

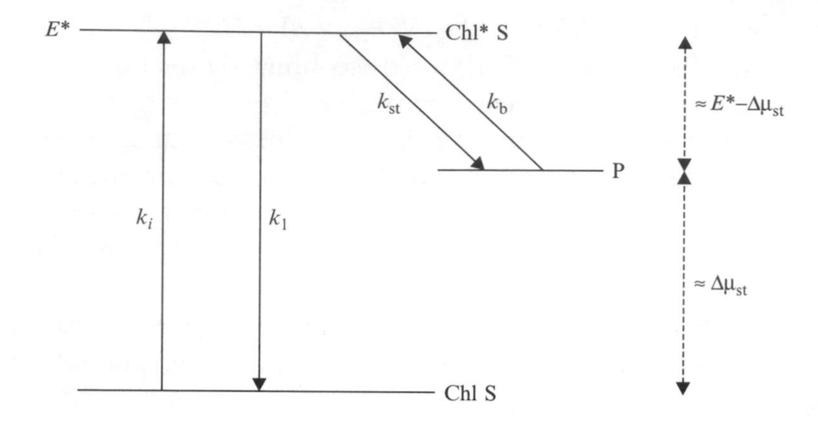

Figure 5.12 Chlorophyll on a substrate S is excited by an energy E^* to an excited state Chl*S. It either returns to the ground state or the energy is stored in P

potential $\Delta\mu_{st} = \mu_P - \mu_{ChlS}$, which can be harvested and put into use. Unfortunately, the Gibbs free energy depends on the temperature. Therefore, the vertical distance between P and ChlS in Figure 5.12, may not be interpreted as the difference in Gibbs free energy at temperature T. That is the reason why at the right of Figure 5.12 we put a sign $\approx$ in front of $\Delta\mu_{st}$.

In a practical case one is interested in the *storage efficiency* ϕ, defined as

$$\phi = \frac{\textit{net} \text{ formation of P per second}}{\text{excitation of ground state molecules per second}} \tag{5.16}$$

> Sweden produces more than 20 percent of its energy needs with re-growable raw materials. The supply is almost twice as much as is now being used. However, power prices have dropped so low in Sweden that construction of new biomass plants is hardly worth while anymore.
>
> *New Energy* 1/2000, 40–42

The *net* formation of P in the numerator is the relevant quantity, as we have to take the storage minus the back reaction. In a derivation of an expression for the storage efficiency ϕ we need the following assumptions ([1], pp 143, 144). (a) The concentration [P] changes only slowly during the day–night–day sequence and can be taken constant for a particular stretch day–night. (b) A steady state has been established. (c) In the dark the chemical potentials μ_P and μ_{Chl*S} are the same (equilibrium).

One then finds

$$\phi = \frac{k_{st}}{k_l + k_{st}} - \frac{k_l}{k_i} e^{-(E^* - \Delta\mu_{st})/kT} \tag{5.17}$$

Here, kT refers to the temperature at which photosynthesis takes place with $kT = 0.025$ [eV]. In the exponent on the right one finds the excitation energy

E^* of the chlorophyll minus the stored exergy. This is indicated, rather imprecisely, on the right of Figure 5.12 as a distance between the Chl*S and P levels. If the exponent in (5.17) is strongly negative, the exponent kills the factor k_l/k_i in front and one is left with the first term.

The first term on the right of (5.17) may be interpreted as the storage efficiency, if there is no reaction back from P. Indeed, once the excited state Chl*S is formed there are only two possibilities: storage in P or return to the ground state. Consequently, the second term gives a correction due to the back rate. As in Equation (5.4) one has competition between the driving 'force' for storage $(E^* - \Delta\mu_{st})$ and thermal disorder, which explains the factor kT in the exponential.

> Algae are a promising source of biomass because they grow extremely quickly. A new process is developed, which requires no drying of algae and cuts the energy needed to supply nutrients.
>
> *New Scientist* 17 July 1999, 17

In a slightly different way one may explain it as follows. According to Figure 5.12 $(E^* - \Delta\mu_{st})$ is related to the energy difference between the storage state P and the excited state Chl*S. The thermal motion is the mechanism, which should drive the back reaction against the energy difference. For constant temperature, if $(E^* - \Delta\mu_{st})$ becomes bigger, the back reaction will have a larger energy barrier to overcome and its probability will go down. That seems useful, but the stored exergy goes down as well, so there is competition between these two effects.

Example 5.3

Plot the storage efficiency ϕ of Equation (5.17) as a function of $(E^* - \Delta\mu_{st})$ for $k_i = 10$, $k_i = 1$, and $k_i = 0.1$.

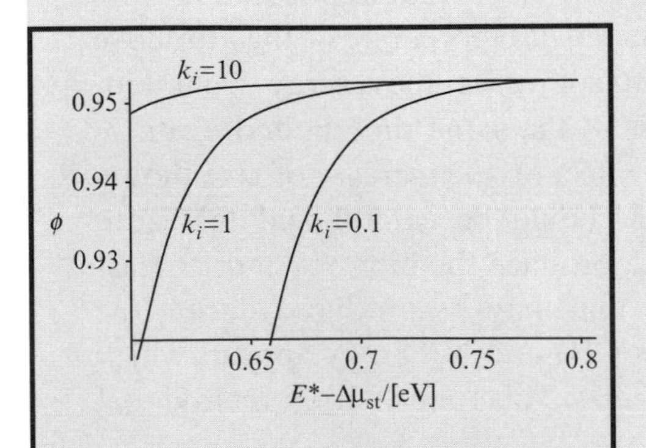

Figure 5.13 The storage efficiency for three values of k_i and as a function of $E^* - \Delta\mu_{st}$

Answer

The rates k_l and k_{st} are found in Table 5.1. The first term on the right of (5.17) then becomes 0.953. This is the maximum possible ϕ, as the second term in Equation (5.17) always subtracts something. A plot procedure from, e.g., Mathematica ® then immediately gives Figure 5.13

It is interesting to look at Figure 5.13. The excitation rate k_i depends on the illumination by sunlight. From Figure 5.13 one may deduce that a storage rate of 95 percent is reached for $k_i = 10$ when $(E^* - \Delta\mu_{st}) = 0.61$ [eV]. For $k_i = 1$ it is reached when $(E^* - \Delta\mu_{st}) = 0.67$ [eV] and for $k_i = 0.1$ when $(E^* - \Delta\mu_{st}) = 0.74$ [eV]. The conclusion is that plants which live at the bottom of a forest with a closed canopy of leaves on top should have a bigger value of $(E^* - \Delta\mu_{st})$ in their photosynthetic system than plants living in open areas. That is needed to store the lower intensity of sunlight at their location. This indeed is the case.

Applications

The main application of biomass is in home heating. The old ways of making a fire have been improved. Instead of open fires, tiled stoves have been widely in use for many years. They are receiving competition from other ways of burning wood, for example in woodchip boilers with a large user convenience. The market for electricity from biomass is still small ([64] and Table 4.4).

> We proved to ourselves and to the world that biodiesel from waste frying oil can power an unmodified vehicle. Not only did we show that this biodiesel powers our Veggie Van, but we showed that there is an alternative to petroleum-based fuel, and ultimately, that there are alternatives to our present fossil-fuelled economy.
>
> *Renewable Energy World* **2** (6), Nov. 1999, 74–86

From the point of view of exergy, the best method is to use all the Gibbs free energy in the biomass, indicated on the right in Figure 5.12. We have already discussed in Equation (4.55) that combustion is not the cleverest way to use the resources of nature. Not much research has been done on this point, as the processing of biomass to liquid fuel for cars, as we discussed in the context of Figure 4.17, is more popular.

As noted before, one of the attractive points of using the exergy from biomass, is that the CO_2 that is liberated is at the same time bound again in new biomass. This seems to neutralize its effect on the increase of greenhouse gases in the atmosphere. However, one should be careful and take into account the fuels and fertilizers needed to produce the biomass and, at least as important, look at greenhouse gases that may be produced during the cultivating of the biomass. In some cases CH_4 and N_2O may appear, which are even stronger greenhouse gases than CO_2. This follows from their global warming potential, depicted in Table 3.1.

5.2 NUCLEAR POWER

In nature there are only a few hundred stable atomic nuclei. That is to say there are a few hundred combinations of Z protons and N neutrons, which do not decay to other nuclei. The major kinds of radioactive decay (α, β and γ) are summarized in Table 5.2. If a nucleus is highly excited it may have other decays as well, summarized in Table 5.4. First we describe the concept of half-life.

5.2.1 Half-life

The decay of a radioactive nuclide follows the simple rule

$$N(t) = N_0\,\mathrm{e}^{-t/\tau} \tag{5.18}$$

where N_0 represents the number of nuclei at time $t = 0$ and $N\,(t)$ the smaller number at a later time t. By differentiation one finds

$$\frac{\mathrm{d}N}{\mathrm{d}t} = -\frac{1}{\tau}N(t) \tag{5.19}$$

This means that the number of decaying nuclei per second at any moment is proportional to the number still present. The constant of proportionality is called the *decay constant* and is given by $\lambda = 1/\tau$. Notice that the exponent in Equation (5.18) should not have a dimension; therefore τ has the dimension of time.

The *half-life* of the nuclide is the time $t_{1/2}$ after which the number has gone

> Chernobyl radioactivity persists in fish.
>
> About 10 percent of the initial peak radioactivity declines with an ecological half-life as long as 8–22 years. This ecological lifetime is smaller than the physical half-life of 30.2 years, as Cs is kept by the soil and dilutes.
>
> *Nature* **400**, 29 July 1999, 417

Table 5.2 Types of radioactive decay from an initial nucleus with Z protons and N neutrons

Decay	Emitted particle	Final nucleus (initial nucleus Z, N)
α	^{4}He nucleus $= 2\text{p} + 2\text{n}$; $Z = 2, A = 4$	$Z - 2$; $N - 2$
β^-	electron, negative charge	$Z + 1$; N
β^+	electron, positive charge	$Z - 1$; N
γ	gamma	Z, N

down to half the original number. Therefore, one substitutes $N(t) = N_0/2$ on the left of (5.18) and finds

$$N_0/2 = N_0\,\mathrm{e}^{-t_{1/2}/\tau} \tag{5.20}$$

Divide on the left and the right by N_0 and take the natural logarithm. This gives

$$-\ln 2 = -t_{1/2}/\tau \tag{5.21}$$

or for the half-life

$$t_{1/2} = \tau \ln 2 \tag{5.22}$$

Note that the half-life is independent of the number of radioactive nuclei present.

5.2.2 *Binding energy*

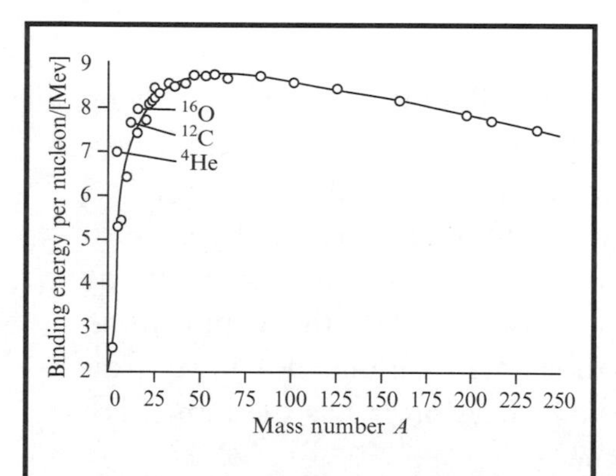

Figure 5.14 Binding energy per nucleon for stable nuclei as a function of mass number *A*

For nuclei, the *binding energy B* is defined as the energy one has to add to separate the $A = N + Z$ nucleons into N free neutrons and Z free protons. Conversely, this same energy is liberated when N free neutrons and Z free protons again form the same nucleus. The principle of nuclear power can be understood if one looks at the curve of the binding energy per nucleon B/A for stable nuclei versus the atomic number A as shown in Figure 5.14.

In Figure 5.14 the maximum binding energy per nucleon B/A is reached when $A \approx 60$. Indeed, ^{56}Fe is one of the most stable elements in the universe. To understand the principle of nuclear power, consider the heavier nuclei ^{235}U and ^{238}U. These two nuclides have the same charge Z and therefore the same chemical properties and are called *isotopes*. Although ^{238}U is not really stable (the heaviest stable nucleus is ^{209}Bi) its half-life of $t_{1/2} = 4.5 \times 10^9$ years is long enough to have survived since the formation of the earth about 4 or 5 billion years ago. In Figure 5.14 it is represented by the data point on the far right. Its isotope ^{235}U only has a half-life of $t_{1/2} = 7 \times 10^8$ years and is still present in a small fraction.

Some of the properties of the isotopes ^{235}U and ^{238}U are shown in Table 5.3. Notice that they both have a value B/A of around 7.6 [MeV]/nucleon. The

Table 5.3 Properties of ^{235}U and ^{238}U

	B/A	Abundance	$t_{1/2}$	$\sigma_f/(10^{-28}\,[\mathrm{m}^2])$
^{235}U	7.59 [MeV]	0.72%	0.7×10^9 years	583
^{238}U	7.57 [MeV]	99.3%	4.5×10^9 years	$< 4 \times 10^{-6}$

abundance of ^{235}U is much smaller than of the heavier isotope, which is not surprising as the lighter one is decaying faster. In the last column the rather mysterious symbol σ_f is the so-called *fission cross-section* for *thermal neutrons*. We will explain these concepts, as they are essential in the understanding of nuclear power.

5.2.3 *Fission*

Thermal neutrons are neutrons in thermal equilibrium with the surrounding matter. Their average kinetic energy therefore is $3\,kT/2$ where kT refers to the now familiar $kT = 0.025$ [eV] at ambient temperatures. When a thermal neutron hits a ^{235}U nucleus it is absorbed and for a very short time forms a ^{236}U nucleus. This is disturbed heavily and quickly splits into two (or sometimes three) smaller nuclei according to the reaction

$$^{235}\mathrm{U} + n \rightarrow \mathrm{X} + \mathrm{Y} + \nu n + \text{energy } (\approx 200[\mathrm{MeV}]) \tag{5.23}$$

This process is called *nuclear fission*. The fission cross-section σ_f indicates the probability that process (5.23) will happen. From Table 5.3 one finds that it is expressed in [m^2] and indeed σ_f defines the cross-section with which the incoming neutron 'sees' the nucleus. If this value of 583×10^{-28}[m^2] from Table 5.3 is compared with the geometrical cross-section of 1.72×10^{-28}[m^2] (Exercise 5.8) one understands that the fission cross-section of ^{235}U is very large indeed. At the same time the cross-section for thermal fission of the heavier nucleus is negligibly small. This is the reason why ^{235}U is used for controlled fission in a power station.

Fission (5.23) may lead to many combinations of final nuclei X and Y. With any combination of nuclei a corresponding number of neutrons ν is liberated, usually two or three. From Figure 5.14 it follows that nuclei X and Y will have higher binding energies than the initial nucleus ^{235}U. Therefore energy is liberated. With a glance at the figure, the position $A = 235/2 \approx 118$ when the nucleus would split into two equal parts, so one would guess a value of 8.4 [MeV]. Then in total an energy of $235 \times (8.4 - 7.6)$ [MeV] $= 188$ [MeV] would be liberated. Of course, the emitted, free neutrons have no binding energy. More precisely, an average value of 200 [MeV] is found. The distribution of fission products may be found in [53], p. 298.

The value of 200 [MeV] for a single fission may be compared with the energy bound by photosynthesis according to Equation (5.13). That value was 4.66×10^{-18} [J] = 29 [eV]. For $C_6H_{12}O_6$ this amounts to 0.1 [eV]/nucleon. Therefore, the energy from fission is seven orders of magnitude larger than in molecular processes like photosynthesis, or combustion. These numerical arguments were the reason why in the past so many physicists were enthusiastic about nuclear power. Their argument was that one would need a much smaller volume of fuel to produce the same power and it would produce a much smaller volume of waste as well.

> More than 450 nuclear devices were exploded at Semipalatinsk nuclear test site in Kazakhstan between 1949 and 1990. Since 1992 the researchers have traced the health records of almost 40 000 people from 53 settlements. Studies based on such data may help to reduce uncertainties in the knowledge of radiation risk. The register is very important in investigating low-dose exposure effects and in re-evaluating the standard criteria of radiation safety based on the Hiroshima and Nagasaki data.
>
> *New Scientist*, 16 October 1999, 12

Although it was well known that radiation might induce cancers and other illnesses, it was felt that a factor of 10 million would buy a lot of safety measures. The other side of the coin is that a nuclear explosion of a certain mass of material will liberate 10^7 as much energy as the same mass of a chemical explosive like dynamite.

Before discussing how commercial nuclear power works in practice, we note that besides fission fusion would also be an option. Suppose a ^2H nucleus (deuterium) with a binding energy of 1.1 [MeV] per nucleon (not shown in Figure 5.14) hits ^3H and forms ^4He and a neutron. From Figure 5.14 one deduces that per nucleon (7.0– 1.1–2.5) [MeV] would be liberated, which is 3.4 [MeV] per nucleon (in reality 3.5 [MeV]). This is even more than for fission and explains why commercial nuclear fusion is still being investigated.

5.2.4 *Nuclear power stations*

In a nuclear reactor the energy liberated in process (5.23) is converted into heat, which drives a heat engine, usually based on steam. In order to understand the operation of a nuclear power plant one has to look in a little more detail at Equation (5.23). The idea is to start the reaction with a few external neutrons and then maintain a *chain reaction*, which keeps itself going while producing useful energy.

The 200 [MeV] of liberated energy is distributed over the kinetic energy of the fission products X and Y (165 [MeV]), the kinetic energy of the neutrons (5 [MeV]) and several types of radiation. The fission fragments X, Y and the

> June 1997 Aleksander Zakhatov made a fatal mistake. He placed a thin shell of Cu on a sphere of highly enriched U. Suddenly, a huge burst of radiation turned the air blue as the U went critical. Zakharov informed his manager, lost consciousness and died from the radiation received.
>
> *New Scientist*, 30 October 1999, 20–21

neutrons will collide with the surrounding material and lose their kinetic energy in the form of heat. From the data given it follows that between 165 and 170 [MeV] of the energy can be taken away as useful heat ([53], p. 81; Exercise 5.9). The number of neutrons in process (5.23) on average equals $\nu = 2.43$, so each neutron will have a kinetic energy of a few [MeV]. At these energies the fission cross-section σ_f is very small; consequently only a small fraction of the liberated neutrons will again produce a reaction (5.23). This fraction is much too small to get a self-supporting chain reaction.

> The health risks associated with radiation from exposures to depleted uranium (the tail of enrichment with 99.8 percent ^{238}U) are relatively low – so low as to be statistically undetectable, with one exception. Radiation doses for soldiers with embedded fragments of depleted uranium may be troublesome. Besides, the chemical toxicity of uranium in the kidneys may be significant.
>
> Bulletin Atomic Scientists, Nov/Dec 1999, 42–45

In order to get a self-sustaining process one has to slow down the neutrons to low thermal energies of about 0.025 [eV] without losing them to absorption. For this purpose one lets the neutrons collide with a *moderator*, usually water (H_2O). In the moderator a neutron with a high speed hits a proton at rest. As proton and neutron have almost the same mass they equally share the kinetic energy of the neutron, which therefore loses half of its kinetic energy. This collision process has to occur many times before the neutron arrives in the thermal region with high fission cross-sections. In the mean time it may get absorbed in the water itself, in the ^{238}U or in the reactor vessel. In this way the number of liberated neutrons per fission event reduces to the so-called *multiplication factor k*.

If the reactor is of infinite extension it will not lose neutrons to the outside world. In this case the multiplication factor is called k_∞ and for a typical reactor it is in the order of $k_\infty \approx 1.10$ to 1.20. Of course this will be larger than k itself. Consider as a thought experiment, a very large spherical reactor with $k \approx k_\infty$. When the radius of the sphere becomes smaller, the loss of neutrons at the surface will increase until at a certain stage the value of $k = 1$ is reached. This is the *critical size* of the reactor.

For another shape, for example that of a cigar, the ratio surface/volume is bigger than for the spherical case. Consequently, one loses more at the surface and the critical size will be larger. For this reason workers with fissile materials are advised never to use spherical shapes for their containers.

For steady state operation, the value of the multiplication factor k should be kept at $k = 1$, so it is clear that k_∞ should be bigger than one. But it must be smaller than the number of neutrons $\nu = 2.43$ liberated by the fission (5.23). The reader should realize that it is a technological *tour de force* that this has been achieved at all.

The trick is to keep the losses as low as possible. For ordinary water and natural uranium with a small (0.7%) abundance of ^{235}U, the multiplication factor is below $k_\infty = 1.1$. The losses of neutrons in water and in ^{238}U are too high to sustain a chain reaction. Therefore at least one of the losses has to be reduced. Sometimes one uses heavy water, in which the ^1H atoms are replaced by the deuterium isotope ^2H; this reduces the absorption of neutrons in water. More often one *enriches* the uranium to about 2.5 percent of ^{235}U.

Figure 5.15 shows a schematic representation of a nuclear reactor. On the left the reactor core is shown, in which fissile materials are present in the form of rods (not shown) and embedded in a moderator to slow down the neutrons. Around the core a reflector reduces the loss of neutrons to the outside. There should be an active mechanism by which the multiplication factor k can be kept constant at $k = 1$. Control rods, indicated in Figure 5.15 and consisting of a material that easily absorbs neutrons are used for this purpose. If need be one can push them a little further into the reactor core or pull them out.

A coolant is pumped through the core to take away the heat of the process (5.23). That may be water under high pressure that exchanges its heat to another water loop where it drives a steam turbine. The coolant has a temperature of about 300 [°C] and consequently the net efficiency is smaller than Carnot's value of 34 percent.

Time dependence and control In order to understand the behaviour of a reactor under a small perturbation, one defines a time l as the time in which a neutron reproduces itself. Neutrons, which are produced in fission (5.23)

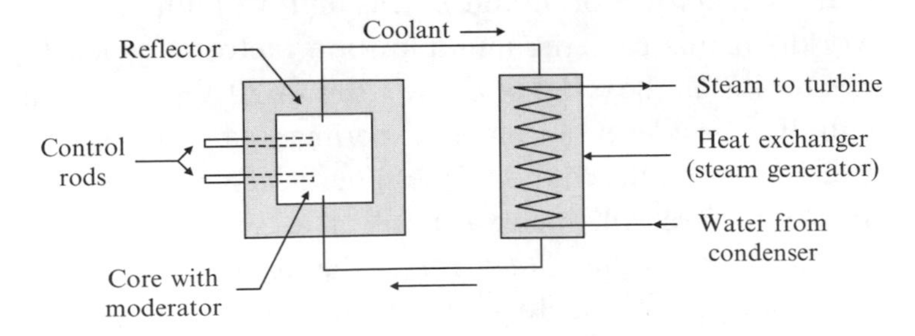

Figure 5.15 Nuclear reactor. The heat from fission (5.23) is taken away by a coolant, which by means of a heat exchanger, drives a steam turbine. A reflector should keep neutrons inside and control rods should keep $k = 1$

The Tokaimura plant lies some 150 [km] northeast of Tokyo. September 30, 1999 technicians were adding U_3O_8 to HNO_3. The two chemicals were to be mixed in an elongated container designed to prevent a critical mass of U coming together. However, they poured it into a tank using buckets. After the 7th bucket the tank contained 16 [kg] of U enriched to 18.8 percent ^{235}U. At 10.35 a.m. it formed a critical mass and started a chain reaction. The process was facilitated by the water, both in the tank and surrounding it, which acted as moderator. Workers had to rush into the building for just minutes at a time to remove the water and stop the reaction.

New Scientist, 9 October 1999, 3–5

will first slow down by collisions and then stay for some time with thermal energies in the moderator before they are absorbed by the ^{235}U of one of the fuel rods. Then they immediately cause another fission. The sum of these two is called l, and approximately $l \approx 10^{-4}$ [s].

Assume that for any reason the multiplication constant increases from $k = 1$ to $k = 1 + \rho$. This means that in the reproduction time $\mathrm{d}t = l$ the number of neutrons increases from N to $kN = (1 + \rho)N$. The increase in the number of neutrons therefore can be written as $\mathrm{d}N = \rho N$. One may put this in differential form as

$$\frac{\mathrm{d}N}{\mathrm{d}t} = \frac{\rho N}{l} = \frac{\rho}{l} N \qquad (5.24)$$

In this equation the derivative $\mathrm{d}N/\mathrm{d}t$ equals a constant times the function $N(t)$ itself. From mathematics it follows that the solution is an exponential function

$$N = N_0\, \mathrm{e}^{\rho t/l} = N_0\, \mathrm{e}^{t/\tau} \qquad (5.25)$$

where $\tau = l/\rho$ is called the *time constant* of the process. For a small perturbation one may take $\rho = 10^{-3}$. The reproduction time $l = 10^{-4}$ [s] leads to a time constant $\tau = 0.1$ [s]. After 10 seconds the number of neutrons has increased by a factor $\mathrm{e}^{100} = 10^{43}$ and as the neutrons originate from fission, the liberated fission energy has increased by the same enormous factor. Of course, before that, the corresponding increase in temperature will have resulted in a meltdown of the nuclear materials, chemical explosions and possibly a blow up of the container building in which the reactor was positioned.

Radioactive food and jewelry from homes near the Tokaimura nuclear plant in Japan are revealing the impact of the accident there last year (1999). Gold is a most sensitive way of evaluating neutron flux. A gold necklace worn by a woman working more than a kilometre from the accident was 30 times as high as normal. It turned out that this was because she returned twice to her house 350 metres from the plant to feed her dogs.

New Scientist, 20 May 2000, 5

With a small time constant like $\tau = 0.1$ [s] one has very little time for counter measures. Fortunately, there is one little mistake in the calculation given above. The neutrons displayed in (5.23) are not all released immediately, but some of them, a fraction $\beta = 0.0065$ result from a decay of fission products X and Y and their daughters and are called *delayed neutrons.* On average, the delay time $t_d = 9$ [s]. If one includes the delayed neutrons in the calculation, the reproduction time becomes l^* with

$$l^* = (1 - \beta)l + \beta t_d \approx \beta t_d \approx 0.05[s] \qquad (5.26)$$

In this case the time constant becomes $\tau = l^*/\rho = 50$ [s], which is long enough to take mechanical counter measures, such as pushing neutron absorbing control rods into the reactor.

However, these control measures only work for small perturbations. For larger values of $\rho(\rho > \beta)$ the *prompt neutrons*, which are released immediately can do the job of exponential increase all by themselves. Still, the presence of delayed neutrons in a conventional reactor helps the designers.

> Cleaning up old nuclear sites often involves removing contaminated concrete. The T. Thiooxidans bacteria should offer a slow but safe way to clean up. The bacteria were mixed with S and cellulose thickener. They then eat through concrete at a rate of 1 [cm] a year as long as the humidity is high. When enough is removed, the gel is allowed to dry out, removed and stored safely.
>
> *New Scientist*, 9 October 1999, 6

5.2.5 *Radiation and health*

The reason to be careful with the operation of nuclear power plants is the well-known health hazard of receiving a surplus dose of radiation. That dose is measured in a unit *gray* (Gy) which corresponds with the absorption of 1 Joule of energy per 1 [kg] of body mass, so in terms of dimensions [Gy] = [J kg^{-1}]. Heavy particles of high energy do much more damage than lighter particles of low energy, even when they dispose of the same amount of energy into the living tissue. Therefore the dose in gray is multiplied by a quality factor Q, resulting in a new unit the *sievert*[56], with the same dimension [Sv] = [J kg^{-1}]. One often meets the old unit rem, meaning radiation equivalent man: 1 rem = 0.01 sievert. In Table 5.4 the most common types of radiation are given with their quality factors Q. The types of radioactive decay were defined in Table 5.2. In order to judge the hazards of radiation, one should know that radiation has always been with us. We receive some radiation from the cosmos, and more from the soil by traces of radioactive elements present there. From these natural sources the average human receives around 2 milli-sievert per year (2 [mSv yr^{-1}]). There is a wide range

Table 5.4 Radiation with quality factors Q

Particle	Energy	Q	Decay type
electron (pos. or negative)		1	beta decay
α		20	alpha decay
n	$E < 10\,\text{keV}$	5	neutron emission $N \to N - 1$; same Z
n	$10\,\text{keV} < E < 100\,\text{keV}$	10	
n	$2\,\text{MeV} < E < 20\,\text{MeV}$	10	
n	other energies	20	
p	$E > 2\,\text{MeV}$	5	proton emission $Z \to Z - 1$; same N
γ		1	decay within the same nucleus
X		1	(medical) Roentgen radiation: change within atomic electron cloud)

around the average, though, depending on the place where one lives and the kind of soil or rocks around.

> Waste should be lowered down boreholes drilled to 4 [km]. The trick is to exploit heat generated by the waste to fuse the surrounding rock and contain any leaking radioactivity. The difficulty will be to ensure that putting heat into the rock does not cause water to start circulating upward by convection.
>
> *New Scientist*, 18 September 1999, 21

Medical applications (roentgen or X-ray diagnostics) and building materials (esp. concrete) add another 2 [mSv yr^{-1}] to the average, which totals at 4 [mSv yr^{-1}]. The question then is what amount of additional radiation is acceptable from other sources? The International Commission of Radiological Protection (ICRP) has given norms [65], which are briefly summarized in Table 5.5. Besides the whole-body dose, the acceptable dose for certain parts of the body are also given. One will notice that the norms for people who are professionally involved with radioactivity are higher than for the general public.

The norms of Table 5.5 are put rather low. An individual who will receive, for example, double the norm, most probably will experience no harm. If a major part of the population were to experiences such a dose, a few would develop cancers or other malign effects. The health hazards therefore are statistical and are found from studying the surviving population of Hiroshima and Nagasaki, survivors from some other accidents, and from animal experiments.

If doses are very much higher than those of Table 5.5, individuals will certainly experience harm. With a sudden dose of 200 [mSv] a human will experience a temporary decrease of white blood cells. For 2000 [mSv] the

Table 5.5 Norms for upper limits of radiation from non-natural and non-medical sources, from ([65], p. 72)

Application	Public/[mSv yr^{-1}]	Occupational/[mSv yr^{-1}]
total effective dose	1	20 (5-year average)
lens of the eye	15	150
skin	50	500
hands and feet		500

resistance against infections will be gone. At 5000 [mSv] nausea and diarrhoea will be acute with death within a few months. A dose of 80 000 [mSv] is lethal within a few hours. These data are taken from a study on the consequences of nuclear war [66]. Civilian nuclear power is planned not to give doses higher than the small numbers of Table 5.5.

5.2.6 *Safety and accidents*

> The ferocious 2000 fire season in the United States saw two American nuclear weapons facilities hit by wildfires in less than 2 months. Although no buildings housing radioactive or chemical materials were burned down, the fires should be viewed as a wake-up call to be better prepared. As frightening as these fires were, they could easily have been much worse.
>
> Bulletin Atomic Scientists, September/October 2000, 10–12

In order to judge the health hazards of nuclear power, the total process from the mining of the uranium to the management of the waste should be analysed. From Table 5.5 one knows that one has to avoid radioactivity entering the biosphere. Let us follow the life cycle of the materials:

- Mining of the uranium produces 'tails' at the mining site. This is a rather local problem, and has to be dealt with locally.
- Enrichment of the uranium to 2.5 percent of ^{235}U. Care has to be taken that in containers no critical mass can be reached. This also holds for some of the points below.
- Production of the fuel elements. When only U isotopes are used, the health hazards are low. But the production is often combined with the handling of used fuel elements with health hazards, see below.

- The nuclear power plant. This will be discussed below.

- Handling spent fuel elements. Here, the leftover uranium is separated chemically from reaction products X and Y of reaction (5.23) and from the heavier nuclides formed from ^{238}U. Most of them are very radioactive. Chemical explosions or fires in these plants may be very dangerous.

- The waste from the spent fuel elements. This will be discussed below together with the waste from decommissioning the plant when it is outdated. After the last fuel elements are taken out, the plant is left for a long time, possibly 100 years, to let the activity of the induced radioactivity in the building and the installations reduce to an 'acceptable' level. Note that, according to Equation (5.18), radioactivity will eventually disappear. But for a long half-life one has to wait a very long time.

The power plant For a power plant, the exponential function (5.25) shows that if anything goes badly wrong with a nuclear power station, it happens fast. The reason to worry about it is the factor 10^7 in energy density compared with a power station based on chemical energy, such as a fossil fuel power station. Even when a runaway nuclear power station has successfully been shut down, one still has to consider the fission products X and Y in (5.23). About 3 [MeV] of the liberated energy of 200 [MeV] is produced by decay of fission products, which continues for an appreciable time.

> A rigorous containment is a principal basis for the further and large-scale uses of nuclear power. Furthermore, future reactors should be passively safe, that is, in an emergency they should not require quick active interventions. There should be a period of days before an action is required for the retainment of the radioactivity in the core.
>
> [67], p. 22,23

Moreover, the absorption of neutrons in ^{238}U, which is a majority constituent of the fuel in current power stations, produces a multitude of heavier nuclei, many with long decay times. One therefore needs mechanisms to cool the reactor core to take the rest of the heat away and mechanisms to avoid the radioactive materials entering the environment. All these mechanisms have been designed and power plants, certainly in the West, have a final containment that only rarely breaks down ([67], p. 22)

Major accidents with power plants occurred in Harrisburg, USA in 1979 and in Chernobyl, USSR (now Ukraine) in 1986. In the Harrisburg accident it is estimated that a single individual may have received a dose of 0.8 [mSv] and the average dose for the population within a circle with a radius of 80 [km] was a factor of 100 less. The Chernobyl accident was much worse in its consequences, due to the fact that the containment badly broke down. There were 135 000 people evacuated who on average had received 100 [mSv], which means that quite a few had received appreciably higher doses. These numbers are substantially higher than the norms of Table 5.5. Still, if anyone of these people develop a tumour it is hard to say whether the origin is the radiation of the accident or something else. At not extremely high doses the effects are statistical in nature, which makes them hard to interpret.

> Passive or inherently safe reactor designs are largely paper studies. It would require at least a couple of decades to demonstrate these designs and deploy the technologies. So far, no utility or consortium of utilities has been willing to take the financial risk of building prototypes for testing, and no government has come up with the financing.
>
> Bulletin Atomic Scientists, July/August 2000, 39–44

Reactor safety is an important aspect of the acceptability of nuclear power. Therefore, scientists and industry are studying designs where an uncontrolled chain reaction is impossible. We mention but one of the many proposals, still in the stage of design. In Accelerated Driven Systems a nuclear reactor is used with a multiplication factor $k < 1$, so it lacks the neutrons for a sustained chain reaction. The required extra neutrons are produced by bombardment of Pb by high-energy protons from an accelerator. If anything goes wrong, the accelerator stops and so does the reactor. In the reactor one uses Thorium, which is abundant in the earth and does not need to be enriched [68].

Nuclear waste Inside a nuclear reactor the ^{238}U will absorb neutrons and form ^{239}U which then decays in two steps to ^{239}Pu. The neutron density is so high that the ^{239}Pu will again absorb neutrons and form ^{240}Pu, which may absorb more neutrons, etc. After some time a multitude of heavier nuclei are present in the reactor, often with a very long half-life. The beginning of the chain of *transuranic nuclei* is shown in Figure 5.16.

The transuranic nuclei build up inside the fuel elements. After some time, in the order of three or four years, the fuel elements are replaced by new fuel rods. The old rods are then stored for some time in order to get rid of short-lived radioactivity. Next they are treated, where the Pu often is mixed with uranium for new fuel elements, as Pu isotopes provide fissionable material as well. In practice, 1/3 or 1/4 of the fuel elements are replaced annually. In nuclear submarines, highly enriched uranium is used that does not need to be replaced for 15 years or more.

The problem with the waste is the presence of many nuclides with very long half-lives. It contains ^{239}Pu, which has a half-life of 24 000 years. ^{239}Pu emits α particles, which are very destructive (see the high Q-value in Table 5.4). So it is very dangerous when dust containing ^{239}Pu is inhaled and reaches the lungs. Other transuranic nuclei have even longer half-lives, so they have to be kept out

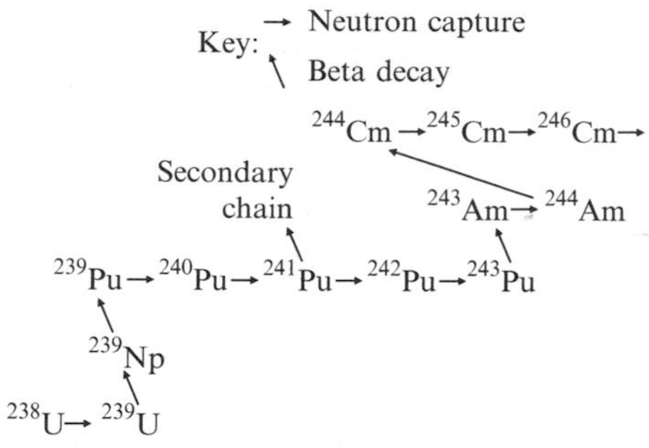

Figure 5.16 Transuranic nuclei produced in nuclear reactors.

> Given the inherent links between the military and the peaceful atom, full realization of an economy powered by atomic energy would require that nuclear weapons no longer be considered legitimate for any country. Fundamentally and in the long run, the problem which is posed by the release of atomic energy is a problem of the ability of the human race to govern itself without war.
>
> Bulletin Atomic Scientists, July/August 2000, 39–44

of the natural environment for a very long time. It stands to reason that a convincing method of safe storage over such a long period of time is very difficult, if not impossible, to find. It is feasible to fission the transuranic nuclei artificially by bombarding them with high-energy particles. Whether at the end of the day one then still has an economically interesting option is a matter for study.

A final point of attention is the production of nuclear weapons. The easiest way to produce a fission weapon is to put a rod of natural uranium in a nuclear reactor. The ^{239}Pu formed is fissionable, as mentioned above. If one keeps the rod only a short time inside, the transuranic chain of Figure 5.16 does not have time to build up, and one has got ^{235}U, ^{238}U and ^{239}Pu. As Pu has chemical properties different from U, they are easy to separate by standard chemical methods. With a critical mass in the order of kilograms, a government or a big terrorist organisation can easily acquire enough Pu to pose a security risk. Of course this risk can be limited by a strong international control mechanism under supervision of the United Nations or a similar organisation. The reader has to judge whether such a control system is realistic in the present international order.

EXERCISES

5.1 Check the statement in the caption of Figure 5.3, that one hour of perpendicular sunshine corresponds to 4.9×10^6 [J m^{-2}].

5.2 A solar collector system is costing 2000 Euro or 2000 US$. In a country like Australia (25° South) it is saving 4000 [kW$_{th}$] a year. Look at the cost of heat on your own energy bill and calculate how many years the installation has to run in order to get the money back. Ignore maintenance and interest costs. Would this installation produce more or less thermal energy for your own local conditions? Note: an industry only makes an investment when the payback time is at most four or five years.

5.3 Show that (2.8) for the simple case of one ideal gas may be written as $p = p_0 \exp(-Mgz/kT)$. Write $m = nMN_A$ where M is the mass [kg] of a single molecule and use $k = R/N_A$.

5.4 Check that equations (5.6) and (5.7) correspond to a total flux of 1000 [W m^{-2}].

5.5 The kinetic flow energy P of wind energy is given in units [J s^{-1} m^{-2}] = [W m^{-2}]. Check this from Equation (5.9).

5.6 Take the wind turbine of Example 5.2. Ignore the vertical velocity differences and take just the velocity at the axis height to calculate the total power comprised in the circle that the blades are covering.

5.7 Explain the difference between the *energy spectrum* depicted in Figure 5.7 and the *photon spectrum* required to discuss photosynthesis. Plot both spectra as a function of energy in [eV] and observe how the peak shifts.

5.8 The geometrical cross-section of nuclei may be calculated with the empirical relation for the nuclear radii $R = r_0 A^{1/3}$, with $r_0 = 1.2 \times 10^{-15}$ [m]. (i) Calculate the geometrical cross-section σ for the nuclei ^{235}U and ^{238}U. (ii) Show that the empirical relation expresses that within all nuclei the nucleons occupy the same individual space. (iii) What is the density ρ of nuclear matter [kg m^{-3}]?

5.9 Assume that during a year in a nuclear reactor 1 [kg] of pure ^{235}U is completely converted by means of process (5.23). What is the thermal power of the reactor in [MW_{th}] and what is the electrical power in [MW_e]?

6 Making up the Balance

Halfway through the book, we will now reflect on the preceding chapters and draw some conclusions. In Section 6.1 we summarize the problems of climate and global change and the fact that renewables or nuclear power do not offer a 'free lunch'. We conclude that we should slow down our rate of environmental change and we reconsider the tasks that science and technology should fulfil for modern society. We point out some methods of achieving that in Section 6.2.

6.1 *GREENHOUSE GASES, POLLUTION AND RESOURCES*

We will summarize the problems which we met in previous chapters concerning greenhouse gases and pollution and add some problems involving finite resources. For the latter we take energy and water as examples. At the end of this section it should be clear that we must reduce our emissions and our claim on natural resources.

6.1.1 *Greenhouse gases*

The most illuminating information is contained in Figure 3.14. There the expected increasing emissions of CO_2 were put down in a wide-curve for a 'business as usual' scenario (IS 92a). The other curves displayed the drastic cut of emissions required to stabilize the CO_2 concentration on a value of 450 [ppm] or 550 [ppm], still much higher than the present value of about 360 [ppm]. We have seen that models were needed to calculate Figure 3.14, with assumptions about the sensitivity of the climate system and the carbon cycle. Still, the real picture will always resemble Figure 3.14 in a qualitative way.

> Since 1976, the world has been warming at a rate of 2 [°C] a century. According to draft IPCC reports this rate is unprecedented in the past millennium. A 3 [°C] rise in Greenland will trigger the irreversible melting of its ice sheet. When that happens we only have ourselves to blame.
>
> *New Scientist*, 20 November 1999, 24

There are methods to sequester the CO_2 by catching the gas at the outlet of industrial installations (Section 4.3). It is not clear yet whether this can be done on a large enough scale and it will have a price tag anyway. Besides, other greenhouse gases with their large global warming potentials, displayed in Table 3.1, are catching up with CO_2 in their global importance. To catch them all and make them harmless seems rather far-fetched.

Finally, one has to distinguish between change and *rate of change*. The climate system has many different timescales built in, from days to thousands of years. On an annual scale we experience the difference between summer and winter and on a longer timescale we know of some very hot summers and extremely harsh winters. We have seen that oscillations in the oceans play a role, possibly in connection with Milankovitch cycles. These variations interact with vegetation and forests with timescales of tens of years. This leads to a conclusion, which we intuitively accept: a temperature change of 2 [°C] in 1000 years is easier to accommodate than a change of 2 [°C] in 100 years. We would expect that a smaller rate of change would lead to less extreme weather events and that nature and human society can more easily adapt to it.

6.1.2 *Pollution*

Pollutants emerge from all human activities. In fossil fuel fired power plants the emissions of NO_x, SO_2 and particles need special attention. In Section 4.3 it was shown that technologies to deal with these emissions already exist and are applied. Because of their high costs these technologies are implemented more frequently in high-income countries than in low-income countries.

Road transport adds to the several kinds of emissions, shown in Table 4.3. Technologies like the three-way catalytic converter in the car exhaust considerably reduce the emissions. The extreme rise in the number of cars on the roads unfortunately nullifies this effect.

For electricity production, which is an important and growing part of the energy conversion field, the alternatives of renewables (Section 5.1) and nuclear power (Section 5.2) were discussed. In these sections we mentioned various environmental aspects and dangers, in particular for nuclear power.

In order to make political choices it is necessary to compare the environmental effects of various methods of energy conversion. This is done in Figure 6.1, which gives an overview of several energy technologies, including the use of fossil fuels and some renewables, which we did not discuss. The entry 'Passive solar energy' refers to using solar collectors for producing hot water. The figure is taken from a solar energy conference [69], where the positive evaluation of solar power must have been popular. A more detailed

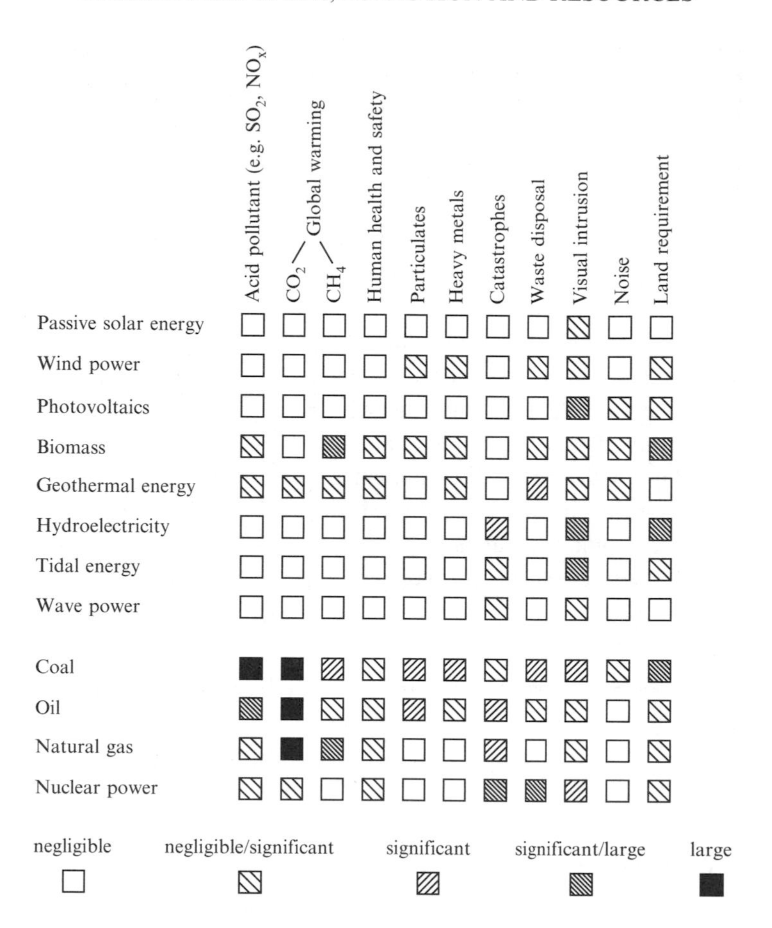

Figure 6.1 Environmental effects of energy technologies. Reproduced with permission of Kluwer Publishers from [69]

but similar summary may be found in [70], pp 36–37. In detail, Figure 6.1 may be open to dispute, but in general it represents a fair judgement.

In the figure fossil fuels rank rather badly; especially coal and oil which may lead to a broad spectrum of emissions. Notice a few other aspects which were not mentioned earlier: almost all options lead to increased land use, some of them lead to noise and all have a non-negligible visual intrusion on the landscape. For a few options, waste disposal has to be organized. The fact that the figure contains implied judgements follows from the waste disposal entry for nuclear power, which is 'significant/large'. On the basis of Section 5.2 we would put that in the highest category as 'large', comparable at least with the acid pollution of coal fired power plants, which in Figure 6.1 was put as 'large'.

It would be interesting to have a last column in Figure 6.1, containing the price of the electricity produced. To do this one could look at the price on the electricity market. Then coal would be one of the cheapest and photovoltaics

> The way in which our so-called highly prestigious Indian Institutes of Technology have only taught the Western models makes me feel that they were keen to make our engineers more the children of that arch-colonialist Lord Macaulay, who wanted all of us to be like the British, instead of making our engineers the children of Gandhi. And it is not just our brown Asian tinkerers who think Western, unfortunately, the whole developing world is madly in love with the highly destructive practices of the West. Getting rid of colonised minds is far more difficult than getting rid of the colonisers themselves.
>
> CSE, 15 February 2000

one of the most expensive. This is the reason why in Table 5.2 one only finds a very small contribution from photovoltaics.

The difficulty is that the market does not include environmental costs, as governments do not charge the power utilities a realistic environmental tax. On the other hand, the cost of nuclear power would be on the low side because of all kinds of government subsidies in the past. There are also government subsidies on fossil fuels [71]. Therefore electricity prices are highly political and we do not give prices for electricity in our main text, although in our appetisers the student will find prices quoted. Reference [71] shows a graph from the International Energy Agency, which predicts decreasing prices for PV electricity, because of economies of scale. They may reach 10 Euroocts/kWh in 2010.

6.1.3 *Resources*

Society is consuming natural resources at an increasing rate. Below, we only point to the resources of energy and water, although other examples could be given.

Energy

In Figure 1.7 a curve for the price of oil as a function of its accumulated use was shown. The data points were scattered around a horizontal line – and in fact that is still the case. The prices of oil, coal and gas do not indicate a quick exhaustion of this resource.

Every year the United Nations publishes a UN Energy Statistics Yearbook ([57]*). In that yearbook one may find the energy use for different fuels over the previous few years. One also will find estimates of resources on the basis of the prospecting and exploration of the big companies. The present authors once used the 1990 information to calculate the number of years the resources would last and came out with about 97 years. A few years later, we found that

*On the internet the most recent, as yet unprinted, data may be found in [54].

> Many will term the extreme drought in Gujarat and Rajasthan a 'natural disaster'. But this is far from the truth. It is a 'government-made' disaster. Over the last one hundred years or so, we have seen two paradigmatic shifts in water management. One is that individuals and communities have steadily given over their role almost completely to the state. The second is that the simple technology of using rainwater has declined. Instead exploitation of rivers and groundwater through dams and tube-wells has become the key source of water.
>
> However, villages that had undertaken rainwater harvesting or watershed development in earlier years had no drinking water problem and even had some water to irrigate their crops.
>
> CSE, 31 May 2000, 3

in 1995 world consumption had increased but estimated resources had increased even more: then they would last for 110 years. The reason is that new resources were found and, perhaps more important, that new technologies were developed which make it possible to get more of the fuels out of the earth. So we do not expect to run out of our global energy resources very soon.

Still, cheap resources of fossil fuel must run out eventually. In the meantime one should not only find ways to reduce the emissions into the environment, but also recall the lesson of Section 4.2 that, by combustion, a lot of the exergy in the form of chemical energy is wasted. There is still room for some clever ideas in this area.

In the energy field we conclude it is pollution, including greenhouse gases rather than the finite resources, which will be the limiting factor in further development. For another finite resource, water, the physical limit of the resource may be reached sooner than we expect.

Water

The essential data for fresh water are summarized in Table 6.1 [72]. Water pouring down as rain is usually suitable for drinking water. Of course, it is the fraction on land that is interesting and this is given in the table (Exercise 6.1). The fraction that rains on the agricultural land, fills the big lakes, seeps down to groundwater basins or flows off the land as rivers is accessible to humans and this estimate is given in Table 6.1 as well.

An important thing to notice is the large fraction of accessible water which is used in agriculture: $3/5 = 60\%$ of the present water consumption and

Table 6.1 Fresh water supply on land/[$10^{12}\mathrm{m}^3\mathrm{yr}^{-1}$] from [72]

Pouring down as rain	113
Accessible to humans	9
Amount of this in rivers	2–3
Total use 1990's	5
Amount of this in agriculture	3

$3/9 = 1/3$ of the available fresh water. It is expected that the growing world population will need more water for animal and human food; also rising wealth puts a heavier strain on the water consumption per person. Therefore, the experts believe that eventually most of the rivers will be used for irrigation. They not longer will reach the oceans but will be diverted to agricultural lands.

As a rough estimate, the water to agriculture may double by irrigation and consequently, food production as well. But a factor of two in population growth and food consumption per person together, is quickly reached. Water resources at this very moment are a source of disagreement and conflict between many countries. The data of Table 6.1 suggest that fights for the scarce resources of water will increase in the future.

In the beginning of this chapter we have already given the conclusion: we must find ways to limit the human strain on the natural environment, and, we should add, people must use peaceful ways to settle their arguments on anything, scarce resources included.

6.2 DEFINING THE TASK

> The role of energy within building varies from 30 percent in OECD countries, 50 percent in non-OECD Europe, to 70 percent in developing countries. Population growth and demand for housing have forced politicians to embark on massive housing schemes without consideration of comfort, energy demand and environmental issues. For example, in the state of Qatar 50 percent of the energy used can be saved by using low energy buildings.
>
> [74], p. 1

Finally, we give some simple examples of what can be done to reduce energy demand by technological and organizational means [73]. We recall the second law efficiency ε, which according to its definition (4.56) compares the useful output of a device with the maximum possible useful output by any system for the same task. Underneath Equation (4.56) the question about usefulness has already been raised. What is useful for one person may be a luxury or waste for another. We do not enter this political debate, but briefly discuss what determines comfort from a physical point of view, and how the physics of building and town planning may reduce energy requirements without jeopardizing comfort [74].

6.2.1 Comfort

At first glance our bodily comfort seems to depend on the air temperature. That is the single most important variable, represented in the weather forecast. Still, one knows from experience that there are other factors as well. An air temperature of 18 [°C] in the sunshine is more pleasant than the same temperature in the shadow. So, radiation exchange is important. When there is wind at that same temperature of 18 [°C], it feels chillier than out of the winds, and of course clothing is an important factor for comfort as well.

Less easily recognized is the need for a *relative humidity* of the air of around 50 percent to feel comfortable. At any temperature, the relative humidity is the amount of water vapour in the air divided by the maximum possible amount of vapour (without condensation) at that temperature. In Example 2.3 the temperature dependence of relative humidity has already been discussed. It was noted that warm air can carry more water than cold air, so the maximum possible amount of water vapour in the air increases with temperature. Consequently, when one is heating the air in a house or building in wintertime one has to add water to the air in order to keep the humidity at 50 percent. In simple dwellings with central heating, one has to put against the radiators a few cisterns, which one daily fills with water.

Heat exchange between man and the environment

The aspects of comfort just mentioned are put in a physical context by studying the heat balance between man and his environment. The instantaneous *energy balance* of the human body can be written as [75]

$$S = M - W_k - E_{sk} - E_r - C - C_k - R \; [\mathrm{W\ m^{-2}}] \qquad (6.1)$$

Here, S is the energy surplus of the body [$\mathrm{J\ m^{-2} s^{-1}}$]. The body keeps the value of S approximately at zero in order to achieve a constant internal body temperature of 37 [°C]. The temperature of the skin may vary more than that of the internal body. The skin temperature may be dependent on the part of the body and is usually somewhat lower than the internal body temperature.

> The formation of goose pimples, occurring at the extreme limit of vasoconstriction because of cold, is a useless action for today's man. It was useful when we were very hairy all over the body surface. The raising of hairs was a good way to improve thermal insulation.
>
> From [75], p. 40

We note that Equation (6.1) is given in [$\mathrm{W\ m^{-2}}$], so for the total balance one has to multiply by the body area A, for which an empirical relation is used:

$$A = 0.202\,(m_b)^{0.425}\,(h_b)^{0.725} \; [\mathrm{m^2}] \qquad (6.2)$$

where m_b is the mass of the body [kg] and h_b is the height of the body [m]. (Exercise 6.2)

The terms on the right-hand side of Equation (6.1) have the following meaning:

- M is the *metabolic rate*, the conversion of internal chemical energy into heat and work (contraction of muscles, which also is converted into heat unless external work is performed).

- W_k is the *mechanical work* on subjects outside the body. At the beginning of Chapter 4 we found that a group of well-trained soldiers was able to produce 69 [W] in potential energy during almost 8 hours. For a body area $A = 2\,[\mathrm{m}^2]$, this would give $W_k = 35\,[\mathrm{W\ m}^{-2}]$.

- E_{sk} is the heat loss by *evaporation* through the skin. Little sweat glands produce water vapour, which moistens the skin. Also, water vapour diffuses from the internal tissues to the outside through the skin layers. Depending on the humidity of the air and the velocity of the air, the water vapour is taken away and replaced by new. For the naked parts of the body these terms will be higher than for the clothed parts. Therefore, the contributions have to be calculated separately, then multiplied by their relative surface area and then added to find the total E_{sk}

- E_r is the heat loss by *respiration* through the lungs. It comprises the latent heat of the exhaled water vapour and the temperature difference between the inhaled and exhaled air.

- C is the heat loss by *convection* from the outside of the clothing to the outside air. We have to distinguish the contributions ΔC for the different parts of the body, corresponding to different types of clothing or skin temperature. From Equation (4.10) one deduces

$$\Delta C = h(T_{cl} - T_\infty) \quad [\mathrm{W\ m}^{-2}] \tag{6.3}$$

where h is the convection coefficient (Table 4.2), T_{cl} the temperature on the outside of the clothing and T_∞ the temperature of the outside air. Formula (6.3) applies to the case where the temperature T_∞ of the outside air is lower than the skin temperature. In that case the heat resistance of the clothing determines T_{cl} and the resulting heat loss. For several kinds of clothing the value of d/k in Equation (4.1), equal to a heat resistance of 1 [m^2], is given in Table 6.2 (Exercise 6.3). The contributions ΔC for each part of the clothing have to be multiplied by their relative surface area and then added to find the total convection loss C in Equation (6.1).

Table 6.2 Mean heat resistance for 1 [m^2] for various clothing. With permission of Elsevier Science from [75], p. 44

Clothing ensemble	(d/k) / [m^{2}K W^{-1}]
Nude	0
Shorts	0.015
Tropical	0.045
Light way summer clothing	0.08
Light working ensemble	0.11
Indoor winter ensemble	0.16
Heavy traditional business suit	0.23

- C_k is the loss of heat by *conduction* to a chair when seated. This is often ignored.

- R is the *radiative exchange* between the body and the environment. As it will depend on the state of clothing, one writes the contribution ΔR from a temperature T_{cl} as

$$\Delta R = \varepsilon\sigma(T_{cl}^4 - T_{surr}^4)\ [\text{W m}^{-2}] \tag{6.4}$$

Here, T_{surr} is the radiative temperature of the surroundings as in equations (4.15) and (4.16). As in Equation (4.16) the emissivity of the part of the clothing enters, usually taken as $\varepsilon = 0.7$ (Exercise 6.4). As before, relative contributions have to be added to find the total R.

> Heat fins are keeping 5 tonnes of elephants cool. African elephants developed broad, thin flaps of ears through which blood is pumped – a cooling mechanism as efficient as that in an air-conditioning unit. The smaller Indian elephants have smaller flaps, and ice-age mammoths and mastodons had small ears, not flaps.
>
> *New Scientist*, 24 June 2000, 48

The body has to keep its energy balance $S = 0$ by manipulating the terms in Equation (6.1). The lowest metabolic activity (data from [75], p. 41) corresponds with sleeping ($M = 40$ [W m^{-2}]). In most climates one keeps the heat losses low by an additional cover. Walking or cooking may amount to $M = 100$ [W m^{-2}], house cleaning is much heavier at $M = 200$ [W m^{-2}] and heavy machine work costs up to $M = 235$ [W m^{-2}]. One of the heaviest activities is competitive wrestling at $M = 500$ [W m^{-2}], but of course one only sustains this for a short time.

The external work W_k at best is up to 20 percent of the metabolic activity M*. For French soldiers it was found to be $W_k = 35$[W m^{-2}]. This demands a metabolic activity of at least $M = 175$ [W m^{-2}] and probably more. From the

*In Exercise 4.2 (c), referring to a fitness centre, a value of 12% was found.

data given in the previous paragraph one may deduce that these soldiers worked very hard for eight hours. Note that heavy house cleaning is seldom done for eight consecutive hours.

Returning to the basic Equation (6.1), the question is how technology can help the body to keep its balance. Appropriate clothing is the first thing. For activities in the open air, one looks out of the window, listens to the weather forecast and decides on the clothing to wear. Inside, modern western men and women like to wear the same light, comfortable clothing all year round. Consequently, homes and buildings have to keep the same internal climate (air temperature, humidity, radiative temperature) all year.

> Modern buildings, clad in glass as a symbol of their modernity, are incongruously dark and require artificial lighting during the day, while the flimsy casing separating them from the outside makes it necessary to use air conditioning all year round, even when the outside conditions are pleasant.
>
> From [73], p. 67–68

What is more, many travellers expect to find the same internal climate everywhere in the world, at all times. Not only inside, but also outside, which is why shopping streets are becoming roofed in many places. Such a task needs a lot of heating and cooling. If, on the other hand, the task is only to subdue extreme climate variations and if one is prepared to adapt the clothing somewhat by putting on a sweater occasionally, cheap and simple technology can be of great help. This is our final topic.

6.2.2 *Building*

We consider two aspects: temperature and ventilation.

> During the first energy crisis in the 1970's the Dutch prime minister advised his people to close the curtains at night. That was about the first time the public at large heard about radiation losses.

Temperature

From equations (6.1), (6.4) and Exercise 6.4, it follows that radiative energy losses are not negligible; they may be of the same order as convection losses to the air. In Equation (6.4) the essential parameter is T_{surr}, the radiative temperature of the surroundings. In the open air one usually takes as an average value $T_{surr} = -10$ [°C]. Inside a house or a building one needs to average the incoming radiation from the walls and from the windows. Although the radiation temperature will be higher than in the open air, at moderate latitudes in wintertime it is usually below T_{cl} with a net radiation loss. In subtropical countries it may cool at night and then a similar radiation loss occurs.

At moderate latitudes a practical method to reduce radiation losses is to build a heating system in the floors and sometimes also in the walls. The heating system may work with water of 50 to 55 [°C], the floors and walls do not become that hot, of course, as they lose energy by radiation and convection. Experience shows that with floor or wall heating the inside air temperature may be 2 [°C] lower than with conventional central heating with the same of comfort.† The result is a lower energy bill for two reasons: the boiler system may work more efficiently at lower temperatures with smaller losses to its surroundings. Besides, with a lower air temperature the losses of the air mass by radiation, conduction and ventilation are smaller as well. An added advantage is that waste heat from nearby power stations or industries can more easily be utilized at the lower temperatures required as one may transport the waste heat over longer distances by a pipeline and still have water that is warm enough for the heating system.

The new Berlin central station will be a huge glass building, housing about 300 kW of photovoltaics in the glass roof. The PV elements are specially designed for glass roof construction. The photovoltaics will serve as shading elements for rail travellers as well as producing solar energy.

Renewable Energy World **2(6)**, Nov. 1999, 34–39

The traditional building in (sub) tropical countries with hot, sunny days and cold nights suggest another possibility. In the Arab country of Qatar, for example, by trial and error, people have found a way to be comfortable at night by building walls from certain materials with a certain thickness d. The physics behind it is that the outside of the wall is heated during the day, the heat wave then slowly passes the wall and reaches the inside at night [76]. Its amplitude becomes smaller as the heat wave loses heat to the wall, but it may still be high enough to feel comfortable inside.

The passage of heat through a wall is given by two equations for which one needs the physics of Chapter 7. They are not derived in this book, but the results are described here. The time t_c to cross a wall with thickness d [m], thermal conductivity k[W m^{-1} K^{-1}], density ρ[kg m^{-3}] and specific heat c_p[J kg^{-1} K^{-1}] equals

$$t_c = d\sqrt{\frac{\rho c_p}{2k\omega}} \tag{6.5}$$

†From WTH company (wth@wth.nl).

where ω is the angular frequency of the day/night variation with $\omega = 2\pi/(24 \times 3600)[s^{-1}]$. The damping depth x_e [m], at which the amplitude is reduced to 1/e of its original value is given by

$$x_e = \sqrt{\frac{2k}{\rho c_p \omega}} = \frac{d}{\omega t_c} \tag{6.6}$$

The dimensions of eqs (6.5) and (6.6) are checked in Exercise 6.5 and we finish with some qualitative remarks.

From Equation (6.5) it follows that a larger thermal conductivity k gives a smaller time t_c in which to cross the wall, which we understand intuitively. Also, a higher density ρ and specific heat c_p increase that time t_c. This may be understood, as the wall will take up more heat while the heat wave is passing. Finally, note from Equation (6.6) that a longer crossing time t_c will lead to more damping.

By choosing appropriate building materials and adapting the thickness d of the walls, one may profit as much as possible from the heat wave at night (Exercise 6.6). In fact, at moderate latitudes this method is also used with modern materials.

In the most classical type of solar heating systems, collector-warmed air is circulated through channels in a massive floor or wall, which then radiates the heat into the room after a time delay of 4–6 hours.

Renewable Energy World **2(6)**, Nov. 1999, 71–73

Discourse with the sun

In many places the incoming sun, passing the windows, is pleasant in wintertime but too hot in the summer. This is true even at moderate latitudes. In such a case the easiest way to achieve this is to have a sunscreen as in Figure 6.2. When the sun is high, in summertime, the screen shields the windows from the direct sunlight. In winter, when the sun is low, it contributes to the heating inside.

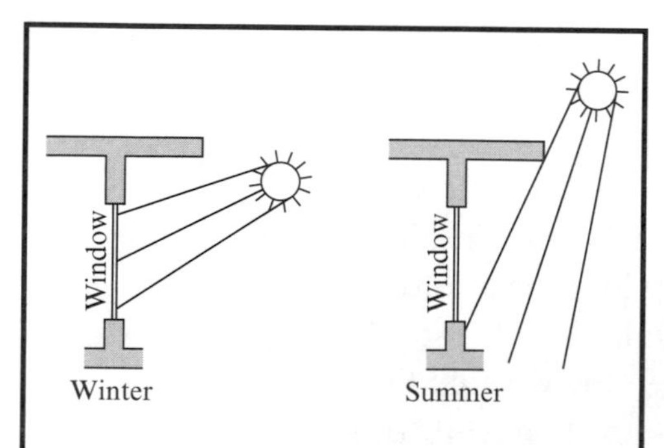

Figure 6.2 A balcony above the window protects the window against direct sunlight in summer, when the sun is high (picture on the right) but allows the sun to enter during winter (picture on the left)

This idea of reacting to the sun can also be applied in other ways. An example is to have double-glazing with blinds in between. In equilibrium the blinds are open. When the sun is shining above a certain intensity, a small panel of PV cells will close it. When the irradiation diminishes a string system pulls back the blinds again.

With a good understanding of physics, much is possible in building. This is very important as buildings have lifetimes of 50 years or more. In designing a building, the architect has to take into account the climate and the course of the sun during summer and winter. He or she should use all possibilities to their advantage. A final example, this time for warm or hot climates deals with ventilation.

Ventilation

In a warm climate good air circulation serves several purposes. It increases the convection coefficient h in Equation (6.3), as shown in Table 4.2; convection thereby takes heat away from the human body. At the same time, the necessary fresh air enters the house. In countries like Iran, one has traditionally used wind towers, as shown in Figure 6.3 [77] The wind velocity increases with altitude (Equation (2.22)) and high-velocity winds pass open windows high up in the tower. The air velocity creates a local low pressure (for physics students: this is Bernoulli's law), which draws up air through the windows downstairs. In other variants one can let the air pass a water surface. The water evaporates, taking part of its latent heat of evaporation from the air, which cools.

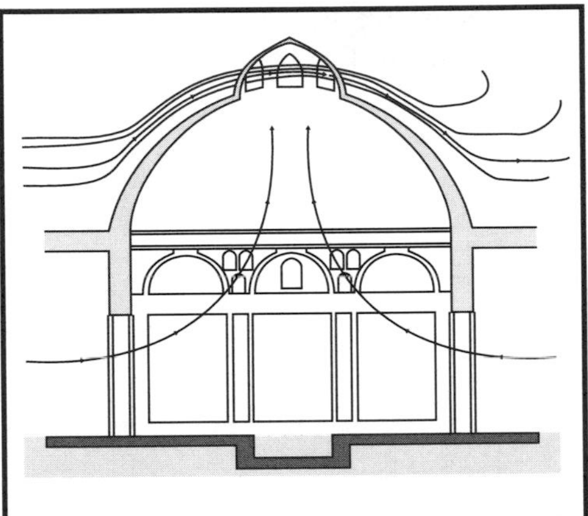

Figure 6.3 Wind tower. The winds pass through the top of the tower, creating a local underpressure. Consequently, air is drawn in through the windows. With permission of Elsevier Science, taken from [77], p. 93

It is important to realize that the concepts of Figures 6.2 and 6.3 do not cost any energy. Once they are built, they function. This is called a *passive use* of the climate conditions. Although the word 'passive' has negative and dull connotations, these passive methods, combined with modern materials and technologies, demand the highest levels of creativity and ingenuity.

EXERCISES

6.1 The land mass covers a fraction $f = 0.29$ of the earth's surface. The annual precipitation on the globe averages $r = 75$ [cm yr^{-1}]. Calculate the rainfall on land and compare with Table 6.1.

6.2 What area do you calculate for your own body from Equation (6.1)? The equation is often simplified for adults as $A = 0.026\ m_b$. What relation between mass and height does this assume? Would that be correct for your own body?

6.3 A human has a skin temperature $T_{sk} = 30$ [°C]; the outside air temperature is $T_\infty = 10$ [°C]. The heat current density q'' [W m^{-2}] through the clothing (from

Equation (4.1)) must equal the heat current density lost to the air (Equation (6.3) with $h = 20\,[\mathrm{W\ m^{-2}\ K^{-1}}]$). Otherwise heat piles up at the interface. For a steady state, calculate q'' for two cases of Table 6.2: heavy winter clothing and a light working ensemble. In both cases calculate T_{cl}.

6.4 Calculate the radiative energy loss for a human when $T_{cl} = 15\,[^\circ\mathrm{C}]$, $T_{surr} = -10\,[^\circ\mathrm{C}]$.

6.5 Check the dimensions of equations (6.5) and (6.6).

6.6 A wall of hollow clay blocks has the following properties $k = 0.36\,[\mathrm{W\ m^{-1}\ K^{-1}}]$, $c_p = 840\,[\mathrm{J\ kg^{-1}\ K^{-1}}]$, $\rho = 1029\ [\mathrm{kg\ m^{-3}}]$. Calculate the thickness d for which the crossing time equals five hours; what is the amplitude with which the heat wave reaches the inside of the wall?

7 Transport of Pollutants

In Chapter 1, pollution was defined as the addition by man of any substance (called pollutant) to the ecosystem that ultimately would cause harm or damage to man.* One of the problems with pollution is that the pollutants do not remain at their position of origin, but disperse through the natural environment. In this chapter we study the way in which pollutants disperse and how to find their concentrations at places distant from the source. Some mathematical equations in this chapter may intimidate some students at first glance, but be assured, they are all simple and easy to grasp.

In Section 7.1 we discuss some basic concepts, based on diffusion as the model for *dispersion*. This dispersion is the spread of pollutants around its source. It should not be confused with the dispersion of light. As dispersion of pollutants happens in space and time we will need a description with both space and time variables. For simplicity, we will restrict ourselves as much as possible to cases with only one space variable, where the mathematics is simple. With more space variables, we will use analogies to find the correct formulas.

Frequently, pollutants are distributed by the winds or by a river. Both are simply described by a flow with a certain, constant velocity. In a first approximation all these aspects can be embedded in a simple Gaussian dispersion model. The result of dispersion is called a *plume* and in the case of a Gaussian it is called a *Gaussian plume*.

When a plume expands and travels with a flow it may encounter substances which take away some of the polluting concentrations. This may be by *absorption* or *adsorption*, together called *sorption*, which will also be discussed in Section 7.1. We will briefly describe the phenomenon of turbulence in connection with Reynolds' number and finish with a discussion of biological aspects, which are not easily modelled physically.

The basic concepts are applied to dispersion in the air in Section 7.2. We will discuss how to find relevant dispersion parameters and see how high a chimney should be in order that pollutants at ground level are below a certain legally required concentration.

*In Chapter 1 we also mentioned energy pollution. This is not discussed in the present text.

Underground nuclear tests produce Pu. The ratio of ^{240}Pu to ^{239}Pu depends on the specific test. It was shown that from the 1968 test the Pu travelled 1.3 km away in 30 years. Probably it was carried by colloids in the groundwater, although other mechanisms remain possible.

Physics Today, April 1999, 19–20

In Section 7.3 we will look at surface waters. We calculate how an accidental high emission at a factory outlet disperses and dilutes, while travelling along with the river flow. This of course is relevant for deciding whether the water may be processed for use as drinking water.

The last Section, 7.4, deals with dispersion in ground water. This water flows much more slowly than water in rivers or the motion of air with the winds. Consequently, it is easier to describe. In fact, the methods and concepts developed by French engineers in the 19th century are still relevant for the description, while in modern times we have computer power to calculate what happens in complicated layered structures.

7.1 BASIC CONCEPTS

Modern computational methods start from the so-called Navier–Stokes equations, which are basic to any kind of flow. These equations cannot be solved analytically. So, approximations are required to find practical solutions. If one is interested in pollution in some average way, for example the total deposition of some substance over a year, a Gaussian shape is often good enough. We will discuss this shape for diffusion and then turn to a few other concepts, mentioned above.

7.1.1 Diffusion

Consider as a thought experiment, a droplet of red ink in a glass aquarium with water at rest. Assume it is done very carefully, so that the drop added does not cause any disturbance of the water and a pipette is used to put the drop somewhere near the centre of the volume. Ink with the same temperature and density as the water is not forced to move from its position. Experience suggests two things. First, that in time the drop of ink will be distributed over the water. Second, that this process will proceed much more quickly when one stirs the water and sets it in turbulent motion. The first process is discussed in the present section; it is called *diffusion*. The second process is called *turbulent diffusion* and will be discussed in Section 7.2.

The single puff

In the example of the drop of ink, the drop may be considered as the instantaneous addition of a polluting substance. This is called a *single puff*, bringing to mind the blowing of a small breath of smoke. Consider the emission of a puff of small size and mass. The puff slowly diffuses into three dimensions, x, y and z. For a small drop in a big aquarium the three x-, y- and z-dimensions may be taken as equivalent. Therefore pick the x-dimension and write down the Gaussian shape (1.13) for emission at the Origin O ($x = 0$)

$$f(x) = \frac{1}{\sigma\sqrt{2\pi}} e^{-x^2/2\sigma^2} \tag{7.1}$$

The fact that the parameter σ appears in the denominator in front of the exponential function guarantees that the integral of $f(x)$ from $(-\infty)$ to $(+\infty)$ is independent of σ. The precise value of the factor in front gives the integral the value one (Example 1.3). It is easy to check that the width of the function $f(x)$ increases with σ. Indeed, the width at half the peak height equals $(2.36) \times \sigma$ (Exercise 7.1).

> To diffuse in a solid, an atom must overcome the potential barrier presented by its nearest neighbours. If the barrier height is E, the atom will have sufficient thermal energy kT to pass over the barrier a fraction $\exp(-E/kT)$ of the time. This gives the temperature dependence of the diffusion coefficients for diffusion in a solid.
>
> [78], p. 545

From experience, one knows that the width of the distribution, and therefore σ, increases with time. From equation, (7.1) one sees that σ must have the dimension of a length, for one can only calculate a numerical value for the exponential function, when $x^2/2\sigma^2$ is a dimensionless number. Without physics or mathematics one cannot go much further, so we just put down the result

$$\sigma = \sqrt{2Dt} \tag{7.2}$$

where D is called the *diffusion coefficient* of the substance in the medium [m^2s^{-1}]. Other words are *diffusivity* or *diffusion constant*. But the word 'constant' is misleading as D is strongly dependent on temperature, and also may depend on location and time.

Equations (7.1) and (7.2) deserve further study. For time $t = 0$ it follows that $\sigma = 0$, so the width would be zero. On the other hand, for finite x, and $\sigma \to 0$ the exponential goes to the limit $\exp(-\infty)$, which is zero as well and one ends up with the ratio 0/0. In mathematical physics there are ways to deal with this odd behaviour*, but the drop has a very small extension and a very high concentration when it starts anyway. So, one may still talk about the emission of a point-like drop at a time a little greater than zero, which for simplicity we put as $t \approx 0$. Remember that we are referring to the single puff, which is the instantaneous emission from a point source. The more complicated emission from a continuous source will follow later.

Diffusion occurs in many circumstances ([78], [79]) but it is a slow process, as is apparent from the small values of the diffusion coefficients in Table 7.1. From that table it also follows that diffusion of a gas in water (at rest) goes much more slowly than the diffusion of the same gas in air (Exercise 7.2). Diffusion therefore may be a good model for dispersion, but the values of the dispersion parameters should be adpated to reality in the following sections.

It is helpful to have a second look at the dimensions of the functions in play. As σ has the dimension of a length [m], the Gaussian shape $f(x)$ in Equation (7.1) has the dimension [m^{-1}] and $f(x)\,\mathrm{d}x$ is dimensionless. As the integral $\int_{-\infty}^{\infty} f(x)dx = 1$, one may interpret $f(x)\,\mathrm{d}x$ as the relative amount of pollutant between x and $x + \mathrm{d}x$. The function $f(x)$ itself may then be interpreted as the relative concentration of pollutant at position x. More generally when an instantaneous point source in one dimension emits Q [kg] at the origin $x = 0$ at time $t = 0$, one would write down for concentration $C(x, t)$

$$C(x,t) = \frac{Q}{\sigma\sqrt{2\pi}} \mathrm{e}^{-x^2/2\sigma^2} \tag{7.3}$$

Equation (7.3) gives the real concentration C with the dimension [$\mathrm{kg\ m}^{-1}$]. Obviously the total emission is found by the integral $\int_{-\infty}^{\infty} C(x, t)\mathrm{d}x = Q$ [kg], for Equation (7.3) is just the Gaussian shape multiplied by a factor Q.

Diffusion in three dimensions

Consider a single puff at time $t = 0$ with mass Q [kg] in three dimensions. A three-dimensional coordinate system (x,y,z) is shown in Figure 7.1. When one considers diffusion in infinite space, the concentration will be a function of the distance r to the origin and look like $C(r)$. If the shape again is Gaussian, it will be found by substituting r for x in the exponential part of equation

Table 7.1 Diffusion coefficients of oxygen gas. From [80], p. 134

In air at 25 [°C]/[$\mathrm{m^2s^{-1}}$]	In water at 21 [°C]/[$\mathrm{m^2s^{-1}}$]
20.6×10^{-6}	2.33×10^{-9}

*One then defines the so-called Dirac delta function $\delta(x)$.

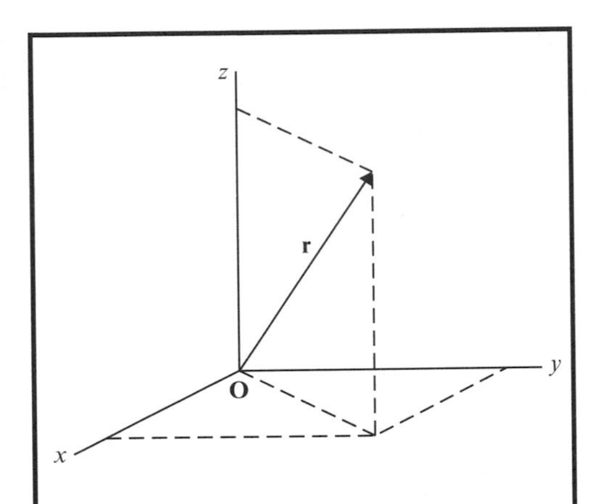

Figure 7.1 A three-dimensional position is indicated by x, y and z coordinates with a distance r from the origin O

(7.3). The integral over the shape again has to be one, but one has to integrate over the three dimensions. One therefore obtains

$$C(r,t) = \frac{Q}{(\sigma\sqrt{2\pi})^3} \mathrm{e}^{-r^2/2\sigma^2} \tag{7.4}$$

The third power in the denominator guarantees that

$$\iiint C(r,t)\mathrm{d}V = Q\,[\mathrm{kg}] \tag{7.5}$$

as the integration (1.14) runs over all space and is performed three times (Exercise 7.3). Note that the time dependence of the concentration C (r, t) is hidden in the width parameter σ according to Equation (7.2). As the volume element dV has dimension [m^3], the concentration C has dimension [kg m^{-3}], as expected in three space dimensions.

Fick's Law

In general, the three-dimensional concentration C will be a complicated function C (x, y, z, t) of the three space variables and time. The thermal motion of the particles and the collisions with their surroundings result in a flow **F** from high concentration to low concentration, where **F** [kg m^{-2}s^{-1}] is the mass of pollutant, which passes 1 [m^2] per second. A relation for the flow **F** is known as *Fick's Law*

$$\mathbf{F} = -D \text{ grad } C = -D\nabla C \tag{7.6}$$

This equation says that the flow is in the direction of diminishing concentration (hence the minus sign) and along the gradient, which is the direction of steepest increase of concentration C (Appendix B, equations (B.10)–(B.12)). There is a factor in between, which is the diffusion coefficient D that we encountered earlier. Relation (7.6) sounds acceptable intuitively and in fact is considered as a good approximation in many cases.

The diffusion equation

The concentrations (7.3) and (7.4) were based on the assumption that, during dispersion, the total mass of the pollutant remains the same at Q

A vertical sea water column may consist of water at different temperatures and salinity's. It will be statically stable at a particular time, if the density increases downward. But that may not remain so. For seawater is a multicomponent fluid and the rates at which heat and salt diffuse molecularly are different. The resulting *double diffusion* may give rise to density changes, resulting in instabilities.

[79], p. 30.

[kg]. This suggests that one may derive an equation for the concentration $C\ (x, y, z, t)$ by describing conservation of mass in mathematical terms. We will not do the mathematics, but directly go to the diffusion equation in its one-dimensional form

$$\frac{\partial C}{\partial t} = D\frac{\partial^2 C}{\partial x^2} \tag{7.7}$$

To give some respectability to this equation, consider the following example.

Example 7.1

Show that the Gaussian shape (7.3) satisfies the diffusion Equation (7.7). Remember that partial differentiation means that one differentiates with respect to that variable, while treating the rest as constants.

Answer

In (7.7) the concentration C occurs both left and right to the power one. Therefore, we may multiply C by a constant, giving

$$C(x) = \frac{1}{\sigma}\mathrm{e}^{-x^2/2\sigma^2} \tag{7.8}$$

We now perform the straightforward differentiations, required to verify (7.7)

$$\frac{\partial C}{\partial t} = \frac{\partial C}{\partial \sigma}\frac{\partial \sigma}{\partial t} = \left(\frac{-1}{\sigma^2} + \frac{x^2}{\sigma^4}\right)\mathrm{e}^{-x^2/2\sigma^2}\frac{D}{\sigma} = D\left(\frac{-1}{\sigma^3} + \frac{x^2}{\sigma^5}\right)\mathrm{e}^{-x^2/2\sigma^2} \tag{7.9}$$

$$\frac{\partial C}{\partial x} = -\frac{x}{\sigma^3}\mathrm{e}^{-x^2/2\sigma^2} \tag{7.10}$$

$$\frac{\partial^2 C}{\partial x^2} = \left(\frac{-1}{\sigma^3} + \frac{x^2}{\sigma^5}\right)\mathrm{e}^{-x^2/2\sigma^2} \tag{7.11}$$

Multiplying Equation (7.11) by D gives (7.9), which obeys the diffusion Equation (7.7)

A diffusion mechanism also is responsible for transport of heat within a substance from a region with a high temperature toward a region with a lower temperature. At the interface between higher and lower temperatures, the particles with the higher temperature (and therefore the higher kinetic energy) experience collisions with those at the lower temperature and share some of their kinetic energy during that process. Again, on average the differences tend to level off, which may be interpreted as a heat flow. Completely analogous to Equation (7.6), this heat flow is described by $\mathbf{F} = -k\nabla T$, with concentration C replaced by temperature T. This is precisely Equation (4.9) in a slightly different notation. Diffusion of heat and of pollutants therefore may be described by the same mathematics, albeit with different physical interpretations and different coefficients.

Point source in a uniform wind: advection

In practice one will often have cases where the source emits its pollution in a flow of air (the winds) or a flow of water (a stream). For simplicity we take a uniform flow in the x-direction with constant velocity u. This is called a point source in a uniform wind, where the word wind comprises water as well. The dispersion of pollutants within the main flow is determined by two effects. In the first place the pollutants will follow the flow, which is called *advection*. In the second place, diffusion will occur, as the concentration gradient still has to level off. These two effects are independent and have to be added.

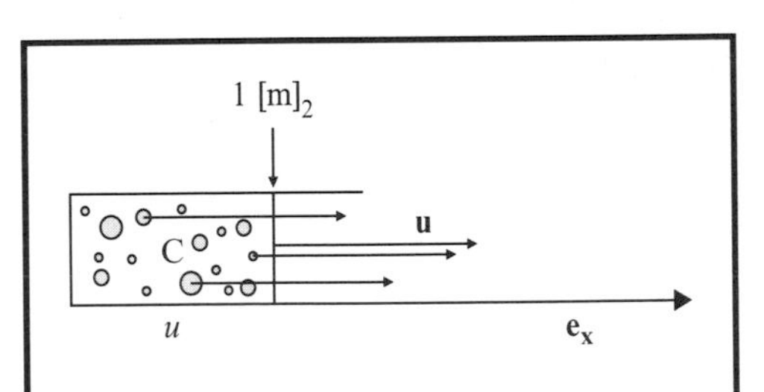

Figure 7.2 Advection. The pollutants in the bag pass one [m^2] in one second, as its length is just the velocity of the main flow

The flow $\mathbf{F}$ still is the mass of pollutants passing a [m^2] per second. The term due to advection may be written down by glancing at Figure 7.2. There, a bag of length u is sketched in the direction of the flow (the x-direction) while the end of the bag on the right has a surface area of 1 [m^2]. As its length is precisely the distance the main flow particles travel in one second, all contents of the bag pass the surface area in one second. This equals Cu [kg m^{-2} s^{-1}]. To obtain the total flow one has to add the contribution (7.6) from diffusion and we get

$$\mathbf{F} = uC\mathbf{e}_x - D\nabla C \tag{7.12}$$

The vector $\mathbf{e_x}$ is the unit vector (a vector with length equalling one) in the x-direction. In Equation (7.12) the x-component of $\mathbf{F}$ has the extra term uC because of the advection; the y- and z-components of $\mathbf{F}$ remain unchanged, as there is no advection in these directions.

Instantaneous emission from a point source in uniform wind Advection implies that for a one-dimensional problem in the x-direction one can no longer have the simple Gaussian (7.3). Instead, the solution to the mathematical equations becomes

$$C(x,t) = \frac{Q}{\sigma\sqrt{2\pi}} \mathrm{e}^{-(x-ut)^2/2\sigma^2} \tag{7.13}$$

The factor $(x-ut)^2$ in the exponent represents a wave travelling with velocity u. At position $x+u$ and time $t+1$ this factor will have precisely the same value as in position x at time t. Therefore it travels a distance u [m] in 1 [s]. The symbol σ in Equation (7.13) still obeys Equation (7.2), so $\sigma = \sqrt{2Dt}$. Therefore, during its travel, the puff widens and the peak lowers. At any time t one may calculate the integral of the concentration $C(x,t)$ from $-\infty$ to ∞, which again gives Q [kg], the total amount of pollutant emitted, which remains constant in time.

In a three-dimensional situation with advection in the x-direction only, there will be diffusion in the y- and z-directions, resulting in

$$C(x,y,z,t) = \frac{Q}{\sigma\sqrt{2\pi}} \mathrm{e}^{-(x-ut)^2/2\sigma^2} \frac{1}{\sigma\sqrt{2\pi}} \mathrm{e}^{-y^2/2\sigma^2} \frac{1}{\sigma\sqrt{2\pi}} \mathrm{e}^{-z^2/2\sigma^2} \tag{7.14}$$

As in Equation (7.4) the factor $\sigma\sqrt{2\pi}$ appears three times in the denominator in order to end up again with a total pollutant of Q [kg].

In relations (7.13) and (7.14) the dimensions of the concentrations are different. As $[Q] = [\mathrm{kg}]$ and $[\sigma] = [\mathrm{m}]$, one has in (7.14) $[C] = [\mathrm{kg\ m^{-3}}]$ and in (7.13) one has $[C] = [\mathrm{kg\ m^{-1}}]$.

Continuous emission from a point source in uniform wind Much pollution is caused by a chimney, smoking in the wind. In its simplest description this resembles a point source in a uniform wind. Such a continuous point source starts at $t = 0$ by emitting a constant amount of pollutant q [kg s^{-1}] from location ($x = 0, y = 0, z = 0$) and continues to do so. One will be interested in the pollution at point P after a very long time $t \to \infty$. Its coordinates are sketched in Figure 7.3.

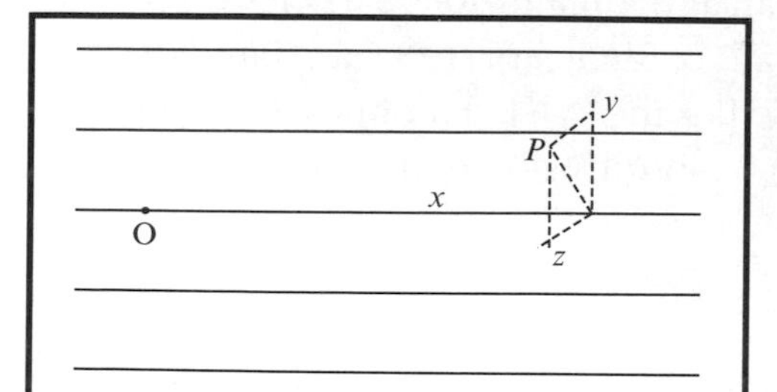

Figure 7.3 Axial symmetry. A point source at the origin and a flow in the x-direction result in symmetry around the x-axis

Notice the axial symmetry around the x-axis. So, the concentration $C(x,y,z,t)$ will only depend on x, t and the distance to the x-axis. However, for the limit of very long time t, a steady state is formed. Mathematically, this means that the time dependence will disappear from the expression.

An approximation, which is valid in many cases, is that advection is a much more powerful means of dispersion than diffusion. This suggests that one ends up with advection in the x-direction and diffusion in the y- and z-directions. Then Equation (7.14) reduces to

$$C(x, y, z) = \frac{q}{2\pi\sigma^2 u} e^{-\frac{y^2 + z^2}{2\sigma^2}} \tag{7.15}$$

where the axial symmetry around the x-axis is expressed by the fact that only the distance to that axis enters the equation by the term $(y^2 + z^2)$ in the exponential. In the x-direction one has advection only, so $t = x/u$ is just the time passed since the puff at (x, y, z) left the origin. This time, therefore, from the familiar expression $\sigma^2 = 2Dt$ is available for diffusion in the y- and z-directions, giving

$$\sigma^2 = 2Dt = \frac{2Dx}{u} \tag{7.16}$$

The dimension of Equation (7.15) can easily be checked from the dimensions $[q] = [\text{kg s}^{-1}]$, $[\sigma] = [\text{m}]$ and $[u] = [\text{m s}^{-1}]$. This gives $[\text{kg m}^{-3}]$, as one expects. The factor $2\pi\sigma^2$ in the denominator of (7.15) guarantees that integration over y and z will each give precisely one, as in Equation (7.1) or (7.4). Therefore

$$\begin{aligned} &\int_{\text{all}} \int_{y,z} C(x, y, z) \mathrm{d}y \, \mathrm{d}z \\ &= \frac{q}{u} \frac{1}{\sigma\sqrt{2\pi}} \int_{-\infty}^{\infty} e^{-y^2/\sigma\sqrt{2}} \mathrm{d}y \frac{1}{\sigma\sqrt{2\pi}} \int_{-\infty}^{\infty} e^{-z^2/\sigma\sqrt{2}} \mathrm{d}z = \frac{q}{u} \end{aligned} \tag{7.17}$$

So, after y- and z-integration only the factor q/u remains.

The factor q/u in Equation (7.15) is due to advection. Consider a slice with thickness $\mathrm{d}x$ and infinitely extended in the y- and z-directions. The pollutants in that slice were emitted during a time $\mathrm{d}t = \mathrm{d}x/u$. This emission took place a time x/u earlier. The total amount of pollutants in the slice are found by integrating over y and z, using Equation (7.17) and multiplying by its thickness $\mathrm{d}x$:

$$\left(\iint C \, \mathrm{d}y \mathrm{d}z \right) \mathrm{d}x = (q/u)\mathrm{d}x = (q/u)(\mathrm{d}x/\mathrm{d}t)\mathrm{d}t = (q/u)u \, \mathrm{d}t = q \, \mathrm{d}t \tag{7.18}$$

This is precisely the amount of pollutants [kg], emitted during a time $\mathrm{d}t$. Therefore, Equation (7.15) is consistent and plausible.

7.1.2 *Sorption*

In environmental science the student will meet the concept of *sorption*, which comprises absorption and adsorption ([9], [51], [81], [82]). *Adsorption* is the attachment of particles or molecules to the surface of a solid or liquid and *absorption* is the uptake of particles or molecules by the total volume of a solid or liquid. Often both processes occur. Charcoal, for example, is a porous material, which may adsorb hydrogen molecules from hydrogen gas. Once the molecules are adsorbed, some of them may diffuse to the inside of the coal and are absorbed. The two processes may be distinguished experimentally from the different timescales on which they operate ([81], p. 690).

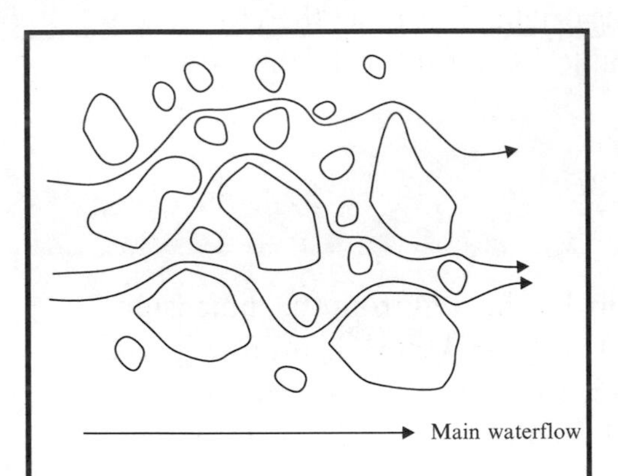

Figure 7.4 Groundwater flow. The main flow is to the right, but the water particles have to crawl through the pores. The solved contaminants may easily be adsorbed

Absorption and adsorption are physical processes, which go back to microscopic forces between atoms and molecules of different substances. If a chemical bond occurs, one uses the word *chemisorption* ([9], p. 117). Other useful concepts are *desorption*, which is the reverse of adsorption or absorption and *partitioning*, which is the process of distribution of a contaminant between a liquid and a solid. Other examples of partitioning can be found in [4] p. 450–1.

In environmental applications, adsorption is very relevant in groundwater problems. Figure 7.4 shows the reason. Although the main flow is from the left to the right, the water particles have to find their way through the pores in the ground, passing many irregular surfaces of the soil particles. These small particles have a large surface and easily adsorb contaminants, which may be dissolved in the water.

Use of adsorption and absorption for cleaning

The physical processes of adsorption and absorption are being used extensively for cleaning ([51], pp 79, 80). If the exhaust from industrial processes, for example, is contaminated by HCl gas, one may lead the exhaust gases through a water bath in which the poisonous HCl is readily absorbed. Similarly, organic liquids may absorb SO_2 and H_2S from the exhaust of fossil fuel furnaces.

Activated carbon is generally in use as an adsorbent of organic contaminants in the air. The pollutants are adsorbed at ambient temperature. Later on the sorption bed is heated and desorption happens. The extracted contaminants can then be treated chemically.

7.1.3 *Biological aspects*

Physical scientists must not close their eyes to biological activities that are not described in the earlier paragraphs. We mention a few points ([83], pp. 15–31).

Accumulation by biological processes

Many organic compounds have a low solubility in water. As a general rule, the solubility of organic compounds in organic solvents like fatty tissue is bigger, the lower their solubility in water. Consequently, the concentration of a poisonous organic compound in fish, like the rainbow trout, may be a factor of 100 to 10 000 higher than in the surrounding water. Not only that, the fish may swim over large distances and transport the pollutant in a way which is not described by the differential equations of physics.

Metal ions entering the food chain

Metal ions may occur in river water, originating, for example, from industrial emissions in the river. When for any reason the water becomes more alkaline (that is, the pH increases), the metals may precipitate on the river floor and become part of the sediments. Or, in another case, sediments may adsorb the metal ions. Small organisms may feed on the sediments, by which the metals enter the food chain, eventually ending up on our plates. The mechanism of entering the food chain is very much dependent on the chemical compounds in which the metals are embedded and, even in the same fish, the concentration of metals in different organs may be quite different.

7.1.4 *Reynolds number*

The behaviour of a flow is often very irregular and turbulent. This happens for high velocities u or for fluids within large physical dimensions l. The classical experiment on this phenomenon was made by Reynolds and indicated in Figure 7.5 [84]. Reynolds used a horizontal glass tube with a flow of

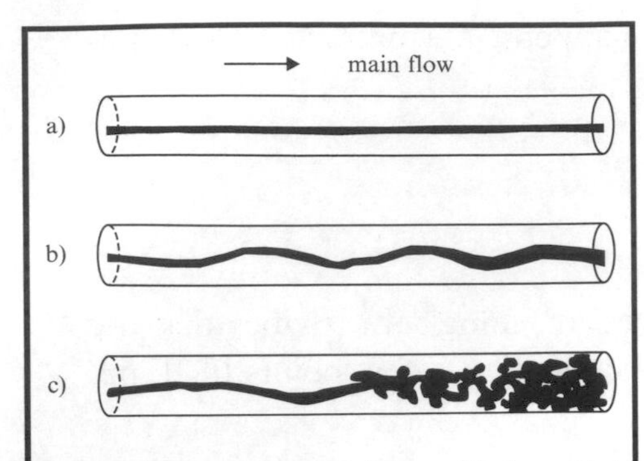

Figure 7.5 Reynolds' experiment with dye added on the left to a flow. In the top (a) the flow is laminar. At higher velocity (middle (b)) the filament starts to waver. At a certain velocity (bottom (c)) it spreads. Reproduced by permission of John Wiley, based on [84], p. 7

It was known in the middle of the 18th century that laminar and turbulent flows represented two distinctive flow types. It was found that results of experiments in open channels and pipes were very different under laminar and turbulent flow, but there was no appreciation of the way in which *viscosity* influences the fluid behaviour. It was the experimental investigations of Osborne Reynolds (1842–1912), Professor of Engineering at Manchester University, which led to a better understanding of the conditions which favour a particular flow type.

[84], p. 6.

water at a certain velocity. On the axis of the tube, on the left in Figure 7.5, a fine filament of dye was introduced. Reynolds slowly increased the velocity of the flow. At a certain velocity the dye filament began to waver (Figure 7.5(b)) and at a still higher speed it suddenly spread all over the tube (Fig. 7.5 (c)). At lower velocities the flow was apparently *laminar*; at a certain higher velocity it became *turbulent*. This transition is determined by a dimensionless Reynolds number *Re*, composed of velocity u, tube diameter l, viscosity μ (a measure for the internal friction within the fluid) and density ρ together. This dimensionless *Reynolds number* reads

$$Re = \frac{\rho l u}{\mu} = \frac{l u}{\nu} \tag{7.19}$$

The ratio $\nu = \mu/\rho$ is called the *kinematic viscosity* and explains the last term in Equation (7.19). The transition from laminar to turbulent motion takes place for a Reynolds number between 2500 and 4000. For the return from turbulent to laminar the point of transition is even better defined at about 2000. For liquids with other viscosity or tubes with another diameter it is always this same value of the Reynolds number that determines the transition.

7.2 *DISPERSION IN THE AIR*

The burning of straw after the grain harvest became a serious nuisance in Western Europe, particularly in England, owing to increased cultivation of cereals and the decrease in animal husbandry. These fires often generate their own cumulus clouds.

[6], p. 87 (photograph).

From common experience one knows that atmospheric motion is very turbulent. At ground level one observes small leaves and pieces of paper picked up by the wind, following very irregular orbits. These may resemble whirls or eddies, which are also observed in flowing streams.

Higher up, one may notice the irregular behaviour of smoke plumes rising from chimneys and picked up by the winds ([6], [85]). We therefore expect that the atmospheric motions correspond to high Reynolds numbers. This indeed is the case as follows from Table 7.2, with some typical scales for horizontal and vertical motions [21]. Table 7.2 also gives the corresponding Reynolds number *Re* from Equation (7.19) with the kinematic viscosity ν of air given in Appendix A.

The large Reynolds numbers *Re* confirms the observations: atmospheric motion is very turbulent. Only in a very thin layer, just above the ground, are the dimensions small enough to produce a laminar flow (Exercise 7.4).

The origin of the atmospheric turbulence is known from Chapter 2: the time varying insolation causes varying evaporation and changing temperatures. These, among other things produce rising air and clouds. The dependence of insolation on latitude gives pressure differences and the main (geostrophic) flows. Because of turbulence, the atmospheric transport of pollutants is complicated and one has to rely on approximations to describe it.

7.2.1 Turbulent diffusion: real plumes and Gaussian plumes

In Figures 2.8 and 2.9 the concept of lapse rate was illustrated as the *decrease* of temperature T with increasing altitude z. More precisely, it is the value $(-\mathrm{d}T)$ divided by the rise $(\mathrm{d}z)$, which gives a lapse rate: $-\partial T/\partial z$. The measured value is almost always positive, hence the minus sign in the definition, although locally at some altitude an *inversion* may occur. Then the temperature increases with height and consequently the lapse rate is negative.

The measured temperature was indicated by the label 'Environment' in Figure 2.9. We also introduced the adiabatic lapse rate as the lapse rate for a parcel of air that does not exchange any heat with the surroundings. The discussion of Chapter 2 may be summarized by the statement that air is stable when the adiabatic lapse rate is bigger than the measured, ambient lapse rate, for then a rising parcel of air soon will be cooler than its surroundings and gravity forces it to go down again.

Industries will get rid of part of their waste by putting up a high chimney. The outgoing plume will be carried away by the winds, which are always blowing high up. Besides, the turbulent motion of the air will quickly dilute the concentrations of unwanted compounds to acceptable levels. This indeed works, as we shall see below, except when the chimney top is below an inversion

Table 7.2 Physical characteristics of atmospheric motion from [21], p. 192

	Speed u/[m s^{-1}]	Length l/[m]	Reynolds number *Re*
Horizontal	10	10^6	10^{12}
Vertical	0.1	10^2 to 10^4	10^6 to 10^8

layer. For then, the rising air is sent down again, unless it is even hotter than the air in the inversion itself. In another example an inversion layer some 100 [m] above ground level may keep the automobile exhaust gases down at ground level, causing health problems for people living in the neighbourhood.

In general, however, a chimney is a proven instrument for diluting wastes. In Figure 7.6 three plumes from a real chimney are shown. The three cases have the same adiabatic lapse rate, dashed on the left. The ambient lapse rate, however, is different and drawn near the adiabatic one. In all three cases the gases leaving the chimney are somewhat hotter than the ambient air. So the plumes will rise a little until they cool by expansion and by the evaporation of the water drops in the plume.*

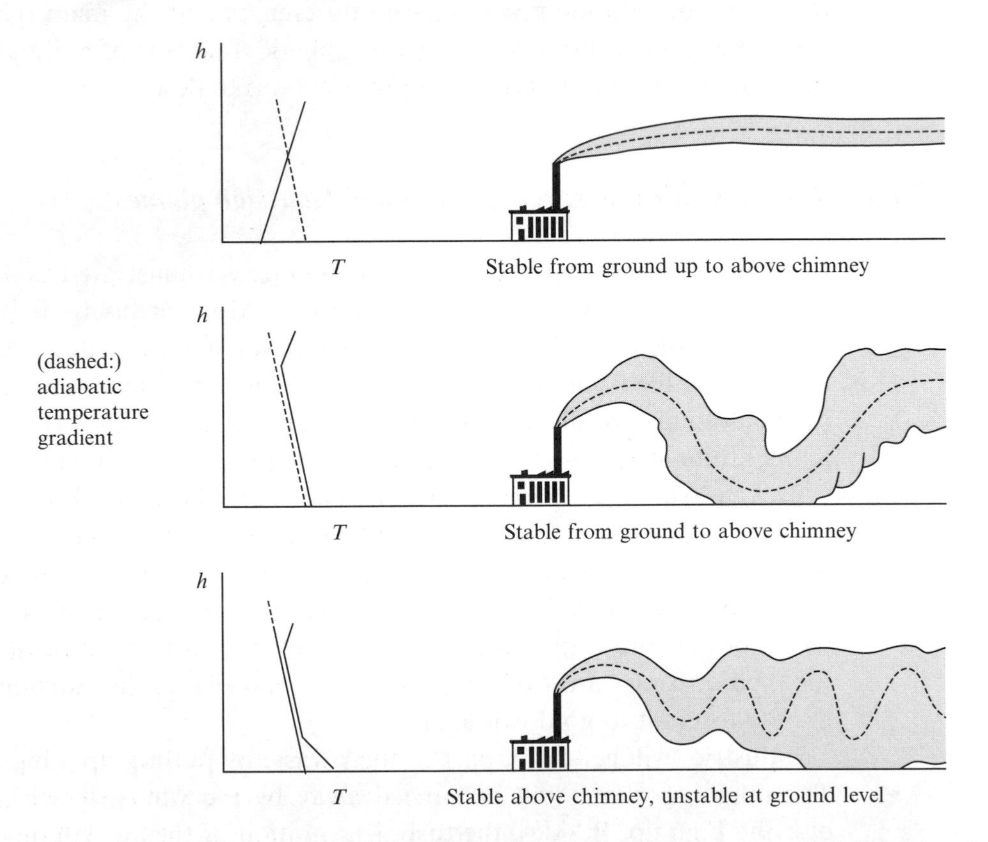

Figure 7.6 Plume shapes of a high chimney (100 [m]) for three ambient lapse rates. The behaviour of temperature T as a function of height h is given on the left of each graph, both for the adiabatic (dashed) case and the ambient (full line) case. The resulting plume is on the right with its axis dashed. The top graph shows an inversion: a high stability gives a narrow plume. In the middle graph the situation is neutral below the inversion, causing a plume which reflects against the inversion and the ground. In the bottom graph, the atmosphere is unstable at ground level, causing strong mixing and a wide plume

*Small water droplets cause the white glare of a plume just leaving a chimney. When they have evaporated, the plume may become grey from small soot particles or invisible when only gases are emitted ([6], p. 49, 51).

When one looks carefully at Figure 7.6 one notices that the plume in the top graph is located in an inversion: the ambient lapse rate is negative. Consequently, the parcels of the plume are forced back to their initial positions close to the dotted centre line of the plume. The middle case is more interesting. There is an inversion somewhat higher than the chimney top. Notice that the up-going plume parcels are forced to go down. Then they continue to go down as below the inversion the ambient lapse rate is adiabatic. At ground level the plume is reflected and goes up again.

> Pollution carried up in large cumulus may be spread over a large area, not by small diffusing eddies but by steady smooth shear which converts a compact mass of air into a very extensive, but still concentrated, sheet.
>
> [6], p. 84 (photograph)

The most interesting case is found in the bottom graph. The inversion above the chimney top forces the plume parcels down. Near ground level the situation is unstable, so parcels with downward velocities continue to go down and parcels with upward velocities continue to go upward. Small perturbations will thoroughly mix the plume, which quickly widens. Note that, in the middle case, the centre of the plume slowly goes up and down, while in the bottom case it is much more pronounced.

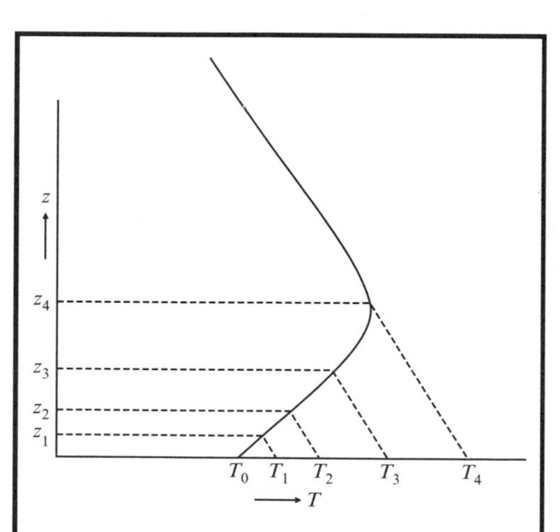

Figure 7.7 Mixing height. The drawn curve shows a hypothetical temperature profile, just before sunrise. When the sun heats the ground, the air above becomes mixed, first at temperature T_1 to height z_1, the mixing height, which is increasing during the day

It is worthwhile to spend a day watching a chimney plume. One may notice a change of direction of the main flow and probably also observe a change in the local ambient lapse rate during the day. What may happen is shown in Figure 7.7. That figure represents a clear night above land during which the lower air has lost a lot of heat by radiation. Around sunrise the temperature distribution may be like the full curve, which displays a temperature inversion up to a height $z = z_4$, possibly a hundred metres ([82], p. 147, [21], p. 131). Above that altitude the temperature roughly follows the dry adiabatic lapse rate.

At ground level the early morning temperature is T_0. The sun starts warming the ground surface and the adjacent layer of air is heated. Air parcels move up by adiabatic expansion, and others move down until the bottom layer is thoroughly mixed with the ground temperature T_1. The mixing results in about the dry adiabatic lapse rate for the lowest layer, as the warm water vapour has no inclination to condense and its mass is only small. So, one observes in Figure 7.7, going upward from $z = 0$, starting from temperature T_1, first the adiabatic lapse rate up to height $z = z_1$, then again the inversion up to height $z = z_4$. The height $z = z_1$ of the lowest, mixed layer is called the *mixing height*. During the day the mixing height increases until at temperature $T = T_4$ the total build-up is adiabatic.

The real story may be much more complicated than described here. Winds from the sea for example, may spoil the simple temperature profile, or clouds may prevent the sun from warming the lower layers adequately. But even in the simplified picture of Figure 7.7 it is clear that the plume from the chimney in Figure 7.6 may encounter a changing temperature profile during the day. Consequently, the plume shape and possibly the plume direction will change.

> The experience of the Chernobyl disaster (April/May 1986) has been very instructive. On the whole, the weather was such that pollution was mainly confined to a fairly definite mixing layer, and because radioactive materials are easily detected it was possible to follow the course of air parcels, emitted at subsequent days. The emissions of the first two days travelled a very tortuous path via the Baltic and Southern Sweden, turning South across Southern Germany, then northwestwards across Wales and England to Western Scotland and Ireland before drifting towards the northern ocean. The locations of the greatest rainout of radioactive particles could not have been determined with accuracy even 12 hours beforehand.
>
> [6], p. 83–85

In order to prepare for a calamity, the owner of an industrial complex would need a well-tested computer program, which is able to simulate the dispersion of poisonous materials in realistic surroundings (buildings, hills, clouds) and all kinds of climatic conditions. For 'normal' emissions at low concentrations one only needs to know the average concentration which reaches ground level from a chimney. Ground level, of course, is the place where people breathe in the air, where crops are accumulating depositions and where pollutants are taken up by surface waters. To estimate the consequence of small irregular pollution over a long period of time, one uses an average over the many possible realistic plumes and assumes such an average to have a Gaussian shape.

A Gaussian plume

For simplicity, one assumes a single dominant wind direction, which is taken as the x-direction. The normal emissions from a chimney are simulated by a continuous point source in a uniform wind, which is taken as horizontal. The expression was written down in Equation (7.15). That equation, however, had the same parameter σ in the y- and z-directions. We take z as the vertical coordinate and y as the horizontal coordinate perpendicular on the wind direction. Clearly, the physics may be and will be different in both directions, resulting in different values of σ_y and σ_z. Therefore, Equation (7.15) is generalized to

$$C(x,y,z) = \frac{q}{2\pi\sigma_y\sigma_z u} e^{-\frac{y^2}{2\sigma_y^2} - \frac{z^2}{2\sigma_z^2}} \tag{7.20}$$

The x-dependence of the concentration $C(x,y,z)$ will be hidden in the values of the so-called *dispersion coefficients* σ_y and σ_z. Diffusion, as discussed in Section 7.1 is only a model for realistic dispersions, which happen in practice. The dispersion coefficients σ_y and σ_z are much larger than would follow from Equation (7.2) and Table 7.1. They are taken from empirical models, which take into account the conditions of the terrain. These models are supplemented by measured data. They are still a measure for the width of the plume and therefore will increase with distance x from the source and soon reach a value of more than 100 [m] ([1], pp 259, 260)

Emissions from a chimney The concentration at ground level from a chimney with height h and emission of pollutants q [kg s^{-1}] is found from Equation (7.20). There it was assumed that the source of the plume is at the origin of the coordinate system with $(x, y, z) = (0,0,0)$. Take the origin of the coordinate system at the foot of the chimney. Then the source is at height h with coordinates $(0,0,h)$, so one has to measure the z-coordinate starting at $z = h$. This gives a concentration

$$C(x,y,z) = \frac{q}{2\pi\sigma_y\sigma_z u} \mathrm{e}^{-\frac{y^2}{2\sigma_y^2} - \frac{(z-h)^2}{2\sigma_z^2}} \tag{7.21}$$

The concentration is highest for $z = h$, as it should, and diminishes when z becomes higher or smaller than h. In fact, $z = h + d$ and $z = h - d$ give the same concentration. Equation (7.21) also gives a concentration for $z < 0$, which is below ground level. This cannot be correct as the plume particles cannot penetrate the ground and one has to assume something about the physics at ground level.

At ground level, the worst-case assumption from the point of view of air quality is that the chemicals emitted from the chimney remain in the air. That means that particles going down to the ground are not deposited but reflected back. The geometry is sketched in Figure 7.8. A particle emitted from a source S is reflected from the ground at $z = 0$ and follows a path upwards again. This may be simulated by a mirror source S′, emitting particles. Although for a plume the particles may not move in a straight line as in Figure 7.8, a mirror source at $z = -h$ will simulate reflection of the plume. Instead of Equation (7.21) one finds

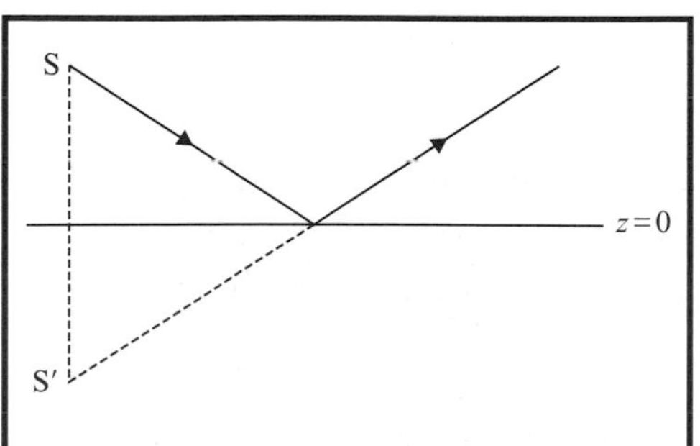

Figure 7.8 Reflection of a source S at the ground, simulated by adding a mirror source S′

$$C(x,y,z) = \frac{q}{2\pi\sigma_y\sigma_z u}\left(e^{-\frac{y^2}{2\sigma_y^2}-\frac{(z-h)^2}{2\sigma_z^2}} + e^{-\frac{y^2}{2\sigma_y^2}-\frac{(z+h)^2}{2\sigma_z^2}}\right) \tag{7.22}$$

This formula only holds for $z > 0$. To protect air quality, the highest concentrations in the air at ground level should be calculated. They are found precisely underneath the axis of the plume ($y = 0$) and as z is only slightly bigger than zero one may put $z = 0$ as well. This results in two identical terms which together give

$$C(x,0,0) = \frac{q}{\pi\sigma_y\sigma_z u} e^{-\frac{h^2}{2\sigma_z^2}} \tag{7.23}$$

Remember that the x-dependence is hidden in the dispersion coefficients σ_y and σ_z.

Example 7.2

Find an expression for the maximum concentration of pollutants in the air at ground level, assuming that the ratio σ_y/σ_z does not depend on the distance from the chimney.

Answer

Rearrange terms in Equation (7.23) and obtain

$$\frac{C\pi(\sigma_y/\sigma_z)uh^2}{q} = \frac{h^2}{\sigma_z^2} e^{-\frac{h^2}{2\sigma_z^2}} \tag{7.24}$$

On the left-hand side the dependence on x, the distance to the source, is only due to to the concentration C. On the right-hand side of Equation (7.24) the x-dependence is hidden in the dispersion coefficient σ_z. The right-hand side is most easily displayed as a function of σ_z/h as in Figure 7.9.

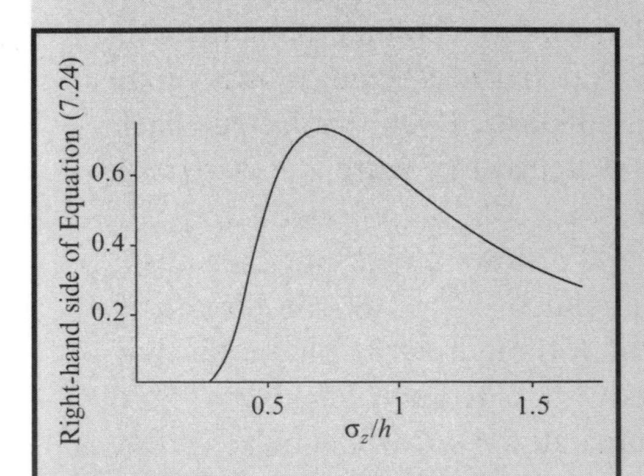

Figure 7.9 The right-hand side of equation (7.24)

In Figure 7.9 there is a maximum of ≈ 0.74 for $\sigma_z/h \approx 0.7$ (Exercise 7.5). Put the left-hand side equal to the maximum value of the right-hand side. Then the maximum ground level concentration becomes

$$C_{\max} \approx \frac{(0.74)q(\sigma_z/\sigma_y)}{\pi u h^2} \tag{7.25}$$

Note that the maximum concentration C_{max} of Equation (7.25) is proportional to the emitted pollution q[kg s^{-1}], as expected. It is inversely proportional to the wind velocity u. This is understandable as for a higher velocity the plume will touch the ground farther away from the source. Finally, it is inversely proportional to h^2, the height squared of the chimney. This is pleasant, as building a chimney which is twice as high, will reduce the ground level concentrations by a factor of 4.

In planning a chimney at a certain site, one has to measure the dispersion coefficients σ_y and σ_z at the selected place for different weather conditions during the seasons. Taking into account the legal maximum concentrations, one then decides on the required height of the chimney.

7.2.2 *Deposition*

In writing down Equation (7.22) it was assumed that, on touching the ground, all particles and gases in a plume will be reflected. This, of course, will not happen in practice. Particles and gases may be adsorbed by surfaces of plants and other objects, or they may be absorbed in seas and oceans. This process is called *dry deposition*. Particles, emitted from chimneys may be large and heavy enough not to follow the slowly expanding plume, but because of gravity may go down earlier. This process is called dry deposition as well.

Wet deposition is a process whereby rain or snow causes pollutants to settle down. The precipitation may be induced by the particles in the smoke, which form nuclei of small water drops. The drops then may absorb the chemicals in the smoke. This process is called *rainout*. Or if rain from some cloud passes the plume on its way down, it may incorporate particles and chemicals from the plume, a process called *washout*.

In most of the cases described, the deposition brings pollutants down sooner than expected in the Gaussian plume model.

Finally, in the atmosphere, many chemical reactions may take place, often under the influence of sunlight. They will change the chemical composition of the plume, sometimes improving the quality of the air, sometimes deteriorating it ([82], p. 291).

7.3 *DISPERSION IN SURFACE WATERS*

Pollution of surface waters and ways to control that pollution are subjects by themselves [86]. Their description is complicated as river flows are very turbulent. This follows from an order of magnitude estimate of the Reynolds number $Re = \rho ul/\mu$, defined in equations. (7.19). For water one has $\rho \approx 10^3$ [kg m^{-3}] and $\mu \approx 10^{-3}$ [Nm^{-2} s] (Appendix A). For a river with $u \approx 1$ [m s^{-1}] and for l a depth ≈ 1 [m] one finds $Re \approx 10^6$, far above the threshold of turbulence.

In Figure 7.10 the usual coordinate system for a river is given. The local direction of the flow is taken as the x-direction, with y the coordinate along the horizontal direction perpendicular to the flow, and z the coordinate along the vertical direction.

> Disposal of untreated organic waste into streams was common in urban and rural areas of the United States until the early 20th century. Besides pathogenic bacteria in the waste, large amounts of organic chemicals and suspended material depleted the dissolved oxygen of receiving waters and killed much of the aquatic biota in some streams.
>
> Also clearing of forests, use of fertilizers and pesticides in agriculture and return water from irrigated fields affect stream water quality.
>
> [51], pp 1465, 1466

In the simplest approximation, an emission quickly mixes over y and z. Then only the x-direction remains and the puff will be modelled by the one-dimensional Equation (7.13). In that equation there was the diffusion parameter obeying (7.2) with $\sigma^2 = 2Dt$. In a turbulent river the diffusion parameter D has to be replaced by an empirical coefficient of longitudinal dispersion, called K.

The values of the parameter K are strongly dependent on the kind of river. It varies from 0.3 [m^2 s^{-1}] for small streams to values of 1000 [m^2 s^{-1}] for a major river ([82], p. 61). In fact, for the river Rhine the coefficient of longitudinal dispersion K increases with the distance from the source, ending up with a value of around 3000 [m^2 s^{-1}] in the Netherlands ([87], p. 48). For a tidal system a value $K \approx 1000$ [m^2 s^{-1}] is measured.

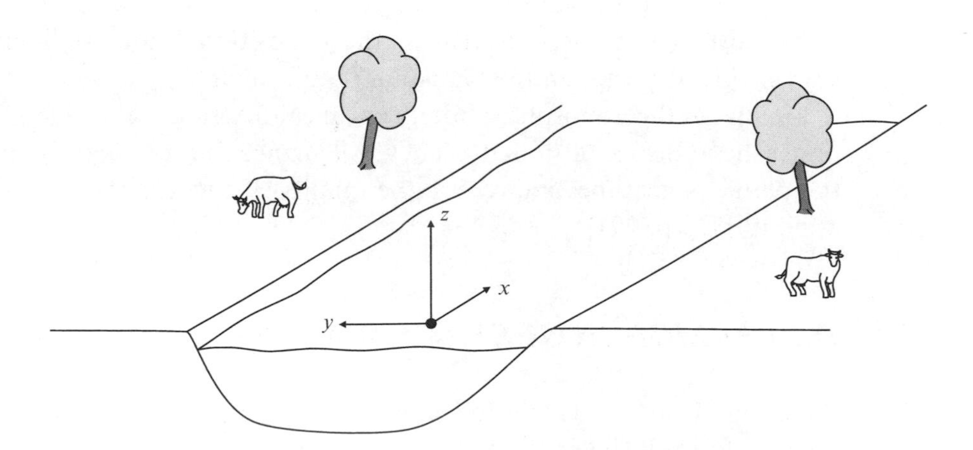

Figure 7.10 Coordinate system along a river

7.3.1 *Calamity: a poisonous discharge into a river*

Industries situated along a river usually have outlets for their waste water. Suppose that by accident a mass Q [kg] of poison is discharged into a river with a flow velocity u. The question then is to calculate the concentrations downstream in order to judge whether swimming should be forbidden or the intake for processing the river water as tap water should be stopped. We use Equation (7.13) with D replaced by K. For simplicity K is taken as constant. In that case (7.13) with the required substitution $\sigma = \sqrt{2Kt}$ gives

$$C(x, t) = \frac{Q}{\sqrt{4\pi Kt}} e^{-(x-ut)^2/4Kt} \tag{7.26}$$

In 1986 during a fire at the Sandoz plant in Switzerland, large quantities of chemical were washed into the River Rhine with the fire extinguishing water. The chemicals caused considerable damage to the ecosystem and led to limitation of the water use by the waterworks. An analysis of this incident proved the inadequacy of the means which were available at that time for predicting the transport of a pollution cloud in the River Rhine.

[87], p. 1

In this one-dimensional case the concentration C has dimension [kg m^{-1}], which is the total amount of pollution in a slice with a thickness of 1 [m] along the river and area the total cross-section of the river. For health reasons one is more interested in the concentration φ [kg m^{-3}] in the water itself. This is easily found by considering the volume of water V passing a cross-section per second [m^{3}s^{-1}]. As the velocity of the water is u, the total amount of pollutants passing a cross-section will be uC[kg s^{-1}]. This is distributed evenly (we suppose) over the volume V [m^{3}s^{-1}] of the water passing per second, so a simple division gives for the concentration φ [kg m^{-3}]

$$\varphi = \frac{uC(x, t)}{V} = \frac{Qu}{V\sqrt{4\pi Kt}} e^{-(x-ut)^2/4Kt} \tag{7.27}$$

As Q, u, V and K have been taken as constant, the only variables in the concentration φ [kg m^{-3}] are x, the distance along the river and the time t. The peak of the Gaussian (7.27) is found where the exponential equals one, so in $x = ut$. Let us suppose that we are 'riding the peak' and move along with the velocity u. Then, with time t the peak gets lower and wider as $\sigma = \sqrt{2Kt}$ is a measure for the width. As shown before, the factor in front of the Gaussian guarantees that the total amount of pollutants remains constant. The lower values of the peak in Equation (7.27) when moving with the flow, imply that the concentration φ decreases in time and the health danger diminishes as well.

In practice, there are several complicating factors. There will be side rivers, discharging into the main river, which increase V. The velocity u may change and the longitudinal dispersion parameter K most certainly changes. Therefore, Equation (7.27) is applied to stretches between side rivers and K may be determined experimentally by tracer experiments. The results are archived and, if a calamity happens, the corresponding calculations can be performed (Exercise 7.8). One might add that some fraction of the pollutants will be removed by the riverbanks or adsorbed by particles which gravitate down to sediments. These pollutants may not stay there forever and may eventually return to the ecosystem.

7.4 *DISPERSION BY GROUNDWATER*

Groundwater is an important resource of drinking water; in the United States for example, some 40 percent of the population depend on groundwater for their drinking ([9], p. 1). One usually takes it for granted that groundwater is free from contaminants, pumps it up and does not spend the 1700 $ or so* necessary to do a complete chemical analysis. Luckily groundwater is still a quite clean and reliable resource and it is worth keeping it that way.

There are many man-made pathways for contamination of groundwater. In Figure 1.4 contamination by a leaking fuel tank has already been shown. In that figure the fuel is apparently lighter than water as it is floating on the groundwater underneath. The vapours from the fuel will rise through the pores in the soil and enter the house above, causing health problems for the people living there. The fuel, of course, might have been heavier than water. In that case it would penetrate the groundwater below.

> How to find water in the underground? Before sunrise, lie down flat in the place where the search is to be made, and placing the chin on the earth and supporting it there, take a look out over the country. Then dig in places where vapours are seen curling and rising up into the air. This sign cannot show itself in a dry spot.
>
> Vitruvirus (25 BC), On Architecture, VIII, I, 1

There are many other sources of contamination of groundwater. Examples are the deicing of roads by spreading chemicals or the utilization of fertilizers by farmers and gardeners. These chemicals will slowly seep down into the soil and dissolve in the groundwater. As another example, rain or snow may contain air pollutants which, after precipitation, may eventually reach the groundwater. Finally, the official and unofficial dumps of industrial and household waste should be mentioned. If they are stored below the groundwater level, the chemicals of the dump will dissolve in the water. If the dumps are stored above that level, rain will seep

*(prices in 1992, [82], p. 11)

down into the dumps, pick up some of its chemicals, dissolve them and eventually reach the groundwater.

Once contaminants reach the groundwater, water will transport them by the processes that were discussed in the previous sections for air and rivers. Even though groundwater motion is slow, advection is still the main transport process in the direction of the flow. But, as shown in Figure 7.4, the flow is creeping and crawling through the pores and will not follow a straight line. Therefore, there is a significant dispersion perpendicular to the average flow. The dispersion of contaminants therefore takes place in plumes of all shapes, just as in the air ([88], Chapter 3), which may be the reason why the same word plume, used for a smoking chimney, is also used for dispersion of contaminants in the groundwater. As was the case for plumes in the air, simple calculations again will only show average behaviour in the dispersion.

> In the 1930's poison baits utilizing arsenic were used in the Midwest USA to counter a grasshopper infestation. Apparently, leftover poison bait was buried when the infestations ended. In 1972 a water-supply well was drilled for a small business. In short order, 11 of 13 employees became ill with arsenic poisoning. Tests showed the well contained 21 [mg L^{-1}] of As.
>
> [9], p. 32

Even more than dispersion in the air or by the rivers, time may cause the plume change in composition. Since, as the groundwater movement proceeds by means of pores, adsorption in the soil particles may remove contaminants from the flow. Also, particularly the upper regions of the soil are many bacteria which induce processes by which contaminants may be changed chemically.

In this section we only describe the main groundwater flow, as it is advection in this main flow which is the major means of dispersion. We start with a description of the top 50 metres or so below the earth's surface and then introduce the hydraulic potential as a means of describing the groundwater motion. We will introduce the equations of Darcy and apply them to horizontal flow. We end with a qualitative discussion of sources and sinks and comment on the cleaning of contaminated groundwater.

> Water has become a serious resource problem since a full two-thirds of the groundwater pumped up in the United States is used to irrigate crops. About one-fourth of this is overdraft, i.e. water drawn at a greater rate than it refills.
>
> [51], p. 8

7.4.1 *The surface layer*

Figure 7.11 is a sketch of how the earth's surface layer might appear. There are *impervious layers*, e.g. clay, through which no water can penetrate and

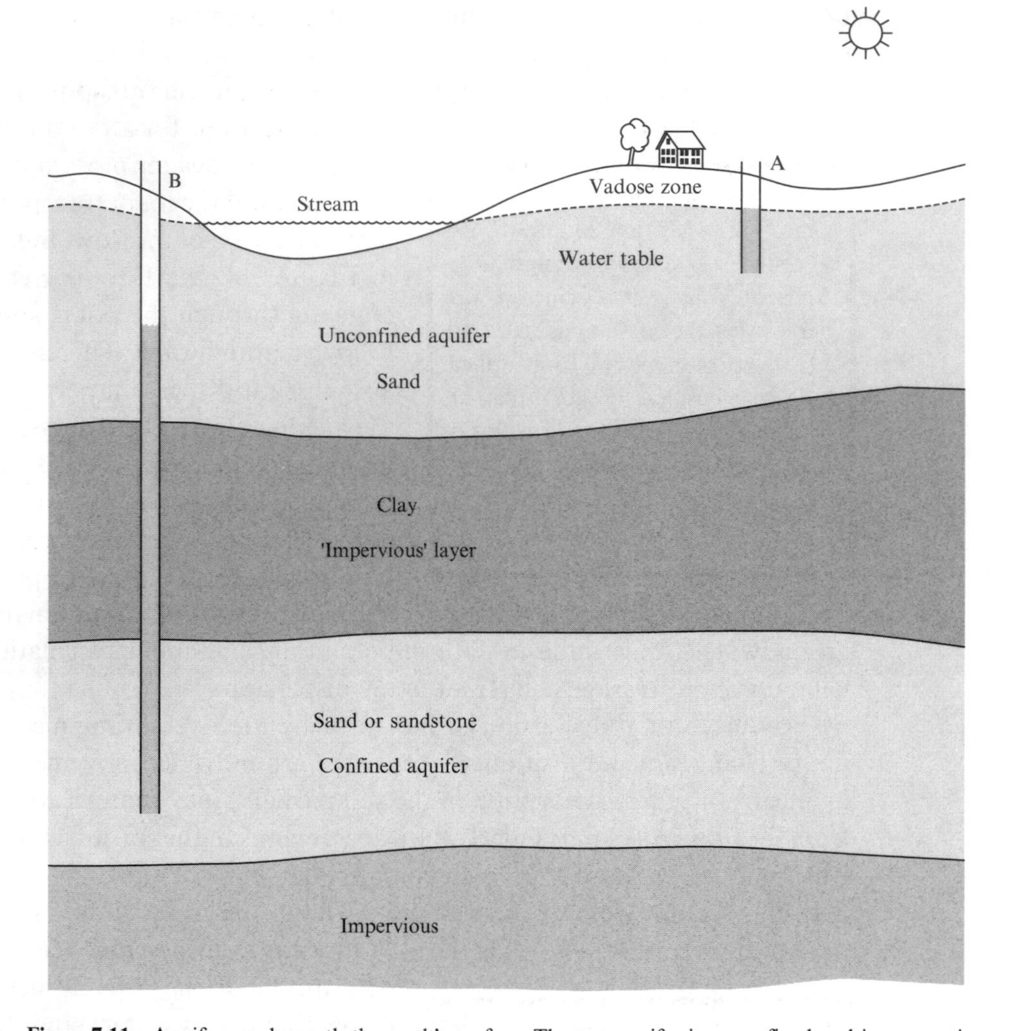

Figure 7.11 Aquifers underneath the earth's surface. The top aquifer is unconfined and in connection with the atmosphere. The lower one is confined between two impervious layers. The open tube at A defines the water table. Between the water table and the surface we have the vadose zone with lots of life. The tube at B shows a rather high water pressure in the aquifer deep down

there are water-carrying layers, called *aquifers*, often consisting of sand or sandstone. Consider Figure 7.11, starting at the bottom, where one notices an aquifer between two impervious layers. This is called a *confined aquifer*. Above the confined aquifer we see an impervious clay layer and above that layer again an aquifer, which is now called an *unconfined* aquifer. The reason for the adjective is that, at the top, the aquifer is in connection with the atmosphere by means of pores in the ground.

At position A in Figure 7.11 one notices a tube, which is open at both sides. On top it is in open connection with the atmosphere and at the bottom it penetrates the unconfined aquifer. In practice, it has a metal casing with little holes pierced in it, so it can penetrate the soil and can still be filled with water. The height to which the water rises is called the *water table*. At that position in the soil one has atmospheric pressure.

In Figure 7.11 the water table is not precisely horizontal but follows the terrain somewhat. Indeed, as shown below, this causes the groundwater flow. Between the water table and the open air one has the *vadose zone*, which may be humid and is also called an *unsaturated zone*. Small capillaries may suck up water into this zone and plants and trees may use the humidity to dissolve their food. In this soil much life is present and many chemical reactions take place.

On the left in Figure 7.11, the water table touches the surface of the stream. This is the steady-state situation. When the surface of the stream goes up, water will penetrate into the sandy banks and the water table will rise, albeit slowly. When the water level goes down, water seeps from the aquifer into the stream.

On the far left, one notices at position B a deep tube, again open at both sides. The tube goes down into the confined aquifer. Apparently, the water pressure down there is so high that the water in the tube rises almost to the water table of the unconfined aquifer. If this happens one may easily withdraw water from the deep, confined aquifer. If the pressure is high enough, a natural source will well up. These pressures arise in *artesian wells*, where the aquifers are sloping down from far away hills. The elevation of the aquifer in the hills is responsible for the high water pressure down in the valleys.

Seeping in the soil

Groundwater moves through the pores between the soil particles, as shown in Figure 7.4. The ease with which groundwater can move must be connected with the properties of the soil, shown in Table 7.3.

The pores taken together may be represented by the *porosity* n, which is the volume of the pores divided by the total bulk volume of a sample. The last column of Table 7.3 gives the porosity for various soils. It appears that the

Table 7.3 Particle size and hydraulic conductivity for various soils ([89], p. 8)

	Particle diameter d/[mm]	Hydraulic conductivity k/[m s^{-1}]	Porosity/% ([88], p. 53)
Clay	< 0.002	$10^{-10} - 10^{-8}$	35–55
Silt	0.002–0.06	$10^{-8} - 10^{-6}$	35–60
Sand	0.06–2	$10^{-5} - 10^{-3}$	20–35
Gravel	> 2	$10^{-2} - 10^{-1}$	20–35

porosities do not vary widely, while the particle sizes do. These sizes influence another property, the *hydraulic conductivity k*, displayed in the centre column of Table 7.3. This property will be defined below, but qualitatively we understand that this quantity will determine the ease with which the groundwater moves.

> If bodies are buried without a casket, decomposition will release organic material. It has been reported that H_2S gas in a well was the result of a 17th century graveyard for black plague victims. Apparently the well had been unwittingly bored through the graveyard.
>
> [9], p. 23

The fact that a clay layer is indicated as 'impervious' in Figure 7.11 implies that the pores in clay must be so small that the pores are very narrow. Many of them must have 'dead ends' blocking the flow. Others provide loopholes for the water but the particles are so small that the water in Figure 7.4 has to travel a very long path (If we approximate the particles by small spheres such as children's marbles and put them in a box, the smaller the marbles the longer the path the water has to take to go from the left to the right. The friction against the soil particles then acts a much longer path against the water movement).

The average flow is the important quantity from the point of view of contamination. It has velocity $\mathbf{u}$, a vector quantity with magnitude u and the direction of the average flow, or main flow. Take a square metre perpendicular to the main flow $\mathbf{u}$ and look at the volume of water [$m^3\,s^{-1}$] passing that square metre. This quantity $\mathbf{q}$ is called *the specific discharge vector* [$m^3\,s^{-1}\,m^{-2}$] = [$m\,s^{-1}$]. It will have the direction of $\mathbf{u}$, but be smaller as only the pores contain water with a fraction $n < 1$ of the volume:

$$\mathbf{q} = n\,\mathbf{u} \tag{7.28}$$

Hydraulic potential

In Figure 7.11 we show two cases of a vertical tube in an aquifer, which is open at both sides. In Figure 7.12 part of the figure is shown with some additional variables. The lower opening of the tube is at position (x, y, z), where z is the height above some fixed horizontal plane, possibly sea level. The height h of the water column corresponds with a hydrostatic pressure $p = \rho g\,h$ at (x, y, z), where ρ is the density of the water (Exercise 7.9). The quantity

$$\phi = z + h = z + \frac{p}{\rho g} \tag{7.29}$$

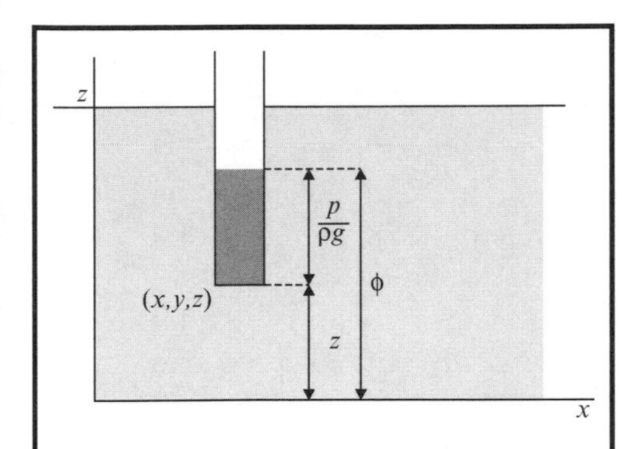

Figure 7.12 Hydraulic potential or groundwater head ϕ. It is the height of the water above some zero level in a double open tube at location (x, y, z), which is located in an aquifer, shaded light grey

is defined as the *hydraulic potential* or the *groundwater head* at location (x, y, z). From the definitions it is precisely the distance of the top of the water in the tube to the $z = 0$ level. Note from Figure 7.11 that the groundwater head for the two aquifers is different. So ϕ may be a function of both the horizontal and the vertical position of the end of the tube. There may be vertical motion, even within an aquifer. For example, when rain seeps down from above. Such cases should be described by a hydraulic potential which depends on the vertical coordinate z.

7.4.2 *Darcy's equations*

One may derive a relation between such measurable quantities as the groundwater head ϕ and the flow $\mathbf{q}$. One finds ([1], pp 216,217)

$$\mathbf{q} == -k \text{ grad } \phi \tag{7.30}$$

The science of groundwater flow originates from about 1856, in which year the city engineer of Dijon, Henri Darcy, published the results of the investigations that he had carried out for the design of a water supply system based on subsurface water. This water was carried to the valley in which Dijon is located by permeable layers of soil and supplied by rainfall on the surroundings.

[89], p. 4

One recognizes the *hydraulic conductivity* k given in Table 7.3. For a given value of grad ϕ the flow $\mathbf{q}$ is proportional to the value of the hydraulic conductivity k. From Table 7.3 it follows that the hydraulic conductivity k for clay is very small. This explains the name 'impervious' layer for a clay layer, but k is not zero. So, eventually some water may penetrate even a clay layer.

The fact that Darcy's equations are written as a single vector Equation (7.30) is more than a mathematical nicety. For, as follows from Appendix B, the direction of the vector $\mathbf{q}$ is perpendicular to the surfaces with constant values of the hydraulic potential ϕ. If one measures the hydraulic potential ϕ one may draw the surfaces of constant ϕ and construct the streamlines of the groundwater flow perpendicular to them. Incidentally, the way in which Equation (7.30) was written also explains the name 'potential' for ϕ. Ohm's Law for the flow of electric charge may be written in precisely the same way (Exercise 7.10).

Example 7.3

From monitoring wells (open tubes) it was measured that over a distance of 500 [m] the difference in hydraulic potential is 2 [m]. (a) Calculate the specific discharge q for a sandy aquifer (inspired by [85], pp. 153, 155) with a thickness of 20 [m]. (b) For groundwater extraction one needs to know how much water is passing a strip with width of 1 [m] over the total depth of 20 [m]. (c) For an estimate of the spread of contamination one needs the velocity u of the main flow. Take some parameters from Table 7.3.

Answer

(a) It follows that grad $\phi = 2[\mathrm{m}]/500[\mathrm{m}] = 0.004$. Table 7.3 gives a range of values for the hydraulic conductivity k for sand. Let us take the largest value $k = 10^{-3}\,[\mathrm{m\,s^{-1}}]$. It follows that $q = 4 \times 10^{-6}\,[\mathrm{m\,s^{-1}}]$. This is a very small number; apparently the official unit of time, the 'second' is not appropriate to discuss groundwater motion. Therefore, take the 'day' and express the discharge q in [m day^{-1}]. One finds $q = 0.35[\mathrm{m\,day^{-1}}]$. (b) The volume of water passing a strip with width of 1 [m] over the total depth of 20 [m] is found by multiplying 0.35 [m day^{-1}] by 20 [m^2], giving 7 [m^3 day^{-1}] passing the strip. (c) the velocity u of the main flow follows from Equation (7.28) with $u = q/n$. With a porosity of $n = 0.30$ this would give $u = 1.2\,[\mathrm{m\,day^{-1}}]$ or 425 [m year^{-1}] (Exercise 7.11).

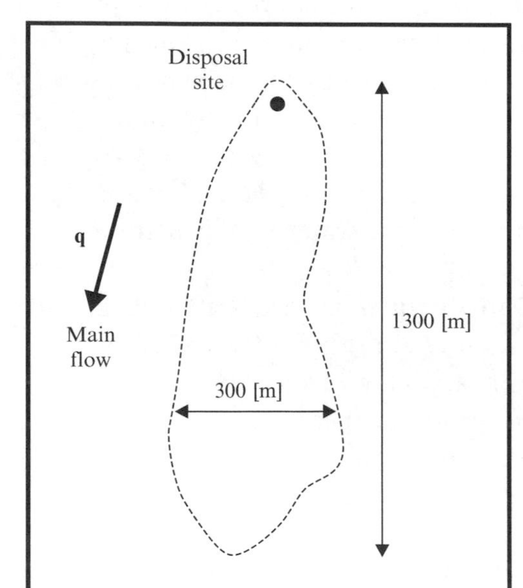

Figure 7.13 Chromium plume, Long Island, USA in an aquifer of sand and gravel after 13 years. With permission of John Wiley from [88], Fig 3–29, p. 79

The order of magnitude, found in Example 7.3 is confirmed by the study of a realistic plume, shown in Figure 7.13. There the plume from a disposal site travelled 1300 [m] in 13 years, which gives $u = 100\,[\mathrm{m\,year^{-1}}]$. The difference with the example above, must be due to a different value of conductivity k or porosity n.

7.4.3 *Sources and sinks*

A source or well is a place where one takes water out of the ground to an amount of $Q\,[\mathrm{m}^3\,\mathrm{s}^{-1}]$. The groundwater is flowing to the source. A sink is a place where one puts water into the ground and the groundwater flows away from the sink. This is described by the same variable Q, which then will be negative. In Figure 7.14 the situation is given for a confined aquifer underneath an impervious layer. Note that the hydraulic potential ϕ is dashed.

For the unperturbed situation the potential $\phi(x, y)$ goes almost horizontal. For a source (left) the potential ϕ goes down towards the well. This intuitively seems right. But it also agrees with the equations. For there can only be a flow $\mathbf{q}$ if the potential changes. From Equation (7.30) it follows that the potential ϕ indeed must go down. For the gradient is in the direction of increasing potential ϕ, which is outwards. The flow $\mathbf{q}$ is in the direction of minus the gradient, so inwards, which is correct.

The right-hand side of Figure 7.14 shows a sink, where wastewater is buried into the ground water. In that case the potential ϕ goes up in the vicinity of the sink.

7.4.4 *Cleaning of groundwater*

If there were a plume as in Figure 7.13, one could drill a well and pump out the contaminated groundwater. In practice a set-up as in Figure 7.15 is used. This example refers to contamination in the upper unconfined aquifer where

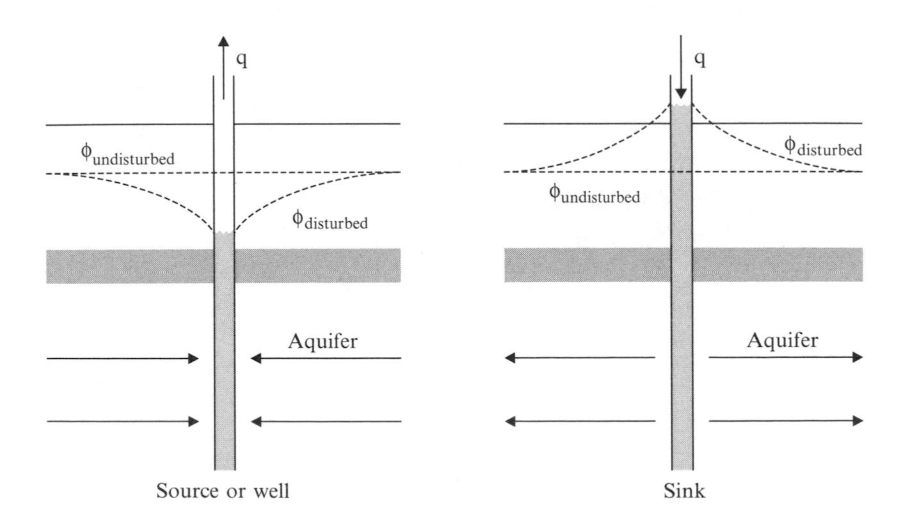

Figure 7.14 A source or well (left) and sink (right) in a confined aquifer. Note the hydraulic potential ϕ, which goes down for the source and up for the sink. Arrows indicate the streamlines in the aquifer

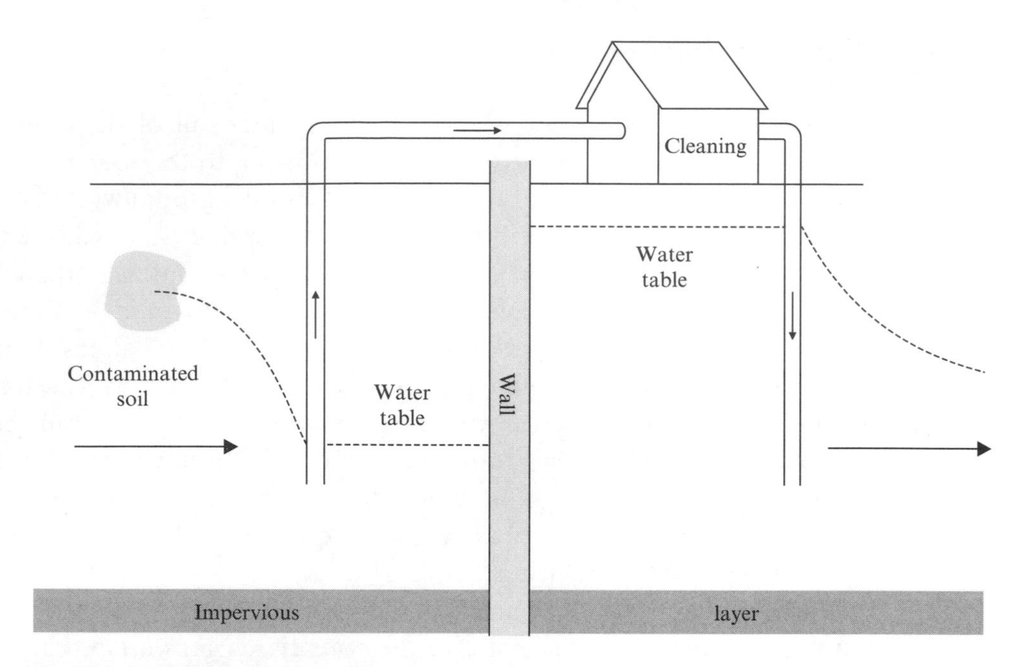

Figure 7.15 Cleaning of an unconfined aquifer. A contaminated plume (left) is cleaned by washing it with the available groundwater flow. A source is drilled on the left, and an impenetrable wall is put down to the confining layer. The water taken out is cleaned above ground and at the other side of the wall is put down again. (inspired by [9], p. 397)

the unperturbed groundwater streams from left to right. Downstream from the contamination one introduces a wall from the surface down into the confining, impervious layer. At the side of the contamination (on the left in Figure 7.15) one takes water out, one cleans it chemically and at the other side of the wall the cleaned water is again put down in the ground. In this way the natural groundwater movement is not disturbed too much.

It is interesting to note the decrease of the water table near the well where one is pumping water out of it, and the rise of the table near the sink, where the water is injected. Between the source or sink and the wall, the table is horizontal as there will be no flow passing the wall. For this unconfined aquifer the groundwater table is identical to the hydraulic potential ϕ.

EXERCISES

7.1 (a) Find the total width at half the maximum for Equation (7.1). (b) (For students with access to mathematics software) reproduce Figure 1.14.

7.2 Calculate from Equation (7.1) for what value of σ the value of $f(x)$ at $x = 1$ [m] equals 1 percent of its peak value. Next, use Equation (7.2) and Table 7.1 to calculate with what time that corresponds for diffusion of oxygen in water and of oxygen in air.

7.3 Show that Equation (7.5) is correct.

7.4 The small layer of the atmosphere near the ground and near ground-based objects, like leaves is described by a velocity $u = 0.3$ [m s^{-1}] and a thickness $l = 1$ [mm]. Calculate the Reynolds number and draw a conclusion as to the turbulence.

7.5 Show that the maximum of the right-hand side of Equation (7.24) occurs at $\sigma_z/h = \frac{1}{2}\sqrt{2}$. Calculate the value of that maximum.

7.6 A power plant with a chimney of $h = 300$ [m] of height emits 55900 [kg] SO_2 per day. Estimate the highest ground-level concentration for two cases: $\sigma_y = 180$ [m] and $\sigma_y = 600$ [m], while in both cases $u = 5$ [m s^{-1}]. Compare the results with the US standards, which are 80 [μg m^{-3}] as an annual average and 365 [μg m^{-3}] as an allowed 24 hour average ([85], pp 273, 320).

7.7 Check the dimensions of the right-hand side of Equation (7.27).

7.8 At $t = 0$ an amount $Q = 1$ [kg] of poison is released at $x = 0$. Study a river with $K = 1500$ [m^2 s^{-1}], $V = 3000$ [m^2 s^{-1}] and $u = 1$ [ms^{-1}]. Plot $\varphi(x, t)$ at $x = 300$ [km], $x = 500$ [km] and $x = 700$ [km] as a function of the time in hours since the release.

7.9 Show that a water column of height h and density ρ exerts a hydrostatic pressure on the liquid underneath $p = \rho g h$ (used in Equation (7.29)).

7.10 Find the similarity between Darcy's equations (7.30) and Ohm's law $V_1 - V_2 = IR$, where the electric current goes from location 1 to location 2. *Hint*. Rewrite the resistance in Ohm's Law in terms of electric conductivity, length and cross-sectional area.

7.11* A sand aquifer has a hydraulic conductivity $k = 7.2 \times 10^{-5}$ [m s^{-1}] and a porosity $n = 0.33$. Grad ϕ is measured as grad $\phi = 0.0043$. Calculate the velocity u of the groundwater flow.

*Inspired by [9], p. 150/151, describing a field experiment in Ontario.

8 Noise

The dictionary describes noise as 'sound that is loud and disturbing'. Many governments have put up regulations to limit noise to 'acceptable levels'. These regulations are usually written in terms of a maximum allowed number of decibels, a unit, which will be defined below.

In this chapter we discuss the various parts of the definition given above. The physics of sound, in Section 8.1, will appear as straightforward science [90]. The qualification 'loud and disturbing' is more difficult to describe, for the concept of 'loudness' will involve the physiological characteristics of the human ear. At certain frequencies, for example, the ear is not sensitive and sound at such frequencies will not be experienced as 'loud'. This point will be elaborated in Section 8.2.

The qualification 'disturbing' in the definition of noise is most difficult to measure with physical equipment. This will become clear by looking at Figure 8.1, which shows the sounds that are entering the room of the man who is reading at the bottom right. There is *direct transmission* of sound from the outside traffic (6 in Figure 8.1), and from the neighbouring room, where his son is making music (1). From that room, incidentally, there is also *flanking transmission* by floor and ceiling (2, 3 and 4). Finally, there is *contact noise* from his little daughter playing upstairs (5). All together this results in a sound level L_0, to be defined in Section 8.1.

The question is whether this sound level L_0 is disturbing. The first assumption has to be that the reading man is not deaf. This means that in official regulations one has to specify the quality of the hearing of the people concerned. As will be seen in Sect. 8.2, the age group of 18–30 years is taken for this purpose. Finally, for a sound to be disturbing it matters where the sound originates. When Fig. 8.1 represent three members of the same family, as suggested above, the man reading will be less annoyed than when the sound is produced by his neighbours with whom he is in trouble. In a physics text human feelings cannot be discussed and we restrict ourselves to what can be measured.*

*It is advisable to have a good relationship with your neighbours. Your music will disturb them less, and if it is too loud and they complain, you will be more inclined to switch down your equipment. This will keep the peace.

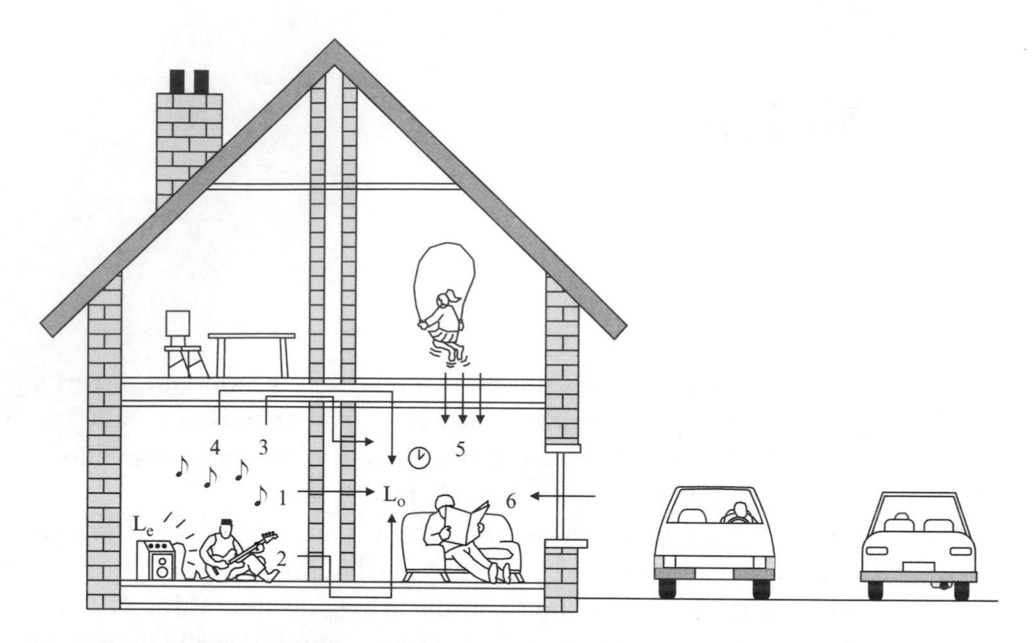

Figure 8.1 The noise level L_0 in a room is determined by direct transmission from an adjoining room (1), traffic noise through the window (6), flanking transmission by floor and ceilings. (2,3,4) and contact noise from upstairs (5)

When a sound level L_0 is too high, it has to be reduced. In Section 8.3 some methods of reducing the transmission of sound are discussed, ignoring the problem of contact noise. We finish with some brief comments on active control of sound, using advanced electronics [91].

8.1 PHYSICS OF SOUND

Sound is a pressure fluctuation in a medium like air, or a concrete wall. It propagates with a velocity† c_0 and is described as a sound wave.

8.1.1 Sound waves

A pressure wave, which is propagating in the x-direction may be written as

$$p(x, y, z, t) = f\left(t - \frac{x}{c_0}\right) \tag{8.1}$$

†The subscript 0 is meant to distinguish the velocity of sound from that of light, which by convention is called c.

Here p is the deviation from the local atmospheric pressure p_0, so the total pressure would be $p_0 + p$. The right-hand side of Equation (8.1) is an arbitrary function of the argument

$$t - \frac{x}{c_0} \tag{8.2}$$

If one takes a time $t + 1$ and position $x + c_0$, the value of Equation (8.2) remains the same as for time t and position x. The right-hand side of Equation (8.1) has its old value and consequently also the left-hand side. This implies that the fluctuation $p(x, y, z, t)$ has travelled over a length c_0 in the positive x- direction during 1 [s], without changing its value. Therefore the velocity of sound will be c_0 [m s^{-1}]. It is convention to take the velocity $c_0 > 0$; if one wants to describe a wave travelling in the negative x-direction, one has to replace $(t - x/c_0)$ by $(t + x/c_0)$ in equations (8.1) and (8.2).

The right-hand side of Equation (8.1) does not contain the variables y and z. At a certain time t only the x-coordinate determines the value of the function f. That value remains the same in a plane perpendicular to the x-axis, as all points in that plane have the same value of x. It follows that Equation (8.1) describes a plane wave, propagating in the positive x-direction.

Harmonic waves

A special case of a plane wave is the harmonic plane wave

$$p(x, t) = A \cos\left\{\omega\left(t - \frac{x}{\omega/k} + \frac{\phi}{\omega}\right)\right\} \tag{8.3}$$

For constant values of ω and ϕ Equation (8.3) has the form (8.1) and consequently represents a plane wave with velocity of propagation

$$c_0 = \frac{\omega}{k} \tag{8.4}$$

Expression (8.3) may be rewritten as

$$p(x, t) = A \cos(\omega t - kx + \phi) \tag{8.5}$$

The cosine function varies between (-1) and $(+1)$ and the pressure function p fluctuates between $(-A)$ and $(+A)$. This value of A is called the *amplitude* of the harmonic wave. The symbol ω is called the *angular frequency* of the wave and k the *wave number*.

For a certain time t, the pressure fluctuation (8.5) has a nice shape which is exhibited in Figure 8.2. From the figure it is clear that a certain value of the pressure fluctuation p at position x_1 appears again for the first time at a length $x_1 + \lambda$. This is called the wavelength λ. Substitution in Equation (8.5) gives

$$p = A\cos(\omega t - kx_1 + \phi) = A\cos\{\omega t - k(x_1 + \lambda) + \phi\} \tag{8.6}$$

The left and the right of Equation (8.6) are equal for all values x_1 if

$$k\lambda = 2\pi \quad \text{or} \quad \lambda = \frac{2\pi}{k} \tag{8.7}$$

Similarly, for fixed position x, the wave (8.5) is periodic in time t. After a *period* T the same result is found for the first time when

$$\omega T = 2\pi \text{ or } T = \frac{2\pi}{\omega} \tag{8.8}$$

The *frequency* f of a wave is the number of complete cosines passing at a certain position x in 1 [s]. This is equal to the number of periods in 1 [s]. Therefore

$$f = \frac{1}{T} = \frac{\omega}{2\pi} \tag{8.9}$$

A frequency is by convention expressed in the unit [Hz] with dimension [s^{-1}]. Using eqs. (8.7) and (8.8) the velocity of sound (8.4) can be expressed as

$$c_0 = \frac{\omega}{k} = \frac{2\pi\omega}{2\pi k} = f\lambda \tag{8.10}$$

Assume for the sake of the argument that Figure 8.2 was taken at $t = 0$. Then the pressure fluctuation at $x = 0$ becomes

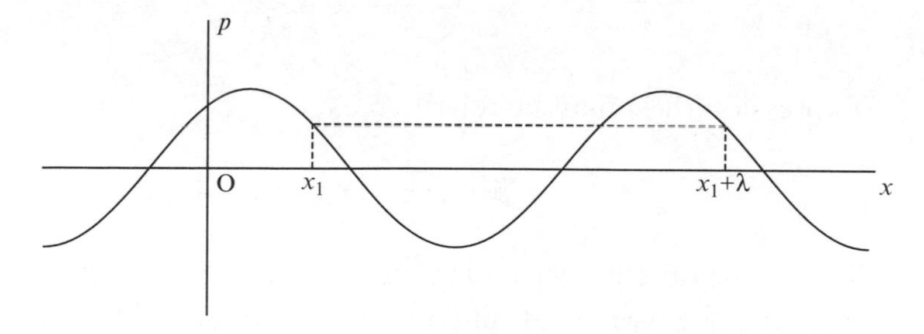

Figure 8.2 Harmonic plane wave (8.5) for a fixed time t. The smallest length after which the graph repeats itself is the wavelength λ

$$p = A\cos\phi \tag{8.11}$$

This determines the *phase* ϕ of the wave. For fixed values of ω, t, A and k, changing the phase ϕ corresponds with shifting Figure 8.2 to the left or the right.

Expression (8.5) may be generalized in order to find a plane wave $p(x, y, z, t)$ propagating in an arbitrary direction. We restrict ourselves to writing down a plane harmonic wave travelling in the positive y-direction with the same A, ω, k and ϕ as Equation (8.5). One finds

$$p(y, t) = A\cos(\omega t - ky + \phi) \tag{8.12}$$

Superposition of waves

Suppose that one has two propagating waves in the same medium, indicated by $p_1(x, y, z, t)$ and $p_2(x, y, z, t)$. Then the pressure fluctuation, which is the deviation from the ambient atmospheric pressure, will be written as

$$p(x, y, z, t) = p_1(x, y, z, t) + p_2(x, y, z, t) \tag{8.13}$$

When both waves have the same frequency but different directions, one will get an interesting *interference pattern.*

Example 8.1

Find the interference pattern of wave (8.5), travelling in the positive x-direction and (8.12) travelling in the positive y-direction. Take $\phi = 0$ and find $p(x, y, z, t)$. Next, take $A = 1$, $t = 0$ and some simple values of ω and k to plot the resulting pressure fluctuation (8.13) as a function of x and y.

Answer

For $\phi = 0$ one finds from (8.5) and (8.12)

$$p_1(x, t) = A\cos(\omega t - kx) \tag{8.14}$$

$$p_2(y, t) = A\cos(\omega t - ky) \tag{8.15}$$

$$p(x, y, t) = p_1(x, t) + p_2(y, t) = A\{\cos(\omega t - kx) + \cos(\omega t - ky)\} = 2A\cos\{\tfrac{1}{2}(2\omega t - kx - ky)\}\cos\{\tfrac{1}{2}(ky - kx)\} \tag{8.16}$$

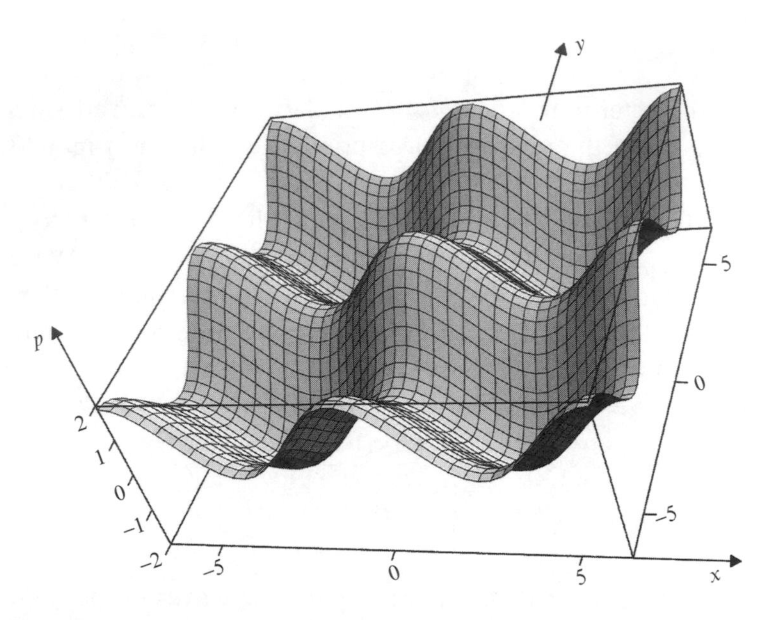

Figure 8.3 Interference pattern from two travelling harmonic waves, one in the x-direction and one with the same frequency and amplitude in the y-direction

For $A = 1$ and $t = 0$ the pressure fluctuation simply becomes

$$p(x, y, t) = 2\cos\left\{\frac{1}{2}(kx + ky)\right\}\cos\left\{\frac{1}{2}(kx - ky)\right\} \tag{8.17}$$

This function is plotted in Figure 8.3 for $k = 1[\mathrm{m}^{-1}]$, $\omega = 1[\mathrm{s}^{-1}]$, $\lambda = 2\pi$ [m] from $-\lambda < x < \lambda$ and $-\lambda < y < \lambda$. Notice the maximum for $x = y = 0$, which reappears after a distance λ in both directions. There are several minima; the deepest one for a distance $\lambda/2$ in both directions. If time increases, it follows from Equation (8.16) that bigger x, y values are needed to give the same pressure fluctuation. Consequently, in time, the interference pattern shifts along the $x = y$ diagonal (Exercise 8.1).

Velocity of sound

A local increase in pressure from the atmospheric pressure p_0 to pressure $p_0 + p$ will be accompanied by an increase in density from ρ_0 to $\rho_0 + \rho$. One may show ([1], p. 267) that the velocity of sound c_0 obeys the relation

$$c_0^2 = \frac{p}{\rho} \tag{8.18}$$

For an ideal gas like air, this equation can be simplified to

$$c_0^2 = \kappa \frac{p_0}{\rho_0} \tag{8.19}$$

where $\kappa = c_p/c_v = 1.4$, a constant, which was encountered earlier in Poisson's Equation (4.41). Indeed, that equation is needed to derive Equation (8.19).

Data for standard air

Standard ambient temperature T and pressure p_0 is defined as ([49], p. 30) $T = 25$ [°C] and $p_0 = 10^5$[N m^{-2}]. From Appendix A one finds $\rho_0 = 1.161$ [kg m^{-3}]. Equation (8.19) gives the velocity of sound in air as $c_0 = 347$ [m s^{-1}]. From a glance at Appendix A it is clear that ρ_0 and c_0 depend on the temperature.

For a certain frequency f, Equation (8.9) gives ω and for known c_0 Equation (8.4) gives k and Equation (8.7) gives λ. Table 8.1 gives the wavelength λ belonging to a certain frequency f in a range, which (as shown in Section 8.2) is relevant for human hearing.

It is interesting that the wavelengths λ are within the range of sizes which human beings may grasp.

8.1.2 *Sound pressure level*

An arbitrary pressure fluctuation may be written as $p(x, y, z, t)$, which at a certain location may be abbreviated by $p(t)$. This fluctuation will be irregular and in the course of time will have alternatively positive and negative values. In order to obtain a single number for a sound pressure level one should not take the time average of $p(t)$ itself, as that most probably will be almost zero. Instead, one defines a *root mean square* value p_{rms} by

$$p_{\mathrm{rms}}^2 = \mathop{\mathrm{Lim}}_{T \to \infty} \frac{1}{T} \int_{-T/2}^{T/2} p^2(t) \mathrm{d}t \tag{8.20}$$

The human ear is able to detect values of p_{rms} in a range of 10^6 or 10^8 between highest and lowest level. Therefore one uses a $\log_{10}$ scale to define a sound pressure level L_p as

Table 8.1 Frequencies and wavelengths in air with $c_0 = 347$ [m s^{-1}]

f/[s^{-1}]	63	82	250	1000	4000
λ / [m]	5.5	4.23	1.37	0.35	0.09

$$L_p = 10 \log_{10} \frac{p_{rms}^2}{p_{ref}^2} = 20 \log_{10} \frac{p_{rms}}{p_{ref}} \text{ [dB]} \tag{8.21}$$

An ice-breaker has two sources of noise: its bubbler and its propellers. The bubbler is a high-pressure air-venting system along the keel that protects the hull by pushing away fragments of broken ice. The continuous bubbling noise has a sound pressure level of 192 [dB] at its origin. Meanwhile the ice-breaker's screw produces cavitation noise of 197 [dB], rising to 205 [dB] when the ship is running at full power.

Some scientists believe that the vessels are so noisy that it is inducing behavioural changes in whales as far as 14 [km] away.

New Scientist, 30 September 2000, 12

In the definition one needs a reference pressure p_{ref} in order to obtain dimensionless numbers of which one can take the logarithm. One takes $p_{ref} = 2 \times 10^{-5}$ [Pa], as at the time the definition was designed it was thought to be the lowest pressure detectable by man. In Section 8.2 it will be seen that, in fact, the threshold is a little bit lower.

It is usual to add decibel or [dB] to the value of L_p, in order to suggest the log scale. For $p_{rms} = p_{ref}$ Equation (8.21) gives $L_p = 0$ [dB]. The human ear will experience pain at a sound pressure level of $L_p = 140$ [dB] (Exercise 8.3). Finally, the subscript 10 in $\log_{10}$ will be omitted, as in this chapter is all logs are to base 10.

Adding independent sound pressure levels

Suppose that one has two independent sources of sound. At a certain location (x, y, z) the first gives a sound pressure $p_1(t)$ and the second one gives $p_2(t)$. The total sound pressure becomes, from (8.13)

$$p(t) = p_1(t) + p_2(t) \tag{8.22}$$

From Equation (8.20) it follows that, still at location (x, y, z)

$$\begin{aligned} p_{rms}^2 &= \lim_{T\to\infty} \frac{1}{T} \int_{-T/2}^{T/2} \{p_1(t) + p_2(t)\}^2 \, dt \\ &= \lim_{T\to\infty} \frac{1}{T} \int_{-T/2}^{T/2} \{p_1(t)\}^2 \, dt + \lim_{T\to\infty} \frac{1}{T} \int_{-T/2}^{T/2} \{p_2(t)\}^2 \, dt \\ &\quad + 2 \lim_{T\to\infty} \frac{1}{T} \int_{-T/2}^{T/2} p_1(t) p_2(t) \, dt \\ &= p_{1,rms}^2 + p_{2,rms}^2 + 2 \lim_{T\to\infty} \frac{1}{T} \int_{-T/2}^{T/2} p_1(t) p_2(t) \, dt \end{aligned} \tag{8.23}$$

If the two sources are independent, $p_1(t)$ and $p_2(t)$ will vary with time independently and their product will be positive as often as negative. It follows that the last term of Equation (8.23) becomes zero and

$$p^2_{\text{rms}} = p^2_{1,\text{rms}} + p^2_{2,\text{rms}} \tag{8.24}$$

Note that, in Example 8.1, the two waves were not independent at all. They had the same single angular frequency ω and in $x = 0, y = 0$, for example, $p_1(t)$ and $p_2(t)$ are identical. Therefore their product is always positive. The argument leading to Equation (8.24) only holds for really independent sources.

For the addition of two independent sources, the sound pressure level L_p follows from equations (8.21) and (8.23) as

$$L_p = 10 \log \frac{p^2_{1,\text{rms}} + p^2_{2,\text{rms}}}{p^2_{\text{ref}}} \tag{8.25}$$

If both sources give the same sound level at location (x, y, z) one gets

$$\begin{aligned} L_p &= 10 \log \frac{p^2_{1,\text{rms}} + p^2_{2,\text{rms}}}{p^2_{\text{ref}}} \\ &= 10 \log \frac{2p^2_{1,\text{rms}}}{p^2_{\text{ref}}} \\ &= 10 \log 2 + 10 \log \frac{p^2_{1,\text{rms}}}{p^2_{\text{ref}}} \\ &= 3 + L_{p_1} [\text{dB}] \end{aligned} \tag{8.26}$$

It follows that the addition of an equally strong source to a source already present, increases the sound pressure level by 3 [dB]. Most people can just hear this difference in sound pressure level. A difference of 5 [dB] is easily observable [92].

If one has two independent sources of which one knows the sound pressure levels L_{p1} and L_{p2}, one has to go back to the sound pressure itself before one may add the pressures. From definition (8.21) it follows that

$$\frac{L_{p1}}{10} = \log \frac{p^2_{1,\text{rms}}}{p^2_{\text{ref}}} \Rightarrow \frac{p^2_{1,\text{rms}}}{p^2_{\text{ref}}} = 10^{\frac{L_{p1}}{10}} \tag{8.27}$$

From Equation (8.25) one finds

$$L_p = 10 \log \frac{p^2_{1,\text{rms}} + p^2_{2,\text{rms}}}{p^2_{\text{ref}}} = 10 \log\{10^{\frac{L_{p1}}{10}} + 10^{\frac{L_{p2}}{10}}\} \tag{8.28}$$

8.1.3 *Acoustic intensity*

The acoustic intensity **I** is the energy leaving a volume per second through 1 $[\mathrm{m}^2]$ perpendicular on the sound wave with dimensions $[\mathrm{J\ m^{-2}s^{-1}}]$. For harmonic plane waves (8.5) one may derive ([1], p. 293)

$$I = \frac{p_{\mathrm{rms}}^2}{\rho_0 c_0} \tag{8.29}$$

where ρ_0 is the density of air and c_0 the velocity of sound. It is easily checked that the dimensions fit (Exercise 8.4). The intensity I has a similar wide range of values like p_{rms}^2. It therefore is customary to express I also in [dB] by

$$L_{\mathrm{I}} = 10 \log \frac{I}{I_{\mathrm{ref}}} \tag{8.30}$$

with $I_{\mathrm{ref}} = 10^{-12}$ $[\mathrm{W\ m^{-2}}]$. There is a simple connection between L_{I} and L_{p}, as

$$\begin{aligned} L_{\mathrm{p}} &= 10 \log \frac{p_{\mathrm{rms}}^2}{p_{\mathrm{ref}}^2} = 10 \log \left(\frac{\rho_0 c_0 I}{I_{\mathrm{ref}}} \frac{I_{\mathrm{ref}}}{p_{\mathrm{ref}}^2} \right) \\ &= 10 \log \frac{I}{I_{\mathrm{ref}}} + 10 \log \frac{\rho_0 c_0 I_{\mathrm{ref}}}{p_{\mathrm{ref}}^2} \\ &= L_{\mathrm{I}} - 0.03\ [\mathrm{dB}] \approx L_{\mathrm{I}}\ [\mathrm{dB}] \end{aligned} \tag{8.31}$$

Obviously, I_{ref} has been chosen such that for common temperatures $L_{\mathrm{p}} \approx L_{\mathrm{I}}$ (Exercise 8.5).

8.2 *HUMAN HEARING*

The sensitivity of the human ear depends on the frequency f of the sound. Experiments have been performed where waves (8.5) with a single frequency were offered to a group of well-hearing people in the age group 18–30 years [93]. The result is shown in Figure 8.4. The lowest dashed curve indicates the sound pressure level L_{p} from Equation (8.21), which can just be heard. One observes that the human ear operates best at $f = 4000$ [Hz]. There $L_{\mathrm{p}} \approx -3$ [dB], which means that even pressures lower than $p_{\mathrm{ref}} = 2 \times 10^{-5}$ [Pa] can be heard.

Experimentally, one next takes a frequency $f = 1000$ [Hz] with a sound pressure level $L_{\mathrm{p}} = 10$ [dB]. Signals at other frequencies which are

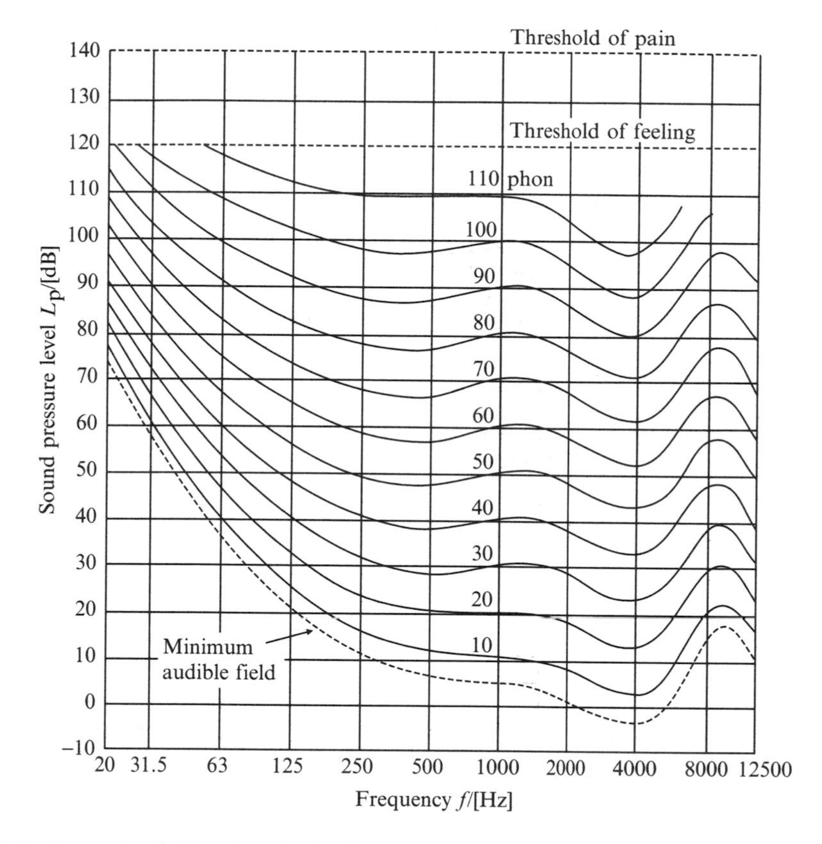

Figure 8.4 Curves of equal loudness for the best hearing sector of the population in the age group from 18 to 30 years. From Annex A, p. 4 of [93] reproduced with permission of the International Organization for Standardization, ISO

experienced as equally loud are then connected by a full curve, called 10 *phon*. In the same way one defines a curve called 20 phon, and so on. The curves of Figure 8.4 are rather complex. At 63 [Hz], for example, 10 [dB] extra brings one from the 10 phon curve to about 25 phon. At 1000 [Hz] it adds, by definition, only 10 phon. Therefore, the extra sound has more effect at the low frequency. See [90], Chapter 11 for information on human hearing.

8.2.1 dBA correction

For environmental regulations, it is human experience which counts and not the uniquely defined sound pressure level L_p of Equation (8.21). Obviously, one has to make a correction based on the curves of Figure (8.4). This is done by smoothing the 'minimum audible yield' curve of Figure 8.4 and inverting it. The result is shown in Figure 8.5. Horizontally, one observes the frequency f and along the vertical axis the *dBA correction* Δ. Some values of

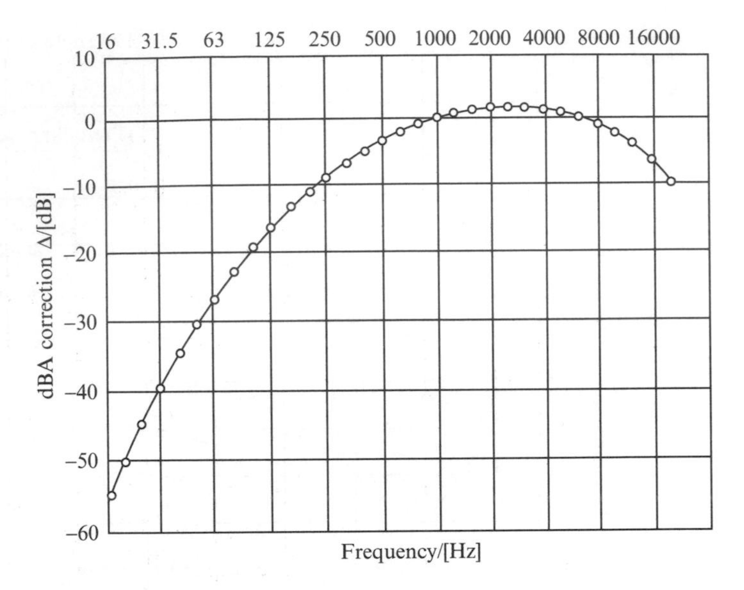

Figure 8.5 dBA-correction curve. The values Δ of the curve at frequency f have to be added to the sound pressure level L_p to obtain the dBA corrected value L_A of Equation (8.32)

Δ are given by circles on the curve. The dBA-corrected sound pressure level L_A is defined as

$$L_A = L_p + \Delta \text{ [dBA]} \tag{8.32}$$

Example 8.2

A signal has a sound pressure level $L_p = 80$ [dB]. Determine the dBA correction Δ for frequencies of 250 [Hz], 500 [Hz], 1000 [Hz] and 2000 [Hz]. Calculate the dBA corrected sound pressure levels L_A. In each case give the correction expected from the 'minimum audible' curve of Figure 8.4.

Answer

$f = 250[Hz] \rightarrow \Delta = -8.6$[dB] $\rightarrow L_A = 71.4$[dBA]. Expected correction $= -11$ [dB]
$f = 500$ [Hz] $\rightarrow \Delta = -3.2$ [dB] $\rightarrow L_A = 76.8$ [dBA]. Expected correction $= -7$ [dB]
$f = 1000$ [Hz] $\rightarrow \Delta = 0$ [dB] $\rightarrow L_A = 80$ [dBA]. Expected correction $= -5$ [dB]
$f = 2000$ [Hz] $\rightarrow \Delta = -1.2$ [dB] $\rightarrow L_A = 81.2$ [dBA]. Expected correction $= 0.1$ [dB]
Clearly, the dBA-correction represents an average (Exercise 8.6)

Table 8.2 Sound levels with permission of John Wiley from [90], p. 282

Level L_A/[dBA]	Source of noise
110–120	Discotheque with roll-n-roll band
90–100	Cockpit of light aircraft
80–90	Truck (64 [km/h] at 15 [m])
70–80	Car (100 [km/h] at 8 [m])
60–70	Vacuum cleaner at position operator
50–60	Light traffic at 30 [m]
40–50	Quiet residential during the day
30–50	Quiet residential at night
20–30	Wilderness

8.2.2 *Regulations*

Governments or local authorities have drawn up regulations on noise nuisance. They have to determine whether certain industries, motorways or airfields are acceptable, or whether certain buildings such as discotheques must have extra sound insulation.

The unit of measurement is the dBA-corrected value L_A. To be precise, one has to measure L_p of Equation (8.21), divided it into the frequencies which the noise contains, apply the dBA-correction (8.32), which is dependent on the frequency, and determine L_A. These procedures are defined by the International Standardisation Organisation; we will not describe them, but just assume that they exist.

> The most important effect in terms of the number of people affected is so-called annoyance. This is strongly connected with the necessity to close windows in order to avoid sleep disturbance or interference with communication, listening to the TV, radio or music. Additionally, there is a number of serious medical effects such as high blood pressure, mental stress, heart attacks and hearing damage.
>
> [94], p.3

The measurements of L_A have to be performed several times during 24 hours, for the nuisance of noise depends on the time when it is produced. From Table 8.2, for example, it follows that the level L_A [dBA] in a quiet residential area during night is about 10 [dBA] smaller than during the day. This is what people appreciate. Also indoors the level should be lower than outdoors.

There are a wide variety of rules and regulations, depending on the country concerned. For practical reasons we therefore will use the proposal of the European Union as an example [94*]. The following definitions are used[†]:

*Further examples may be found in [90], Chapter 12. For the USA consult [10], Chapter 12 or the EPA website.

†In the text we summarize. Member States may deviate from the definitions given.

- L_{day} is the average sound level during the day (7–19 hours)
- $L_{evening}$ is the average sound level during the evening (19–23 hours)
- L_{night} is the average sound level during the night (23–7 hours)

As a next step the sound levels are multiplied by the number of hours they refer to and divided by 24. This average may be calculated by a generalization of Equation (8.28), where two sound pressure levels were added. This gives ([94], Annex I)‡

$$L_{den} = 10\log\left\{(1/24)\left(12 \times 10^{\frac{L_{day}}{10}} + 4 \times 10^{\frac{L_{evening}+5}{10}} + 8 \times 10^{\frac{L_{night}+10}{10}}\right)\right\} \quad (8.33)$$

One notices that day and night get a higher weight by an extra 5 [dBA] or 10 [dBA] in the exponent (Exercise 8.7). In urban areas in the European Union the main sources of noise are traffic noise and aircraft noise. At least 25 percent of the population of the EU is experiencing a reduced quality of life ([94], p. 3).

A study for the World Health Organization [95] puts the advisable steady-state noise level inside bedrooms at 30 [dBA] and for single noise events at a maximum 45 [dBA]. During the night, outdoors, the level also should be, at most, 45 [dBA] in order that people may sleep with open windows. During the daytime, outdoors, the advised maximum noise level is 50 [dBA]. Most countries accept higher levels. For example, a Dutch regulation puts as acceptable outdoors in rural areas 55 [dBA] and in cities 60 [dBA]. In France one may get insulation grants when the outdoor noise level from road traffic exceeds 60 [dBA] (Ref [95] p. 229).

8.3 *REDUCING SOUND LEVELS*

The sound level in a room such as the one in Figure 8.1 with the man reading is, for the main part, determined by transmission through the walls and windows. Figure 8.6 shows a simplified picture of the same room. The window on the right will be discussed later. First the concept of noise reduction is defined with the example of the wall on the left.

8.3.1 *Noise reduction*

An acoustic intensity I_i is hitting the wall from the left and an intensity I_t is transmitted. The *noise reduction* of the wall is defined as

‡In another definition L_{den} is the highest of the three numbers L_{day}, $L_{evening} + 5$, $L_{night} + 10$ (Exercise 8.7)

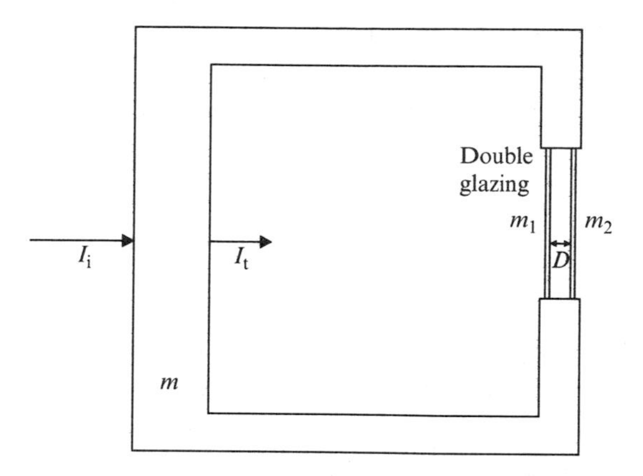

Figure 8.6 Transmission of sound. The wall on the left receives an acoustic intensity I_i and transmits an intensity I_t. It has a mass m [kg m^{-2}]. The double glazing on the right consists of two glass sheets with masses m_1 [kg m^{-2}] and m_2 [kg m^{-2}] with distance D [m]

$$R = 10 \log \frac{I_i}{I_t} \quad [\text{dB}] \tag{8.34}$$

Ref. [90], Chapter 12 uses the symbol *NR* for this quantity. Ref[1] uses the term 'air insulation'.

Let the acoustic intensity entering the wall be called L_i and the intensity transmitted by the wall be L_t. From Equation (8.30) the value of (8.34) is equal to the reduction in acoustic intensity by the wall:

$$R = 10 \log \frac{I_i}{I_t} = 10 \log \frac{I_i}{I_{ref}} - 10 \log \frac{I_t}{I_{ref}} = L_i - L_t \ [\text{dB}] \tag{8.35}$$

From Equation (8.31) it follows that this is also equal to the reduction in sound pressure level by the wall.

A massive wall

The incident acoustic wave from the left in Figure 8.6 will put the wall, with a mass of m [kg m^{-2}], into vibration. The wave passes through the wall and on the right puts the air into vibration causing the pressure wave with sound level L_t. It is understandable that the mass m enters into the equation for noise reduction. We do not give the complete formula ([1], p. 312) but just the approximation for $mf > 500$ [kg m^{-2}s^{-1}]. This gives, for a massive wall

$$R = 20 \log m + 20 \log \frac{f}{500} + 12 \quad [\text{dB}] \tag{8.36}$$

Here, the mass m is expressed in [kg m^{-2}] and the frequency f in [Hz]. In practice there are losses from flanking transmission. A practical mass law for a real wall therefore becomes (Exercise 8.8)

$$R = 17.5 \log m + 17.5 \log \frac{f}{500} + 3\,[\text{dB}] \tag{8.37}$$

A doubling of the frequency means that one replaces f by $2f$ and the noise reduction R gets an additional $17.5 \log 2 = 5$ [dB]. It also follows that moped noise, which contains rather high sound frequencies is more easily reduced by a massive wall than the lower frequencies of heavy trucks.

It should be mentioned that a small leak strongly reduces the noise reduction R. For example, if $R = 50$ [dB] and a small hole is drilled with a relative area of 10^{-4}, it is reduced by 10 [dB] ([1], p. 310). So, if one is worried by noise from the neighbours, it is advisable to look for small holes in the walls.

Double glazing

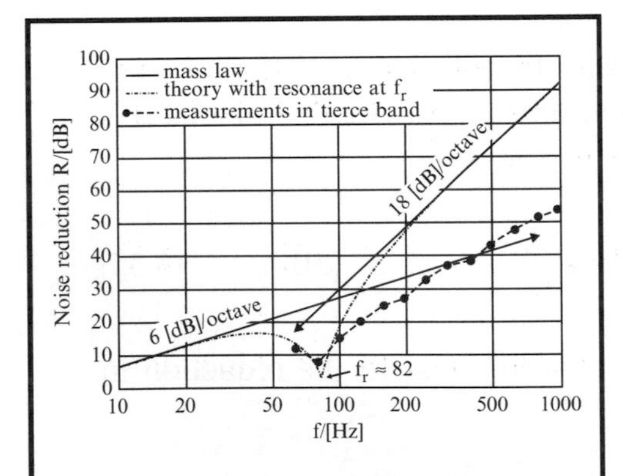

Figure 8.7 Noise reduction R as a function of frequency for double glazing with masses of $m_1 = 20$ [kg m^{-2}] and $m_2 = 10$ [kg m^{-2}] and a distance $D = 0.08$ [m]. Reproduced by permission of Delft Technological University from [96], Figure 3.21)

On the right of Figure 8.6 a window is shown with double glazing and air in between. Figure 8.7 shows the measured noise reduction R (bold points) for values $m_1 = 20$ [kg m^{-2}] and $m_2 = 10$ [kg m^{-2}] at a distance $D = 0.08$ [m]. With these values the dashed curve and the experimental data show a minimum at $f_r = 82$ [Hz]. The dashed curve represents a theoretical formula ([1], p. 314) and the straight lines are approximations of that formula at low and at high frequencies. We do not derive the formulas here, but try to make the result acceptable.

In the limit of low frequencies it follows from the theoretical formula, that a doubling of frequency f (called octave) gives rise to an additional 6 [dB] in noise reduction. This corresponds with the theoretical mass law (8.36) for a massive wall. Apparently, glass and air together with a thickness of about 0.1 [m] experience the incident wave as a single mass. Which suggests that the wavelength of the sound wave is much larger than 0.1 [m]. Indeed, from Table 8.1, it follows that a frequency smaller than 63 [Hz] has a wavelength larger than 5.5 [m].

In the limit of very high frequencies, the formula gives 18 [dB] for a doubling of frequency, which looks like three separate walls. Anyway, for the frequencies shown the wavelength becomes less than 1 [m], so in the high-

The accomplishment of noise abatement objectives requires that control requirements be imposed on noise producers. Noise control is best achieved at the source. For new products, noise emission standards can be specified and designed into such products as aircraft engines, motor vehicle tires, industrial machinery and home appliances.

[10], p. 408

frequency limit, the three media which the wave has to pass may be treated as independent. The experimental data lie considerably lower than this limit, which is ascribed to the mounting of the glass at the edges, where effectively it can be considered as a single sheet [96].

Most interesting is the minimum at $f_r = 82$ [Hz]. This frequency happens to be the *eigenfrequency* of the air between the two glass sheets. Therefore, when the incident wave has this frequency, the air will easily pick up the sound wave, resonate with it, reinforce it and compensate for the losses in the walls.

The following formula holds for the resonance frequency f_r of double glazing with masses m_1[kg m^{-2}] and m_2 [kg m^{-2}] for the glass and a distance D [m] in between:

$$f_r = \frac{1}{2\pi}\sqrt{\frac{\kappa p_0}{D}\frac{m_1 + m_2}{m_1 m_2}} \approx 60\sqrt{\frac{m_1 + m_2}{m_1 m_2 D}} \qquad (8.38)$$

where $\kappa = c_p/c_v \approx 1.4$ is the Poisson constant and p_0 the atmospheric pressure.

It immediately follows that, for a larger distance D, the resonance frequency goes down. The usual thermal double glazing would have a distance D of about 6 [mm]. With $m_1 = m_2 = 10$ [kg m^{-2}] one would find $f_r = 346$ [Hz], which means that traffic sounds with approximately these frequencies will not be reduced (Exercises 8.9 and 8.10). Usually distances of 16 [cm] between the glass sheets gives such a low resonance frequency that one is far to the right of the resonance in Figure 8.7. That will eliminate the noise of most traffic.

Example 8.3

Make Equation (8.38) plausible by using an analogy from simple mechanics.

Answer

The easiest example of a resonance is shown in Figure 8.8 (a). An elastic spring with mass μ is displaced a little bit to the right over a distance x. The empirical Hooke's law says that the resulting force to the left may be written as

$$F = -kx \qquad (8.39)$$

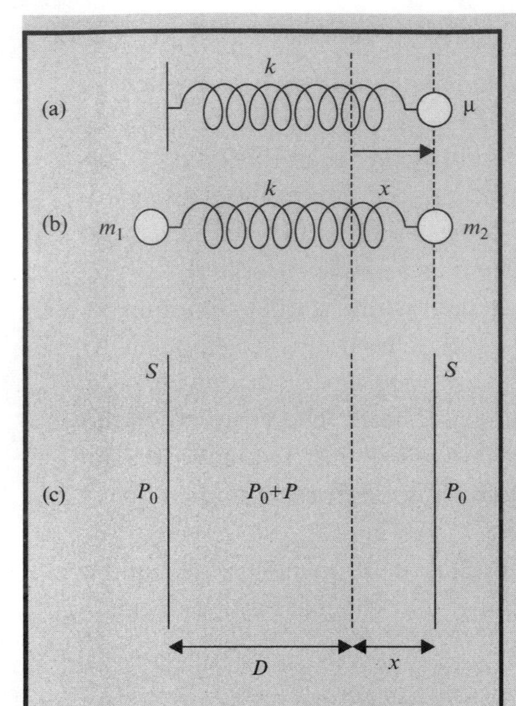

Figure 8.8 Simple vibrations as a model for the air between double glazing

where k is a constant of proportionality [N m^{-1}]. The resulting frequency of the spring appears to be

$$f = \frac{1}{2\pi}\sqrt{\frac{k}{\mu}}\ [\mathrm{s}^{-1}] \qquad (8.40)$$

This is the eigenfrequency of the string and when any external periodic force is applied with this frequency, the spring will resonate.

Figure 8.8 (b) shows two masses m_1 and m_2 connected by the same spring. The mass on the right is displaced over the same distance x and again Equation (8.40) will give the resonance frequency. In this case one needs an effective mass μ to reproduce Equation (8.40). It is given by

$$\mu = \frac{m_1 m_2}{m_1 + m_2}\ [\mathrm{kg}] \qquad (8.41)$$

Figure 8.8 (c) shows the double glazing with surface area S. The glass on the right is displaced to the right over a distance x. The pressure between the glazings $p_0 + p$ is a little bit lower than p_0 at the outside (so $p < 0$). Figure 8.8 (c) is completely analogous to Figure 8.8 (b) if one puts

$$k = \kappa p_0 / D \qquad (8.42)$$

where κ is Poisson's constant. One may prove that Equation (8.42) and (8.41) together give the resonance frequency, when substituted in Equation (8.40). The masses in Equation (8.41) then have to be the masses in [kg m^{-2}].* Here, we only check the dimensions of Equation (8.42). In Equation (8.41) the masses have dimension [kg m^{-2}], therefore the effective mass μ is also in [kg m^{-2}] instead of in [kg]. Poisson's constant κ is dimensionless, the pressure p_0 is in [N m^{-2}], D in [m] and therefore k is in [N m^{-3}]. Underneath the square root in Equation (8.40) the dimension becomes [N m^{-3}]/[kg m^{-2}] = [kg m s^{-2} m^{-3}]/[kg m^{-2}] = [s^{-2}], where we used [N] = mass × acceleration. The square root gives the required dimension [s^{-1}] = [Hz] for the resonance frequency f_r.

*To be precise, equations (8.41) and (8.42) are correct for a sheet with surface area $S = 1$[m^2]. Otherwise both have to be multiplied by S, which of course gives the same result (8.40) (Exercise 8.11)

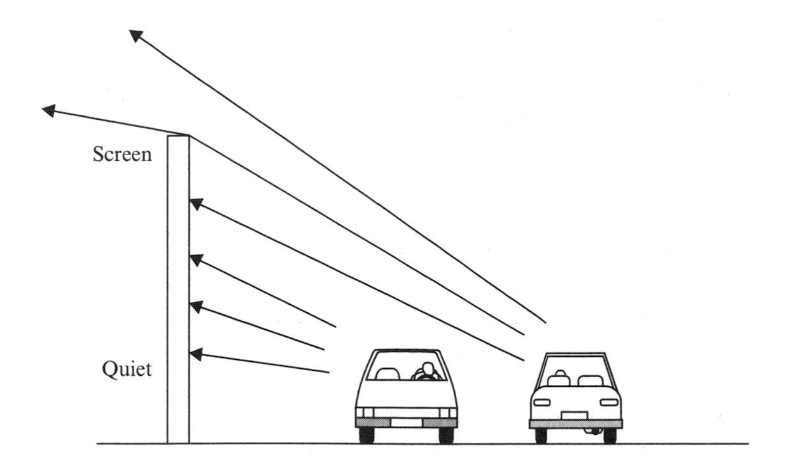

Figure 8.9 A noise screen blocks the direction of the traffic noise towards the residential areas. Note that the propagation of the wave is slightly curved

Noise screen

Massive walls and double glazing do not help much against traffic noise in the garden, or in a residential area. The solution chosen there is to put a noise screen between traffic and people, as shown in Figure 8.9. For the frequencies which dominate traffic noise between 100 [Hz] and 150 [Hz], the wavelength will be in the order of 3–4 [m] (Table 8.1).

Like all waves, sound waves do not travel in straight lines but disperse a little over distances, the size of a wavelength. The screen therefore loses some effectiveness over larger distances; indeed one may notice a sound reduction when walking towards the screen.

Active control of sound

Inside an aircraft or a car, the noise of the engines, resulting in a fluctuating pressure $p_1(x, y, z, t)$ may be rather unpleasant. To reduce that, one makes use of the principle of superposition of waves, written down in Equation (8.13). With one or more computer controlled speakers one creates a second sound wave $p_2(x, y, z, t)$ such that the sum (8.13) is close to zero ([91], Sect. 10.15). A reduction of sound pressure level in the order of 10 [dB] may be achieved.

EXERCISES

8.1 For students having access to simple mathematical software. Reproduce Figure 8.3. Take a few steps $t = T/4$ and $t = 2T/4$ to find out how the interference pattern shifts.

8.2 Reproduce Table 8.1

8.3 For the threshold of pain ($L_p = 140$ [dB]) calculate p_{rms} from Equation (8.21). Compare with the atmospheric pressure.

8.4 Check the dimension of Equation (8.29)

8.5 Check numerically the 0.03 [dB] in Equation (8.31).

8.6 Check the dBA-correction of Figure 8.5 for frequencies of $f = 125$ [Hz] and $f = 8000$ [Hz] against Figure 8.4

8.7 Ref [1] defines L_{den} as the highest of the following three numbers: $L_{day}, L_{evening} + 5$ and $L_{night} + 10$ [dB]. The EU proposes Equation (8.33). Discuss which definition will give the highest value and check your argument by inserting some numerical values.

8.8 Find the noise reduction for brick at a frequency of $f = 500$ [Hz], with a thickness of 22 [cm] and a density of 1600 [kg m^{-3}], where on both sides one has got 12 [mm] plaster with a density of 1500 [kg m^{-3}]. For engineers, an acceptable result for a wall separating two apartments would be 48 [dB].

8.9 (a) Check the number 60 on the right in Equation (8.38) and find the resonance frequency for the values given in Figure 8.7. Compare with Figure 8.7.

8.10 Take double glazing with glass of 4 [mm] and density 2500 [kg m^{-3}]. Calculate the resonance frequency (a) for a distance between the glass sheets of 6 [mm] and (b) for a distance of 16 [cm]. Note that traffic noise has frequencies of 100 [Hz] and higher.

8.11 (For students who like a little algebra). Prove Equation (8.42), by applying Poisson's relation (4.41) to Figure 8.8 (c) and take a sheet with area S, which moves a little to the right.

9 Environmental Spectroscopy

Physics has a tradition of accurate experimentation [97]. Physicists are trained to perform experiments in a controlled manner. In this process they learn how to set up an experiment, how to measure the correct variables, not disturbed by uncontrolled variations in other parameters. They learn to analyse their results in terms of a physical model or law and to apply statistical methods that allow them to estimate the accuracy with which they have obtained their result. This is the reason why the outcome of a typical physical experiment is always reported as an interval in which the value of the measured parameter will be found. The temperature of a classroom, for example, may be given as 20.3 ± 0.2 [°C].

> Cavendish, an extremely wealthy eccentric, carried out some of the most meticulous researches on the fundamental problems of electricity and of chemistry. His most famous experiment was a direct measurement of the force of gravity acting between leaden balls hung on torsion wires. More modern measurements have scarcely improved on the value he obtained 200 years ago.
>
> [97], p. 72

By performing careful measurements, physical scientists may contribute to the understanding, mitigation and possible solution of environmental problems. For example, it is known that very small concentrations of certain chemicals in food may cause cancers. Environmental spectroscopy provides a means to find out whether these compounds are present in a given sample and with what concentrations. Spectroscopic methods also may be used to find out what gases are present in the atmosphere and their concentrations. This gives a method to find out how the concentrations of very sensitive greenhouse gases develop over time (see Table 3.1). It also gives a way of measuring the ozone concentrations at various altitudes.

Spectroscopic methods have in common that they use the interaction of electromagnetic radiation with matter and the resulting change in the electromagnetic spectrum, hence the name spectroscopy. In the present chapter we therefore discuss the absorption and emission of electromagnetic radiation qualitatively* in Section 9.1. We proceed in Section 9.2 with a discussion of

*A more quantitative discussion requires the use of Quantum Mechanics and is given in Ref. [1], chapters 2 and 7.

the adventures of the solar spectrum while passing through the atmosphere. In Section 9.3 we turn to some experimental techniques, finishing with remote sensing. We apply some of them to the ozone problem in Section 9.4: too little ozone high up in the stratosphere and too much low down in the troposphere. We will explain the appropriate measuring technique called lidar.

Other measuring techniques with other applications will be discussed in Chapter 10. Some simple experiments for a students' lab are given in Appendix D. It has to be understood that most or all of the experimental methods described in this book have applications which are much wider than the environmental field. The need to measure very small concentrations of certain compounds also arises in industry, where small additions may improve or deteriorate their production processes. The scope of the two chapters therefore is wider than the examples, given here, are suggesting.

9.1 BASIC SPECTROSCOPY

Early experiments in spectroscopy were performed by measuring the light emitted by a flickering candle (even today the *candela* is used as a unit that measures the brightness of a light source). That light was then transmitted through a material or a solution and detected through a piece of coloured glass by the eye. By careful measurement the elementary laws that describe the transmission of light, the spectral content of light and the elementary laws of spectroscopy were discovered and tested.

> Newton, in 1671, performed one of the first systematic investigations of the spectral composition of light. He let a sunbeam from a little hole in the window blind pass through a lens and a prism. The result was a rainbow on the wall.
>
> [97], p. 19,20

The principles of spectroscopy may be illustrated with the common experience of the shades of green, observed in a forest. Plate 2 shows a colour photograph with many shades of green, while Figure 9.1 represents part of that photo in black and white; there the shades of green are replaced by shades of grey. The interesting feature is that all leaves receive the same sunlight, but the outgoing light shows so many different colours, or in physical terms, so many wavelengths. The physical explanation of this phenomenon is the following.

The white sunlight impinges on the leaves, and penetrates the first few layers of molecules, where it will be scattered. Chlorophyll molecules in the leaves absorb the red light, while carotenoid molecules absorb the blue light. Only the green light is not absorbed and is able to 'escape'. The observed variations are due to small differences in chlorophyll or carotenoid

Figure 9.1 The photograph shows the rainforest on the island Palawan in the Philippines. The rich vegetation is manifested by the amazing variety in shades of green. Plate 2 shows a colour photograph, where the variety of shades show up more clearly

concentration, which change the reflection and scattering properties of the leaves. Furthermore, the chlorophyll and carotenoid composition varies among plants giving rise to slight spectral variations. Since the human eye is a very sensitive detector of colour differences all these variations are observable to us.

The preceding paragraph contains several concepts which deserve some attention: white sunlight, absorption, difference of behaviour between molecules in their interaction with radiation and reflection of light. We discuss these concepts below, starting with the 'white' sunlight.

9.1.1 The electromagnetic spectrum

In Figure 1.8 the wavelength composition of the sunlight was shown, arriving at the earth surface. In the visible region, between 0.400 [μm] and 0.800 [μm] all wavelengths are represented with almost the same intensity. To our eyes sunlight looks 'white' and also a refrigerator looks 'white', which means that all components of the visible spectrum are reflected from its surface with about the same intensity.

Units

For historical and practical reasons, the units used in spectroscopy are not the official SI units. The electronvolt or [eV] is the energy unit used in spectroscopy; it is the energy an electron gains while decreasing its potential energy by 1 [V]. In Appendix A it is expressed in [J].

In Chapter 8, Section 8.1, sound waves were discussed. Electromagnetic waves have a similar behaviour, but they propagate even in vacuum and have a much higher velocity c. Still, the same equations apply although the convention in spectroscopy is a little different. The frequency is called ν instead of f and the *wave number* $\bar{\nu}$ is defined as

$$\bar{\nu} = \frac{\nu}{c} = \frac{\omega}{2\pi c} = \frac{k}{2\pi} \tag{9.1}$$

which differs a factor 2π from definition (8.4) of the wave number of sound waves, which was called k. Officially the wave number $\bar{\nu}$ should be expressed in [m^{-1}], but spectroscopists still use [cm^{-1}] as unit. Of course, 1 [cm^{-1}] = 100 [m^{-1}].

The complete electromagnetic spectrum

The connection between the frequency ν of an electromagnetic wave and the energy of the light quanta is given by Planck's relation (1.3) with h as Planck's constant. In this way an electromagnetic spectrum may be characterized by one of the physical quantities: frequency ν [s^{-1}], wave number $\bar{\nu}$ [cm^{-1}] or energy E [eV]. The complete electromagnetic spectrum in practical use is given in Figure 9.2, although for most environmental spectroscopy only the

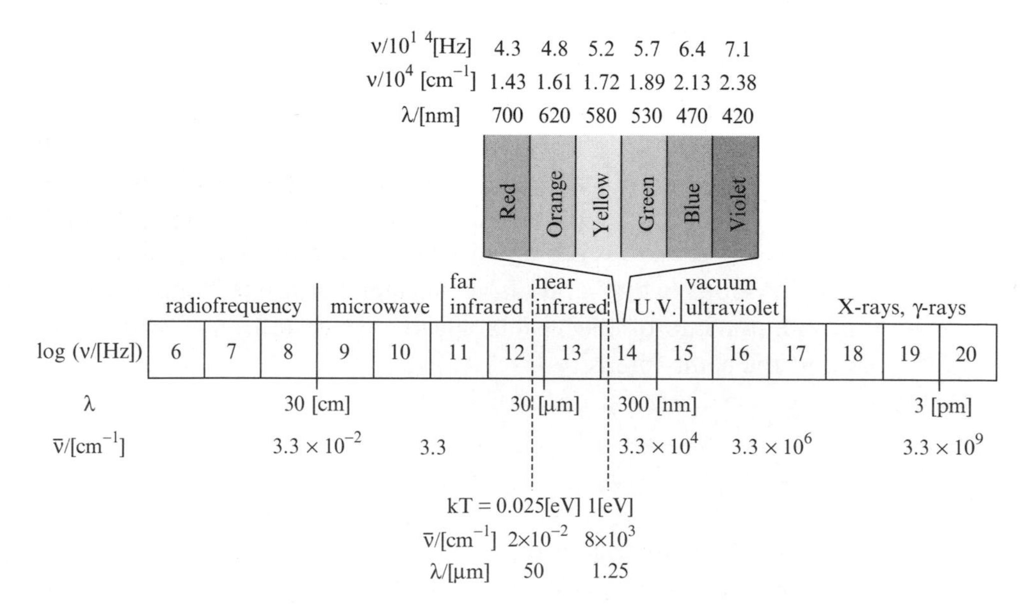

Figure 9.2 Electromagnetic spectrum and its spectral regions. Note the frequency ν, the wavelength λ and the wave number $\bar{\nu}$. Reproduced by permission of the author from [49], Figure 18.1

region between wave numbers 1 [cm^{-1}] and 100000 [cm^{-1}] is relevant. A characteristic wavelength in the orange is 680 [nm], which corresponds with 1.8 [eV] of energy (Exercise 9.1).

9.1.2 Transition between states: absorption and emission

It follows from quantum mechanics that atoms and molecules can only be in states with a well-defined energy E. The more complex the atom or molecule, the denser the states as a function of energy. For an arbitrary molecule the allowed energy levels are sketched in Figure 9.3. At a certain temperature T the thermal interaction between the molecules will result in a Boltzmann distribution over the lowest levels with a spread of about kT. We have already met this phenomenon in the discussion of Equation (5.4).

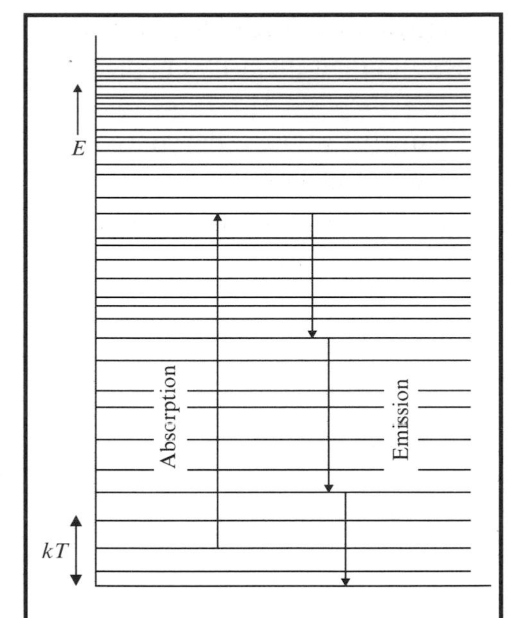

Figure 9.3 Energy levels of an arbitrary molecule. Absorption occurs from the lowest occupied states with energies up to kT above the lowest state to higher levels. For emission the molecule must first be excited to a higher state and then, in a few steps, decrease in energy

In a typical *absorption experiment* a detector monitors the amount of light transmitted through the sample, while the frequency of the light source is swept over the relevant frequency range. An atom or molecule may absorb electromagnetic radiation from an initial state with energy E_i to the final state with energy E_f if the following condition is met:

$$E_f - E_i = h\nu \tag{9.2}$$

In an *emission experiment*, the electromagnetic radiation emitted by the sample is detected as a function of the frequency of the emitted radiation. Of course, first the molecule has to be excited, then it may descend by a few intermediate states. Since in principle the same energy levels are involved in the absorption and the emission process, the same spectroscopic information is obtained from each type of experiment. Generally, emission experiments are more sensitive since the emitted light is detected versus a dark background. On the other hand, there may be so few emitted photons that the signal becomes hard to detect.

Absorption and emission spectra consist of a, sometimes large, number of lines, corresponding to transitions between the various possible energy levels E_i and E_f. Some of these lines may be intense, others weak, or expected lines may even be totally absent. In addition, absorption and emission lines, certainly of molecular transitions, are generally very much broadened.

Spectroscopic methods are widely applied since each atom, molecule (small or large), molecular aggregate or even complete plant is uniquely characterized by a specific set of energy levels. Transitions between these levels occur by the absorption or emission of electromagnetic radiation and result in spectroscopic features, which are highly specific for the atom, molecule or organism concerned. This has already appeared from Figure 3.9, where the absorption of five gases between 3 [μm] and 30 [μm] was displayed. A spectrum may act as a '*bar code*' identifying a certain molecule, in the same way as in a supermarket it identifies the articles one buys. That may be observed from the absorption spectrum of H_2O vapour in the yellow region, as shown in Figure 9.4*. In the small wave number region between 16 850 [cm^{-1}] and 17 050 [cm^{-1}] the positions of the lines and their intensity uniquely determine the molecule.

In our description of the shades of green in Figure 9.1 we used the concept of reflection. This essentially is a macroscopic phenomenon, where interaction with the top few layers of molecules is added, resulting in outgoing waves. The concept is most easily described for a mirror, where an incident

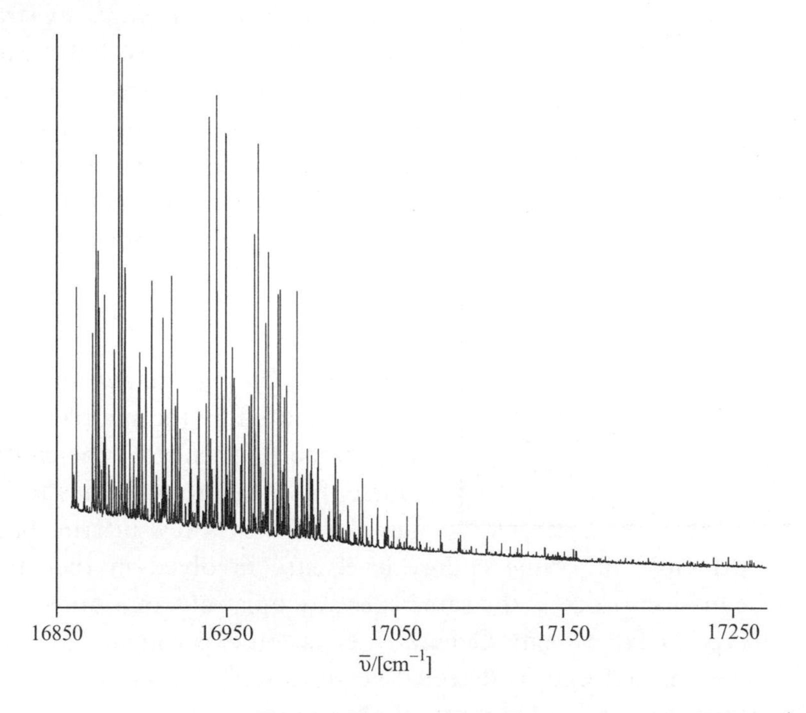

Figure 9.4 Absorption spectrum of H_2O gas in the wave number region around 17000 [cm^{-1}]. Reproduced with permission of Academic Press from [98]

*This spectrum was measured by Dr. Hans Naus and Dr Wim Ubachs in the Laser Facility of the Free University, Amsterdam, Netherlands [98]. They kindly put it at the disposal of the authors.

plane wave results in an outgoing plane wave, while the angle of incidence equals that of reflection.

9.2 THE SOLAR SPECTRUM AT GROUND LEVEL

The earth and life upon it depend critically on the intensity and spectral composition of the solar light that reaches the earth's surface. In Chapter 3 we have already discussed how the delicate balance between the inflow of solar energy and the outflow of the earth's infrared radiation determines the temperature of the earth's surface. This interplay is determined by the presence of greenhouse gases in the atmosphere.

At ground level part of the incoming solar energy is captured by photosynthetic organisms (Section 5.1). The solar spectrum at ground level needs to have the right wavelengths, which can be absorbed by these organisms. To check this, Figure 9.5 shows how the solar spectrum indeed covers the major absorption bands of photosynthetic organisms. After absorption, the organisms convert solar energy to chemical energy.

Photosynthesis has, at a very early stage of evolution, allowed life to persist in the absence of an external chemical energy source and was essential for life to develop into the rich diversity of forms known to us today. Photosynthetic organisms are at the basis of the earth's food chain and, last but not least, the energy stored by photosynthesis has been the source of all fossil fuels which

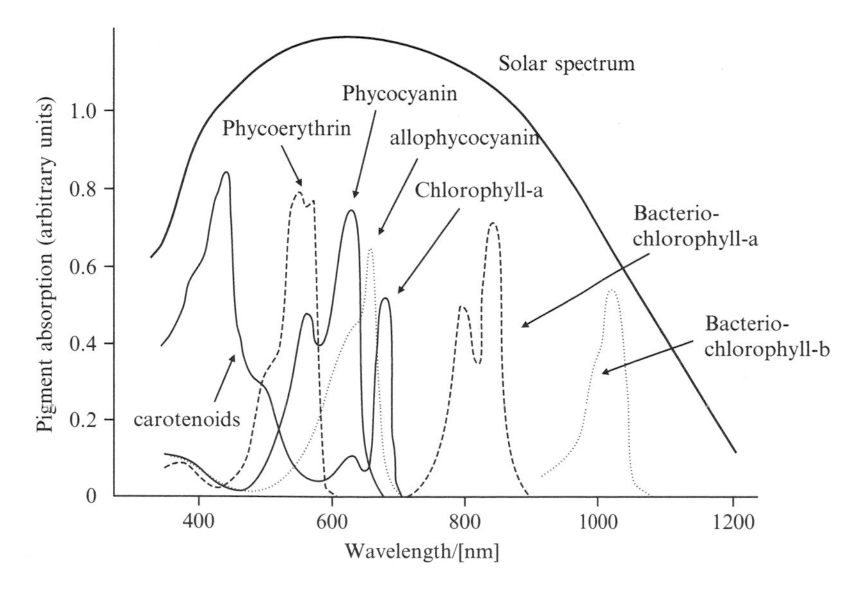

Figure 9.5 Overlap of the solar spectrum with the absorption spectra of a variety of photosynthetic organisms. Notice that they operate in different parts of the spectrum. On the chlorophyll curve a 'shoulder' at 500 [nm] is indicated, which originates from the carotenoids in the sample

are being used at an increasing rate (see Chapter 11). In a few decades we may have to cope with the world's energy crisis and need to develop artificial photosynthesis using some of the engineering designs of the original biological process. This is one good reason to study how the spectral composition of solar light is changed by passage through the atmosphere.

A second good reason is that not all solar radiation has benign effects on life. In Section 9.4 we will discuss how radiation with wavelengths smaller than about 0.295 [μm] or 295 [nm] may destroy biomolecules, such as DNA and proteins, needed for the survival of life. In Figure 1.8 it was indicated on the far left that ozone absorbs the solar radiation in that region. A decrease in the ozone layer implies that bio-molecules, which originally had their absorption bands just to the left of the solar spectrum, and therefore were well protected, may become increasingly vulnerable to UV damage in the future. For all these reasons it makes sense to describe how sunlight changes while it is travelling to the earth.

> Keeping sunburn at bay may not be enough for parents wanting to protect their young children from the sun. A study of oppossums has shown that doses of UV-B lower than those needed to cause sunburn in their young offspring, can lead to widespread melanoma in later life.
>
> *New Scientist*, 11 December 1999, 6

9.2.1 *How the solar spectrum changes when travelling to the earth*

Figure 1.8 gives an impression of the absorption of the incoming solar radiation by the atmosphere as both the incoming radiation at the top of the atmosphere and the radiation at sea level are shown there. Absorption by the molecules H_2O, CO_2, O_2 and O_3 are indicated in that figure. They cause 'dips' and structure in the final spectrum.

In order to know the solar spectrum when the composition of the atmosphere is changing, one needs models in which the experimental information contained in spectra such as Figure 9.4 for H_2O vapour is used as input. One also needs the solar spectrum as a function of altitude in order to get an idea of the atmospheric chemistry that may take place.

Using the physics of the absorption and scattering processes, it is possible to construct a model that describes how light arriving from the sun is transported through the atmosphere and eventually arrives at the earth's surface. Basically, the atmosphere is divided into thin layers and the amount of radiation leaving each layer is balanced against the amount flowing into it using the physical laws of absorption and scattering. Using such a model, the amount of solar light penetrating our atmosphere at every height can be calculated as a sum of direct and diffuse radiation.

The result of such a calculation is given in Figure 9.6. It shows the solar spectrum as it enters the atmosphere (top) and at sea level. The curve at sea level represents direct radiation only. Direct radiation, as the name suggests, is the attenuated solar radiation itself. The indirect radiation is the result of all kinds of scattering processes. They can be calculated, but are experimentally not distinguishable. Plate 3 can give more detailed information than Figure 9.6, as it is in colour. On that plate the two spectra, just mentioned are shown, but also spectra at 8.8 [km], the level of Mount Everest. There the higher 8.8 [km] curve is the summed direct and diffuse radiation and the lower 8.8 [km] spectrum represents the direct radiation only.

In Figure 9.6 the structure of the incident solar radiation is evident. It represents the gases which are present in the outer solar atmosphere and there absorb the outgoing spectrum. The Ca atoms in the solar atmosphere, for example, absorb at 393 [nm] and 397 [nm], giving dips in the spectrum. Likewise, one may note the Na line in the solar spectrum at 589 [nm].

One also notes that the ozone in the upper atmosphere absorbs all radiation with a wavelength below 295 [nm]. In Figure 9.6 this can be seen for sea level. But in Plate 3 the same appears to be true for a height of

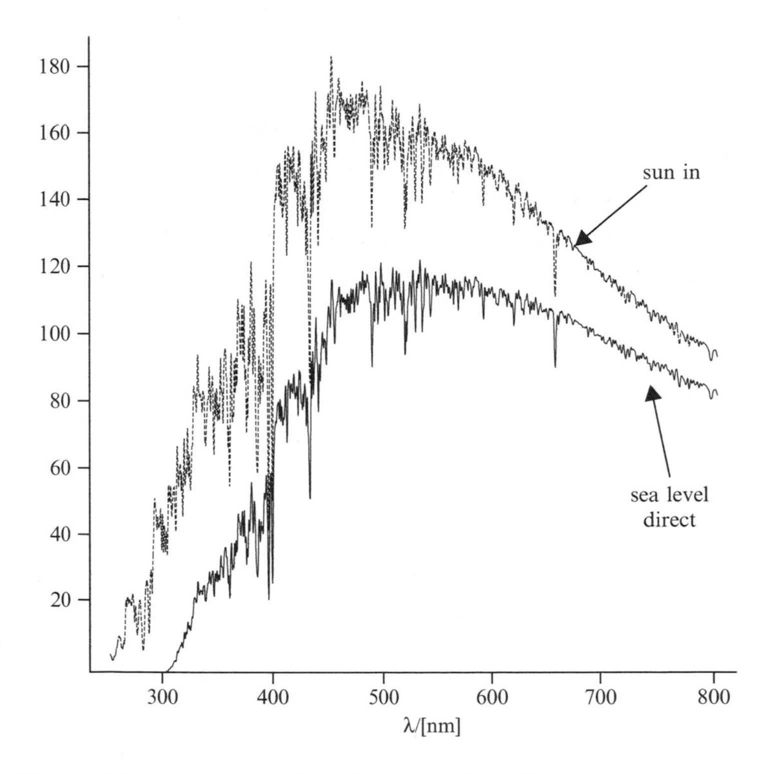

Figure 9.6 Voyage of the solar spectrum through the atmosphere. The top curve represents the incoming solar spectrum on top of the atmosphere. The bottom curve shows the direct radiation at sea level. These spectra as well as those at 8800 [m] are shown in Plate 3. With kind permission of Dr. Kylling taken from [99]

8800 [m]. The bulk of the absorbing ozone, therefore, must be still higher in the atmosphere.

9.3 *EXPERIMENTAL TECHNIQUES IN ENVIRONMENTAL SPECTROSCOPY*

Spectroscopic techniques are very common in environmental science. These techniques are well established in physics and in physical and analytical chemistry. Commercial standardized equipment is available, although often specific adaptations must be made. This is needed, for instance, to perform experiments from a satellite, or at extremely low concentration or when using a star as a light source. Finally, the development in *laser technology* has led to completely new techniques such as *lidar*, which will be discussed in Section 9.4.

The most important spectroscopic techniques are connected with absorption and emission, as sketched in Figure 9.3. *Absorption* occurs when a state with a low energy is excited to a state with higher energy. In *absorption spectroscopy* the decrease of light intensity is detected when the light is passing through a medium. Molecules or atoms in the medium can be identified by the presence of certain absorption lines, just as in a bar code. Since the position and shape of these lines depend on the precise environment, these spectra can also be used as a quantitative monitor of the molecular surroundings, such as the local temperature.

Emission is the reverse of absorption: an atom or molecule goes from a state with higher energy to a state with lower energy. In general, the incident light beam first excites the sample under consideration, the light then is absorbed and re-emitted in a direction and with a frequency which may be different from that of the original excitation beam.

9.3.1 *General overview of spectroscopic methods*

In all absorption and emission *spectrometers* the frequency or wavelength of the light is swept over the frequency range of interest. The transmitted or emitted light is detected and compared with a reference signal. We start our overview with absorption and note that each frequency, shown in Figure 9.2, will require its own instruments.

In Figure 9.7 the schematic outline is shown of a general *spectrophotometer*, absorbing in the visible (VIS) and the ultraviolet (UV). For this VIS-UV spectrophotometer the source is a so-called broad band light source which emits in a broad range of frequencies. A tungsten/iodine lamp will cover the

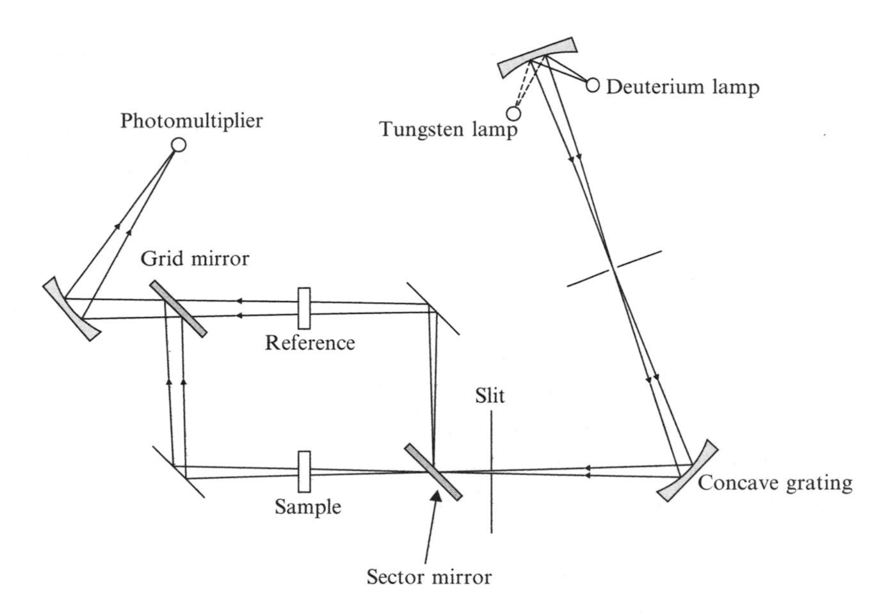

Figure 9.7 Layout of a UV-VIS absorption spectrophotometer. A concave grating disperses the incoming light beam and the light is focussed on the entrance slit. The desired colour (wavelength) is selected by rotating the grating. A motor driven sector mirror sends the beam alternating to the sample, or a well-known reference. Then the two beams are recombined and focussed on a photo-multiplier, which detects an AC signal (unless both beams happen to have the same intensity). The AC signal is transduced, amplified and converted to a ratio I/I_0

400–1000 [nm] wavelength range; a deuterium or xenon discharge lamp is generally used in the near UV (down to a wavelength of 180 [nm]). Deep UV photons with a wavelength shorter than 180 [nm] cannot be generated with lamps. Tunable lasers are a (relatively expensive) way of generating these short wavelengths.

Another wavelength region is the infrared (IR). Infrared radiation is usually generated by heating a quartz surface or a ceramic filament. With the correct composition of the material the emission from these light sources can be made to closely mimic that of a black body. NMR and ESR spectroscopies use microwave radiation that is typically generated by an electronic device called a *klystron*.

In all spectrophotometers an element is included to separate the frequencies present in the radiation source so that the variation of the absorption or emission with frequency can be detected. For the VIS-UV spectrophotometer shown in Figure 9.7 the light, after passing the sample, is dispersed by a grating or a prism. A *grating* is a ceramic or glass plate, usually with an aluminium coating, in which lines have been cut at a distance of about 500 to 1000 [nm]. Constructive interference between light waves reflected from different parts of the grating depends on the angle and, as a consequence, a spectrum is produced.

The detector in Figure 9.7 can be a single wavelength detector like a photodiode or, more appropriately, a diode array or a charge coupled device (CCD), which allows the recording of a total spectrum in one shot. In the IR one applies small band gap semiconductors or integrating spheres. To detect microwaves one uses a crystal diode consisting of a tungsten tip in contact with a semiconductor.

We now turn to emission. In *emission spectroscopy* the sample is typically excited with a monochromatic light beam and the emitted light is detected under some large angle with respect to the direction of the incident beam, through a filter or a monochromator for frequency selection. Both the excitation and the detection frequency can be varied. To increase the signal-to-noise ratio in emission experiments one often uses a technique called *single photon counting*, which allows a precise statistical analysis of the data.

There may be a time delay between the excitation and the subsequent emission, called the *fluorescence lifetime*. Experiments which measure this lifetime allow for further discrimination, even between identical emitters, since the fluorescence lifetime is often sensitive to the molecular environment. The fluorescence of many relevant molecules may be quite variable, dependent on the species and the precise conditions. A simple experimental set-up is shown in Figure 9.8 and Plate 4.

9.3.2 *Some examples*

In this subsection we discuss a few spectroscopic experiments that are used in environmental research. The experiments that will be discussed are all based on the absorption, emission or reflection of light.

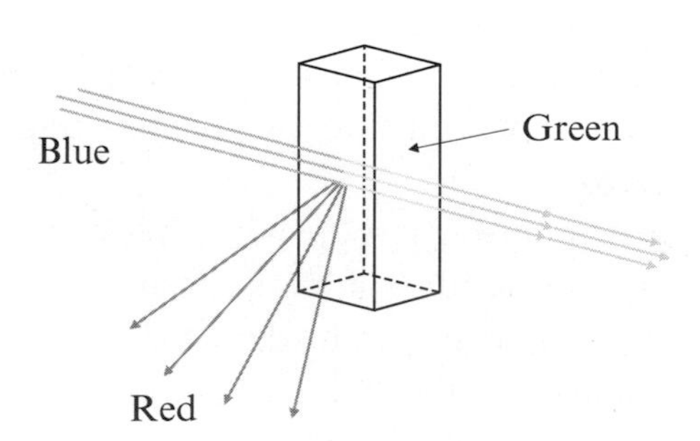

Figure 9.8 Fluorescence of a chlorophyll *a* solution. From the left a blue 488 [nm] beam, produced by an Ar^+-laser is entering a cuvette. This contains a chlorophyll *a* bound to protein (the green material). The absorbed photons excite chlorophyll molecules that rapidly decay to their lowest excited state from which they emit red fluorescence with a maximum of about 670 [nm]. Plate 4 gives a colour presentation

The UV-dose spectrophotometer

As illustrated later on (Figures 9.17 and 9.18) the negative health effects of UV radiation increase with smaller wavelength. In order to judge the health effects of UV radiation, which is present at a certain location, one therefore needs to measure the UV radiation at a few different wavelengths. A simple instrument [100] to do this consists of three channels, one to measure the amount of UV radiation at 305 [nm], a second channel to record UV at 320 [nm] while a reference signal is measured at 340 [nm]. The light is detected with photodiodes through filters with a half-width of about 10 [nm]. The biologically effective UV dose rate [mW m^{-2}] is calculated from a linear combination of the irradiance in the three channels. Note that a high dose is not only bad for humans, but for plants and animals as well.

Figure 9.9 shows an instrument installed on the Tibetan plateau to measure the local UV dose rate as a function of time. The Tibetan plateau is the highest plateau on earth, about 4000 [m] above sea level. Due to the strong solar radiation caused by the shallow atmosphere and the specific properties of the surface, the Tibetan plateau is an important factor in the heating process of the atmosphere, with possibly important effects on the local,

Figure 9.9 Three-channel filter instrument installed on the roof of the Institute of Tibetan Plateau Atmospheric and Environmental Science Research in Lhasa, Tibet. Reproduced with permission of the authors from [100]

regional and global atmospheric circulation, in particular during the monsoon season.

As mentioned before, the ozone in the higher atmosphere absorbs UV radiation and acts as an ozone filter. The effective thickness of that filter is measured continuously by satellite and expressed in so-called Dobson Units, to be defined in Sect. 9.4. In order to check whether the ozone filter indeed directly influences the UV dose, it is interesting to compare the direct measurements in Tibet with the thickness of the filter at the same time.

The amount of light detected at the three wavelengths mentioned above, 305 [nm], 320 [nm] and 340 [nm] is, of course, not only a function of the ozone filter, but also of the cloud cover and the solar zenith angle, which is the angle between the sun and zenith.* By measuring at different zenith angles and wavelengths one is able to correct for these effects.

Figure 9.10 shows a time series of the UV dose rate calculated from measurements as described above and the corresponding height of the total ozone column as measured by the Earth Probe TOMS satellite during the same period 24 June–1 December, 1996. Use a plastic triangle to check that the two signals correlate: a thick layer corresponds to a low UV rate Ref. [101] also gives the afternoon measurements. It is worthwhile consulting it on the web.

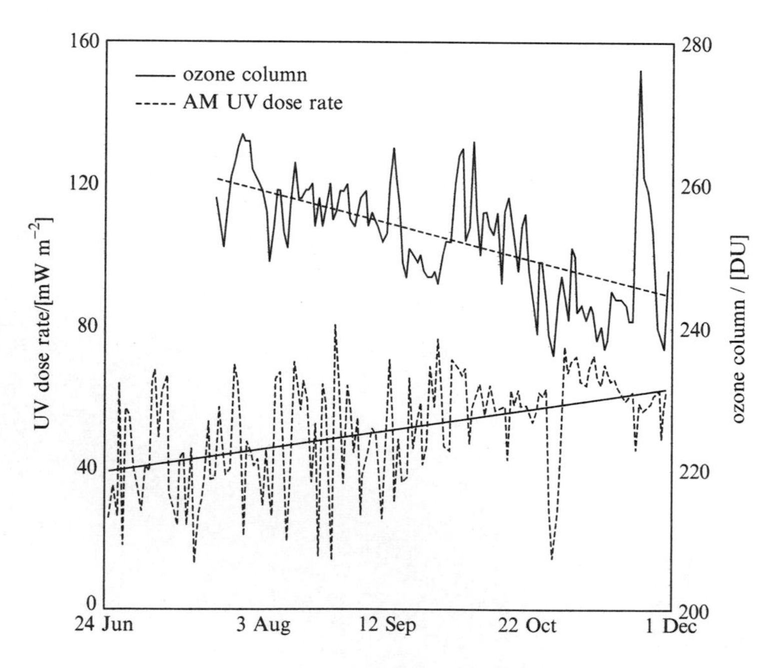

Figure 9.10 Time series of the biological effective UV dose rate from Lhasa, Tibet, in the period 24 June to 1 December, 1996. The graph AM (dashed, lower curve) refers to morning measurements at a solar zenith angle of 60°. The straight lines are simple fits to the measurements. The total ozone column as measured by Earth probe TOMS is also shown (higher curve). Taken with permission of the authors from [101], Figure 1

*This angle determines the length of the path through the atmosphere.

Fluorescence detection of carcinogens

Polycyclic Aromatic Hydrocarbons (PAHs) form a group of chemicals consisting of two or more fused benzoic rings, composed only of carbon and hydrogen. If oxygen, nitrogen and sulfur atoms are also part of the rings, one uses the term *Polycyclic Aromatic Compounds* (PACs).

In general PAHs and PACs have very similar physico-chemical and eco-toxicological (poisonous) properties, but their precise behaviour critically depends on their specific structure and chemical composition. PAHs and PACs are natural constituents of oil and occur in many petrochemical products. PAHs and PACs are further produced during the incomplete burning of fossil fuels or other organic matter.

PAHs and PACs have been recognized as potent *carcinogenic* compounds. One of the compounds that was identified is benzo(a)pyrene (BaP), a five-ring aromatic hydrocarbon. The general structure of BaP is shown in Figure 9.11. For instance, BaP is one of the many constituents of cigarette smoke. Also in automobile exhaust fumes PAC/PAHs were shown to be responsible for the carcinogenic activity, although BaP occurs only as a minor constituent. Nevertheless, the BaP concentration in a sample is often used as an indicator for its total carcinogenic potential.

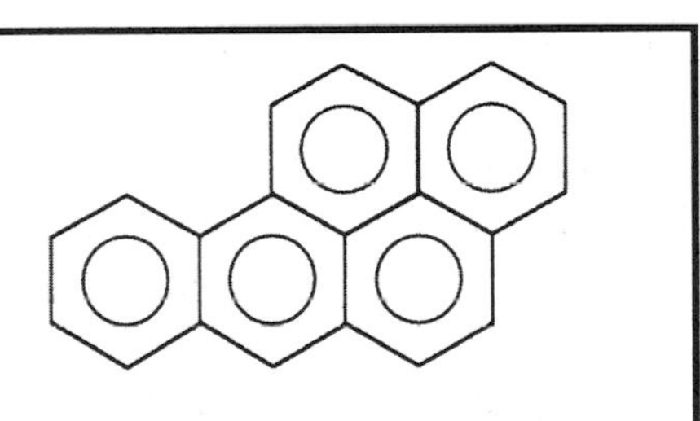

Figure 9.11 The structure of benzo[a]pyrene (BaP), one of the best studied Polycyclic Aromatic Hydrocarbons (PAHs)

Following their uptake in the blood stream of mammals, these compounds are changed by the digestive process (metabolized) and often reactive species are formed that may interact with DNA and form so-called DNA adducts. In specific cases these *adducts* may be extremely carcinogenic. To identify the pathway and mechanism by which these adducts are formed one must be able to identify and distinguish highly similar and isomeric PAHs with high sensitivity and high accuracy.

Optical spectroscopic methods can be used to identify which type of adduct has been formed and to quantify the total amount of adduct, usually relative to some standard. Fluorescence techniques are particularly suitable because the signal is measured against a dark background. However, in case one simply illuminates an ensemble of very similar molecules with a broad-band light source, one excites all the molecules in the ensemble non-selectively and consequently one obtains a broad emission spectrum that contains only very little information about the individual molecules and their environments.

Fluorescence line narrowing (FLN) One can overcome the problem of a broad spectrum by using a narrow band laser to excite the fluorescence.

This is called *fluorescence line narrowing* (FLN) and is illustrated in Figure 9.12. This figure demonstrates the effect of fluorescence line-narrowing for a chlorophyll-protein complex (called PS2 RC). Note that the horizontal variable is the wave number $\bar{\nu}$ [cm^{-1}]. Figure 9.12 compares an emission spectrum obtained with broad band excitation, with a spectrum excited in the red edge of the major chlorophyll *a* transition around 680 [nm]. All the sharp peaks resolved in the spectrum correspond to molecular vibrations of the chlorophyll molecule.

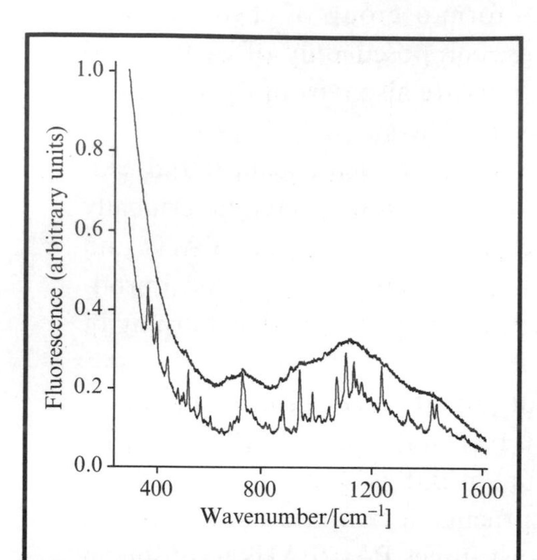

Figure 9.12 FLN on protein PS2 RC. The wide peak shows the emission spectrum following broad band absorption. The horizontal axis represents the energy difference between the exciting and emitted frequencies. The narrow peaks show up on laser excitation at around 680 [nm]. The peaks represent internal vibrations of the protein and the chlorophyll molecule [102]

FLN is also an excellent tool used to distinguish the various adducts that are formed between DNA and metabolized benzopyrene products. Figure 9.13 shows FLN spectra of mouse skin DNA obtained at different time points (4 hours to 7 days) after the addition of the carcinogen BaP to the mice [103]. The BaP was applied to shaved areas on their backs and small amounts of skin were removed at specific times for later study. The FLN spectra were obtained with a wavelength of 356.78 [nm] for excitation, which is selective for certain BaP-DNA adducts.

Looking at the spectra, one observes two types of contribution. The first are sharp lines, which represent molecular vibrations of the benzopyrene piece. The second is a broad feature around 381 [nm]. This one specifically originates from BaP-adducts that were inserted into the DNA helix. This particular type of inserted DNA-adduct reaches a maximum concentration after 1–2 days and is then slowly removed. After that time the adduct apparently is recognized as a hostile element. Most of the other adducts cause a much more significant perturbation of the DNA helix and are often recognized much faster by the cellular DNA-repair machinery. These adducts are then quickly removed from the body.

Optical remote sensing of water quality

Coastal water or the water of inland lakes may show a variety of colours. Clear water is blue; water rich in aquatic humus is yellow and water containing large amounts of photosynthetic algae or cyanobacteria can vary from blue-greenish to red. The colour of surface water is of course directly related to the substances dissolved or suspended in it and the micro-organisms living in it.

Monitoring the colour of the water from a satellite or an aeroplane is called *remote sensing** [104]. It yields important parameters that can be used to study environmental processes such as the production of biomass or the distribution of (polluted) suspended water. Special applications include the mapping and trapping of oil slicks, ice cover and floating layers of algae.

The colour of water is determined by the light scattered out of the water and the light reflected from the surface of the water. Light that originates from below the water surface shows the characteristic influences of diverse coloured components present in the water. One can distinguish two contributions to a reflectance spectrum measured by remote sensing:

(1) light absorption, which diminishes the light intensity and changes its spectral composition and
(2) light scattering, which changes its angular distribution in a frequency dependent manner.

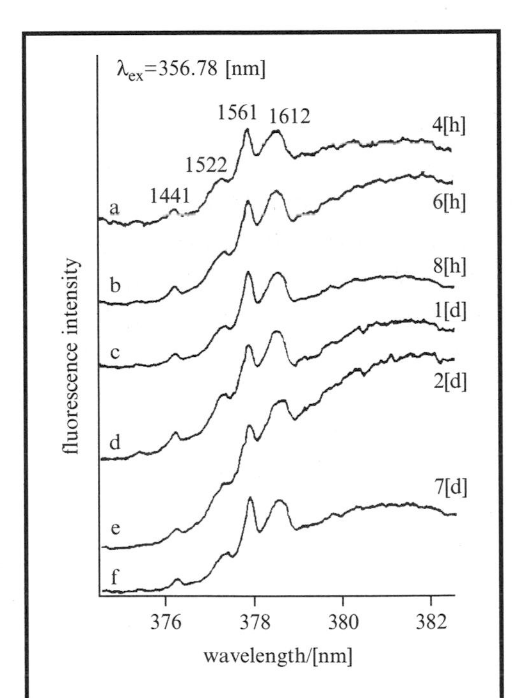

Figure 9.13 FLN spectra of mouse skin DNA obtained at different times after application of BaP. The broad peak at 381 [nm] builds up slowly until the adduct is recognized as 'enemy' and removed. Reproduced with permission of Oxford University Press from [103], Figure 7

The intensity of the reflected light, as detected by a satellite, decreases due to absorption in the atmosphere and increases with the amount of scattering taking place. In general, after the subtraction of atmospheric influences, the measured reflectance spectrum is corrected for the contribution of the water surface to obtain the subsurface reflectance. This subsurface reflectance is then related to an optical model for water. As one of the many examples† of the application of remote sensing we discuss an experiment performed by Dekker and co-workers [105]. Here, an aeroplane flew over the lake area south east of Amsterdam in the Netherlands. They could record both continuous spectra from 435 [nm] to 805 [nm] or image the area via selected spectral bands.

Figure 9.14 shows the irradiance measured at 1000 [m] above three of the lakes. Note that the top curve refers to the lake Vuntus, from which much light is reflected. From the position of the peaks in the spectrum one may deduce that it has a yellowish or green-brownish colour. The bottom curve refers to a 'water-supply' lake, which is dark

*One may distinguish between remote sensing where one uses the direct and indirect solar radiation as a light source or remote sensing where a laser or radar beam is used as a source. In this text we are discussing the first, using the sun.
†Consult http://earthobservatory.nasa.gov for the most recent satellite pictures of the earth.

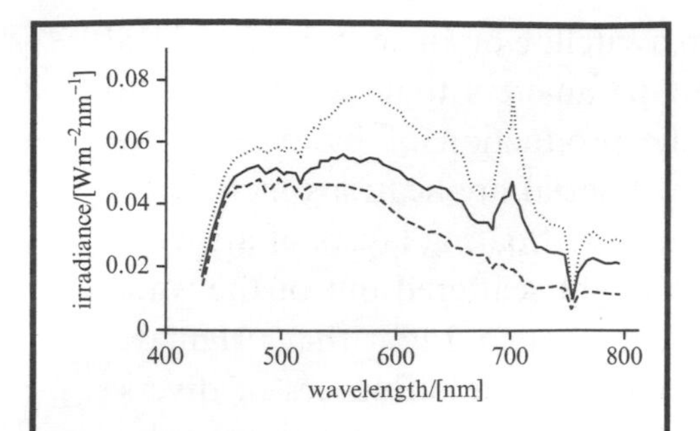

Figure 9.14 Irradiance from three lakes, measured from an aeroplane at 1000 [m] altitude. The top curve refers to 'dirty' lake Vuntius, the lower one to a clean water-supply lake and the middle one to a third lake (Oost Loenderveen). Reproduced with permission of the authors from [105]

and green to brown. The Vuntus is a popular sailing lake, very eutrophic‡ with blooms of cyanobacteria in late summer, whereas the water-supply lake is forbidden for water recreation and its water is meant as drinking water for the town of Amsterdam. In all three spectra the light level below 500 [nm] is relatively high, probably due to direct reflection from the water surface.

The chlorophyll concentration in water is a measure of water quality as a large concentration suggests the presence of organic materials, which may oxidise according to Equation (1.1) and then take away the oxygen required for fish. It also suggests the presence of other chemicals, which may be unhealthy (see Section 1.1). The small dip in the irradiance at 680 [nm] due to chlorophyll suggests a straightforward way to detect the chlorophyll concentration in surface waters. One could measure the irradiance in a window around 680 [nm] and divide it by the irradiance measured in a window around 710 [nm]. Small values of this ratio indicate high chlorophyll *a* concentrations whereas high values indicate clear water.

As an example, we show in Figure 9.15 and Plate 5 the deduced chlorophyll concentrations in a small river, called the Vecht [106]. The black-and-white Figure 9.15 is processed such that dark grey indicates a low chlorophyll concentration and light grey, a large concentration. This is much better seen in colour Plate 5, where dark colours indicate a low chlorophyll concentration, whereas yellow and red indicate a high chlorophyll concentration. The river in the picture streams to the North (upward in the figure). One notices halfway along the graph, a canal with a relatively high chlorophyll concentration crosses the river, upon which the water quality in the river deteriorates for a while.

9.4 *SOLAR UV, OZONE AND LIFE*

In this section the impact of solar UV on life on earth will be discussed. Most of the UV-radiation entering the earth's atmosphere is removed by the ozone layer in the stratosphere and for that reason the intensity of solar UV at the earth's surface decreases rapidly for wavelengths below 300 [nm], as was shown in Figures 1.8 and 9.6. This is precisely the region where harm can be done to important biomolecules like DNA and proteins.

‡'eutrophic' means rich in minerals and food for plants, especially algae.

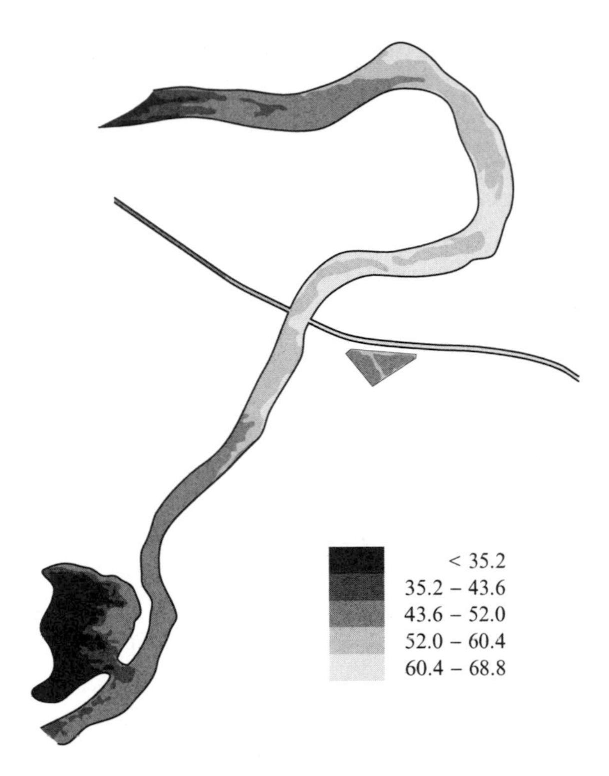

Figure 9.15 Chlorophyll *a* flows North (to the top of the figure) into the river from the canal crossing the graph with permission of Rijkswaterstaat from [106]. See also Plate 5

During the last decade, mankind has been confronted with a decreasing thickness in the ozone layer, dramatically manifested by the occurrence of the ozone hole above Antarctica. These days, weather reports in New Zealand and Australia not only provide expected temperatures, wind or rainfall, but also give UV-warning for skin exposure, reflecting the status of the ozone layer on that day. One of the major causes for the observed destruction of the ozone layer is now known to originate from chlorofluorocarbons (CFCs), such as freon, which was typically used as a coolant in refrigerators. CFCs consist of chlorine (Cl), fluorine (F) and carbon (C).

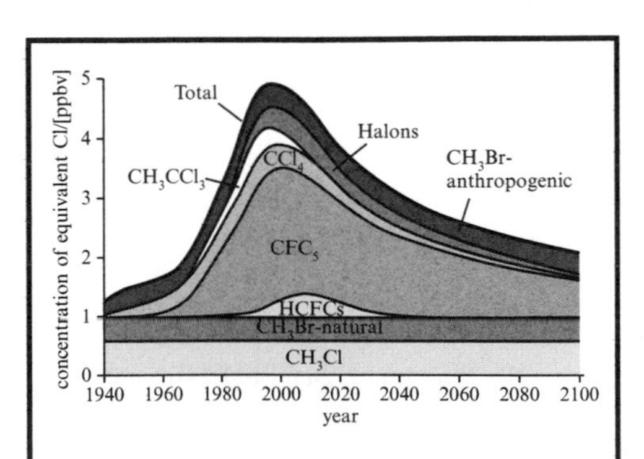

Figure 9.16 Known and projected concentrations of ozone depleting chemicals in the atmosphere. Reproduced with permission of the American Association for the Advancement of Science from [107], Figure 1

Since the seventies, about 10 million tons of CFCs have been released into the atmosphere. With lifetimes between 50 and 100 years, Figure 9.16 dramatically demonstrates how their presence will be felt deep into this century, even with the most rigorous methods adopted to stop their production. The world will be confronted with a significant increase in solar-UV irradiance and it will be of crucial importance to find out the impact of this change on plants, animals and humans.

However desirable ozone might be in the stratosphere, too much ozone close to the earth's surface is another serious threat to life. Ozone is a very oxidative species that very effectively damages biological material. Nitrogen oxides (NO_x) and hydrocarbons C_xH_y, formed during the burning of fossil fuels catalyse the formation of ozone by UV-light in industrialized areas. In this section, both effects will be discussed: too little ozone high up and too much ozone low down in the atmosphere.

9.4.1 *Solar UV and biological molecules*

Solar light as it enters the earth's atmosphere contains a significant amount of radiation in the wavelength region 200–350 [nm] (Figures 1.8 and 9.6). Light with a wavelength below 220 [nm] is largely absorbed by the molecular oxygen and nitrogen in the atmosphere. Light in the 220–300 [nm] region is absorbed by the ozone in the stratosphere, as was indicated in Figure 1.8. The removal of these parts of the solar spectrum is essential since the interaction of solar UV with biological cells may lead to severe damage of the genetic material (*mutagenesis*) or even to the killing of cells. In complex, multicellular organisms, exposure to solar UV may lead to damage of essential parts of the organism and, in the case of humans, to skin cancer.

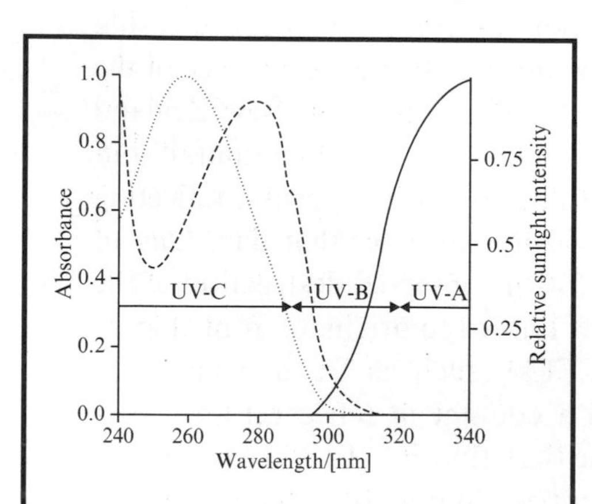

Figure 9.17 Absorption spectra of DNA (dotted curve) and α-crystallin (dashed curve). The solid curve indicates the shape of the solar spectrum at the earth's surface in the wavelength region up to 340 [nm]

Figure 9.17 shows the absorption spectra of two important biomolecules: *DNA*, the carrier of the genetic code, and *α-crystallin*, the major protein of the mammalian eye lens. In the figure the definitions of the three ranges of UV radiation are also indicated. These are UV-A (or near-UV), UV-B (or mid-UV) and UV-C (or far-UV). The absorption of the two biological molecules in the region $200 < \lambda < 400$ [nm], only overlaps with the solar emission spectrum in the wavelength region 290 [nm] $< \lambda <$ 320 [nm], the UV-B region.

Absorption does not always cause damage. In photosynthesis, for example, the absorbed energy is utilized. In the UV-B region, however, absorption is harmful. The prime targets of UV light in biological cells are the *DNA bases*. The DNA bases, guanine (G), thymine (T), cytosine (C) and adenine (A) absorb between 200 [nm] and 300 [nm] with, in particular, the G base extending its absorption to the longer wavelengths. Absorption of a UV photon by one of the bases may lead to its oxidation and thereby destabilize the DNA structure, thus giving rise to an incorrect reading of the genetic code.

For a typical biological cell, the absorption is 90 percent dominated by nucleic acids and 10 percent due to proteins. Proteins may also carry specific cofactors. For instance, the protein in our blood that transports oxygen from the lungs to the tissue, *haemoglobin*, contains a *haem* group which is active in the binding of oxygen. It is well known that haemoglobin is responsible for the red colour of blood.

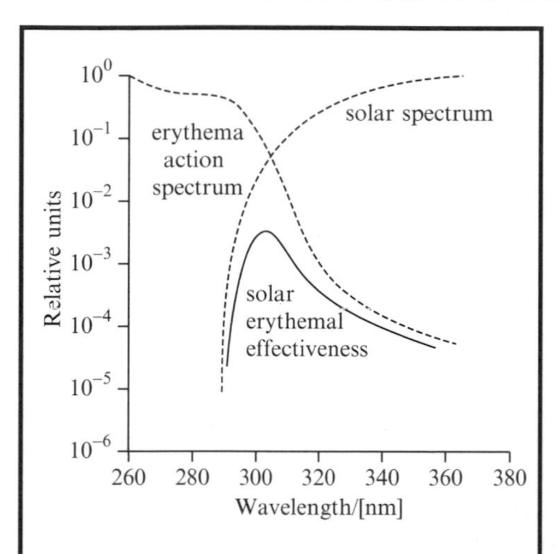

Figure 9.18 Erythema (sunburn) action spectrum $E(\lambda)$, plotted together with the solar spectrum $I(\lambda)$ at the earth's surface. Their product $E(\lambda)I(\lambda)$ is the damage done: the solar erythemal effectiveness. Reproduced by permission of Plenum Press from [108], Figure 8.4, p. 119

The interaction of solar UV with biological cells may lead to severe damage of cellular proteins of the genetic material and cause the death of cells. Figure 9.18 shows the *action spectrum* $E(\lambda)$ for sunburn as an example. This action spectrum reflects the damage done by a unit of irradiation of a certain wavelength λ by producing sunburn (*erythyma*) in human skin. Although the action spectrum clearly extends above 300[nm], implying a contribution of skin pigments and haemoglobin to the action spectrum, the major part originates from the direct absorption of solar UV by DNA. This is even more pronounced as the vertical in Figure 9.18 is drawn on a logarithmic scale. On the same scale the figure shows the UV-tail of the solar irradiance $I(\lambda)$. The damage done by radiation of wavelength λ is therefore given by the product $E(\lambda)I(\lambda)$, which is also shown in Figure 9.18 as the full curve. Note how the efficiency of producing sunburn increases dramatically in the region between 320 and 280 [nm], precisely where the solar-UV irradiance on earth (still) collapses.

9.4.2 *The ozone filter*

Ozone (O_3) forms a thin layer in the stratosphere, with a maximum concentration at between 20 [km] and 26 [km] above the earth's surface. The fact that ozone forms a very thin shield indeed, is probably best demonstrated by the fact that the amount of ozone in the atmosphere corresponds to a thin layer of about 3 [mm] at standard temperature and pressure. Figure 9.19 shows the absorption spectrum of ozone between 240 [nm] and 350 [nm]. Ozone has a strong absorption band in the UV with a maximum around 255 [nm] and effectively absorbs all the incoming solar radiation below a wavelength of

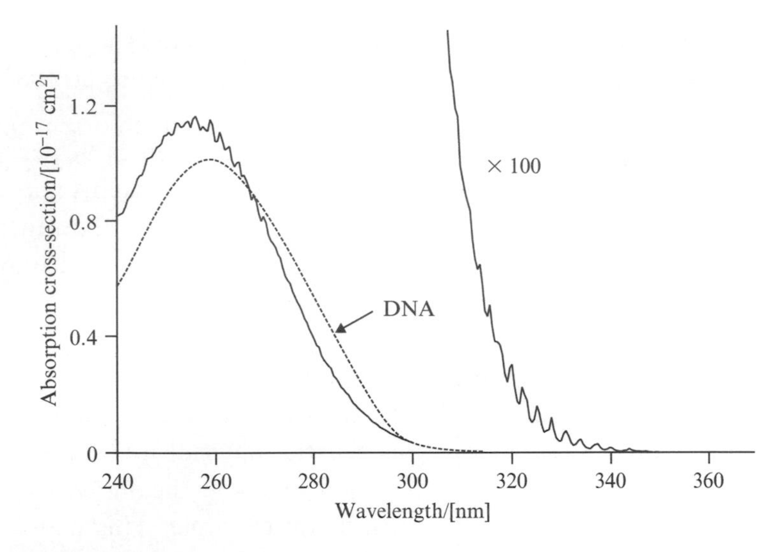

Figure 9.19 Absorption spectrum of ozone in the wavelength region 240–300 [nm] and on the right, on a 100-fold enlarged scale from 300 [nm] to 350 [nm]. The cross-section was kindly made available by Dr. A. Kylling, who used data going back to [109]. For comparison the DNA absorption spectrum from Figure 9.17 is added, normalized to the same scale

295 [nm]. In the maximum the absorption cross-section is about 10^{-17}[cm^2], half of the maximum absorption at the low-energy side is at about 275 [nm], while at 290 [nm] the absorption is down to 10 percent of its maximum. We have added the absorption spectrum of DNA (already shown in Figure 9.17) to that of the ozone filter and the remarkable overlap may not be coincidental.

Ozone is permanently formed in the atmosphere in the 100 [km] altitude region by the combination of atomic oxygen (O) and molecular oxygen (O_2). The oxygen atoms are the result of photo-dissociation of O_2 by UV photons of wavelength shorter than 175 [nm]. The efficiency of ozone formation depends on a multitude of factors, amongst which are the availability of O_2, changes in the stratospheric temperature, the presence of certain chemicals and of dust originating from volcanic eruptions.

Most of the ozone production takes place above the equator, where the amount of incident solar UV light is maximal. Ozone formed at these latitudes then diffuses towards the poles where it is 'accumulated'. The effective thickness of the ozone is conventionally expressed in *Dobson units* [DU]. A layer of 100 [DU] corresponds with 1 [mm] of thickness under standard temperature and pressure. The ozone layer may increase from effectively 3 [mm] at the equator to 4 [mm] above the poles at the end of the winter. The ozone concentration shows strong daily and seasonal fluctuations and tends to be highest in late winter and early spring. Remember that Figure 9.10 gave the thickness of the ozone layer above Tibet as about 250 [DU].

The breakdown of ozone also takes place in the stratosphere and a simplified scheme is shown in Figure 9.20. Nitrogen dioxide (N_2O), essentially a product of the biological nitrogen cycle, or chlorine compounds such as chloric oxide (ClO) or chlorofluorcarbons (CFCs), are released in the troposphere, where they are very stable. Slowly a small fraction 'leaks' to the stratosphere, where the CFCs may be converted to nitric oxide (NO) or free chlorine atoms under the influence of UV light. The thus-formed free radicals then react with O_3 to form O_2, and in this process the original compounds are reformed and the cycle can be repeated. One chlorine atom may catalyse the destruction of as much as 500 000 ozone molecules. 'Natural' chlorine originating from sea salt exists in the atmosphere in concentrations of about 0.6 parts per billion volume air [ppbv]. The man-made CFCs have increased the amount of free and bound chlorine by a factor of four to about 2.5 [ppbv].

In 1979 a small ozone hole was detected above Antarctica during early spring, that disappeared again during the summer. The size of the ozone hole has since then grown dramatically* During the polar winter special conditions occur in the upper atmosphere above Antarctica, where CFCs and NO_x are absorbed by stratospheric ice clouds, and released as active chlorine compounds and $HNO_3.2H_2O$ crystals. In early spring, solar UV converts these compounds into free chlorine atoms that completely deplete the ozone layer above Antarctica.

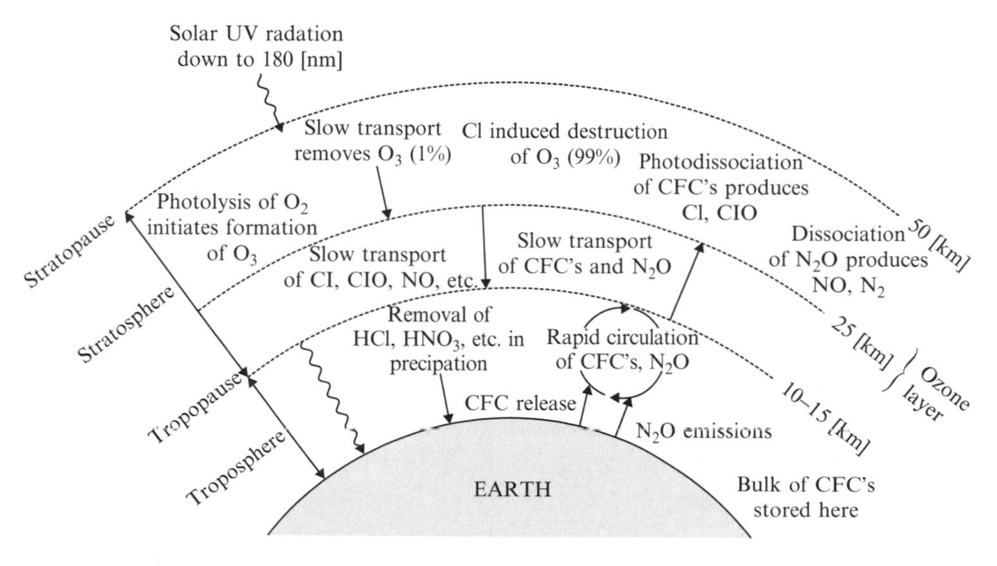

Figure 9.20 Processes that determine the concentration of ozone in the stratosphere (reproduced with permission of National Academy of Sciences from [110]; reproduced in [111], Figure 9.3, p. 146)

*see http://toms.gsfc.nasa.gov/ or http://earthobservatory.nasa.gov

To a less dramatic extent ozone is also depleted at lower latitudes. Figure 9.21 shows the result of a series of ozone measurements made since 1926 at Arosa, Switzerland at an altitude of 1850 [m] using a version of the spectrophotometer described in Figure 9.7. Since 1970 the total ozone has reduced by about 0.8 [DU] per year, which amounts to a total decrease of about 10 percent today. In late autumn 1999, also above North-west Europe a large area of very low ozone concentration was detected, less than 165 [DU], where the safe value is 200 [DU].

Figure 9.22 shows the total ozone decrease and the total skin cancer increase between 1972 and 1988 in North America. As discussed above, a decrease in ozone leads to an increase in UV-B due to two effects: more UV at a certain wavelength and a shift of the transmission to shorter wavelengths,

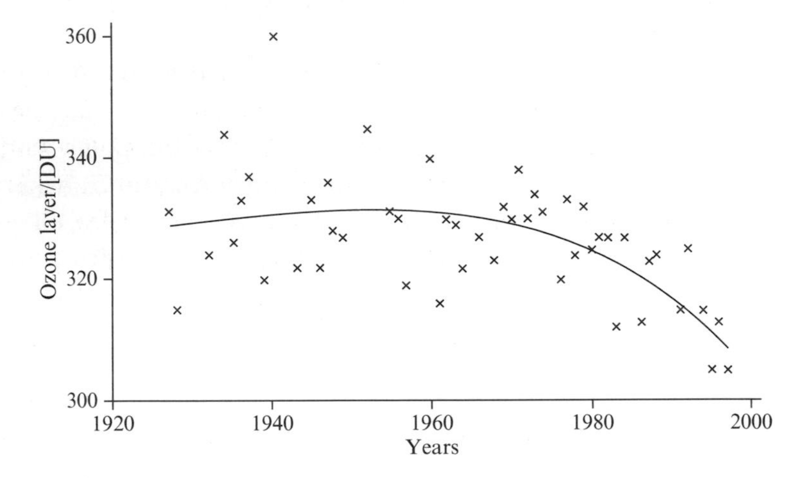

Figure 9.21 Annual average total ozone at Arosa, Switzerland between 1926 and 1997. Reproduced from [112] by courtesy of Dr. J. Staehelin, Institute of Atmospheric Sciences, Zürich, Switzerland

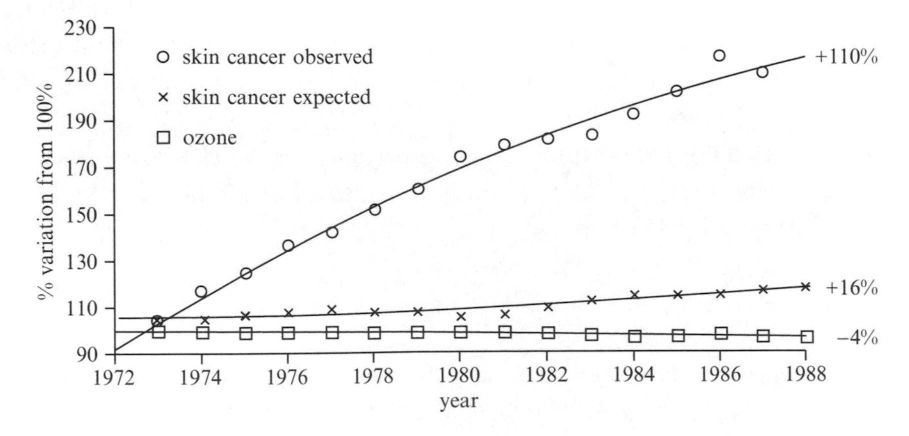

Figure 9.22 Total ozone decrease and skin cancer increases for North America. Reproduced from [112], Figure 8, which was based on [113] with permission of the Royal Meteorological Society

by which the overlap of the solar irradiation spectrum with the absorption spectrum of important biomolecules increases. As shown in Figure 9.22, dermatologists have noticed an increase during the last decade of melanoma and non-melanoma skin cancers, with a relatively higher frequency occurrence for people living in the temperate climates. But it is difficult to separate the contribution of an increased UV-B irradiance from those of increased sunbathing, increased tourism to sunny places and the application of sun-protection creams. From a careful analysis of the available data it was estimated that per % of ozone depletion, UV-B dose increases by about 2 percent resulting in a subsequent skin cancer increase of 4 percent. This explains only 16 percent of the increase in skin cancers, shown in Figure 9.22. The rest probably is due to the other effects mentioned.

9.4.3 *Lidar as a technique to measure stratospheric ozone*

Long-term studies of atmospheric changes have become more and more important due to increasing pollution. Key issues are the ozone problem, the detection of smog and the detection of dust particles or aerosols. To detect changes in the stratospheric ozone, a detection network was established to establish a reliable set of data. *Lidar* (Light-Induced Detection And Ranging) is a laser technique that is one of the methods employed in such a network.

Lidar methods have been used since the 1960's. Pulsed lasers are employed and a frequently used method is DIAL (Differential Absorption Lidar), which is based on the application of two laser beams of different colours (wavelengths). Lidar uses pulsed lasers, which are aimed vertically upwards. The light pulses, when travelling upwards through the atmosphere are gradually absorbed or scattered. The amount of absorption or scattering depends critically on the type and number of particles and molecules that are encountered by the beam, and their absorption and scattering cross-sections at the wavelength of the laser.

In a lidar experiment, the number of photons which are scattered back down to the earth is detected as a function of the difference in time between the moment of the firing of the laser and the arrival of the back-scattered photon at the detector. To make the method selective for specific particles present in the atmosphere, two laser wavelengths are applied in DIAL: the first wavelength is absorbed and scattered by the molecule one wishes to study, while the second wavelength is only scattered. In Figure 9.23 this situation is schematically represented by the 'on' and 'off' wavelengths, respectively.

In the following we give a qualitative description of a DIAL experiment. Suppose that $S_\lambda(t, \Delta t)$ is the signal that measures the amount of light of

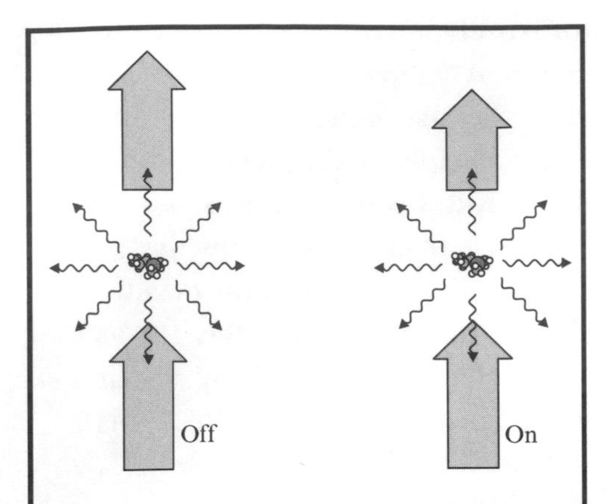

Figure 9.23 Principles of the DIAL process. The left shows the 'off wavelength': scattering only. The right shows the 'on' wavelength: the target molecules absorb and scatter radiation. Note that scattering happens in all directions, including the backwards direction

wavelength λ back-scattered on the detector between a time t and a time $t + \Delta t$ after the laser was fired. This signal corresponds to the scattering by molecules in the volume extending from R to $R + \Delta R$ in the direction of the beam. The relation between distance R and time t is given by the fact that light with a velocity c has to travel to the scattering point and back again, resulting in a total path length of $2R$. This results in

$$R = c\frac{t}{2} \tag{9.3}$$

and

$$\Delta R = c\frac{\Delta t}{2} \tag{9.4}$$

It is assumed here that the timespan τ of the laser pulse is short compared to the sampling interval, $\tau \ll \Delta t$. In practice, excimer lasers are often used with a typical pulse length of 10 [ns], which corresponds to a 3 [m] extension in space. In real experiments signals originating from a depth of at least 150 [m] (corresponding to 1 [μs] time difference) are averaged in one channel. So, the condition $\tau \ll \Delta t$ is fulfilled.

Note that all molecules between the earth and the distance R will contribute to scattering and absorption. Therefore, there is a complicated relation between the signal $S_\lambda(R, \Delta R)$ and the concentration $N_{\text{abs}}(h)$ where h indicates the height h where the absorbing molecules are present. By measuring $S_\lambda(R, \Delta R)$ as a function of R, the concentration $N_{\text{abs}}(h)$ can be found.

A typical result of a stratospheric ozone measurement using lidar is shown in Figure 9.24. This experiment was carried out by the Jet Propulsion Laboratory (JPL) in the USA. The 'on' wavelength was 308 [nm], the 'off' wavelength 353 [nm]. At 308 [nm] ozone absorbs weakly, while at 352 [nm] it does not absorb at all (see Figure 9.19).

Typically, 1500 laser pulses are required for the measurement of one data point in Figure 9.24. To illustrate the difficulty in obtaining good statistics one should realize that, from a pulse with 10^{19} photons that is scattered from a layer of 600 [m] thick at an altitude of 40 [km], one receives maybe two (!) back-scattered photons. This thickness layer corresponds with a time difference of 4 [μs] between light coming from the top and the bottom of the layer. In Figure 9.24 one may note the maximum in the ozone concentration at about 25 [km] of altitude. Figure 9.24 was measured with different

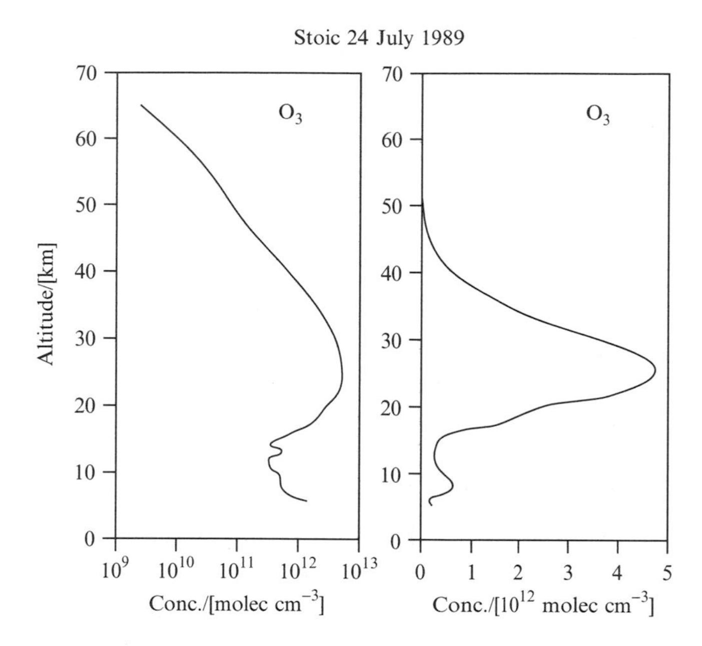

Figure 9.24 Ozone concentration as a function of altitude. The left-hand graph has a logarithmic concentration scale, while the right-hand graph has a linear scale. Reproduced by permission of Dr Stuart McDermid from [114]

measurement devices. The measurements of the ozone concentration versus height for the many devices agree very well.

9.4.4 *Ozone in the troposphere*

Following the discharge in a thunderstorm one can smell the sweetish odour of ozone formed by the electric discharge. During World War II Los Angeles citizens reported eye irritation from 'smog', a thin white fog that spread out from the butadiene factories in the city. Many farmers reported damage to their crops. In contrast to SO_2 air pollution, which is a chemical reductant, this 'smog' was an oxidant and ozone was proposed to be its main constituent. It was found that nitrogen dioxide plus almost any volatile organic compound produced ozone in the presence of sunlight. Furthermore, automobile exhaust when exposed to sunlight produced ozone. When brought into contact with plants, such UV-illuminated automobile exhaust reproduced the damage induced by smog.

Smog formation is a free radical chain reaction in which nitrogen oxides and water-based free radicals act as catalysts. The chain reaction is initiated by solar UV light by which radicals are formed and terminated by NO_2 that consumes the free radicals. The UV-induced free HO_x radicals ($x = 0$, 1 or 2) can either lead to the production or to the destruction of ozone, dependent on the NO to ozone ratio. In the presence of high amounts of NO, ozone is formed, while with low amounts of NO, ozone is consumed. Smog reactions need a fuelling compound, for instance CO, which is in that case converted into CO_2.

Tropospheric ozone plays a detrimental role in the human, animal and plant environment due to its strong oxidation potential. The maximal concentrations are frequently found at more than 1 [km] above sea level and thus cannot be monitored by ground-based ozone sensors. Tropospheric ozone is likely to contribute to, for example, forest damage at these altitudes in the northern Alps.

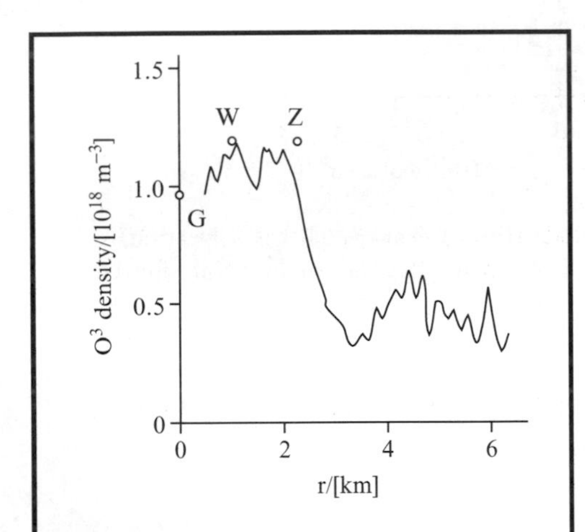

Figure 9.25 Ozone density by lidar as a function of height in Garmisch-Partenkirchen, 11 April 1991. Local ground-based stations are denoted by G (Garmisch), W (Wank) and Z (Zugspitze). Here, r is the altitude above Garmisch at 0.7 [km]. Reproduced by permission of Lambda Physik from [115], Figure 17

Experiments employing a lidar technique may be able to reveal the complete temporal and spatial evolution of tropospheric ozone. However, tropospheric ozone concentrations are generally much lower than those in the stratosphere. Therefore, for significant absorption, shorter wavelengths than those typically used in a stratospheric lidar experiment ($\lambda = 308$ [nm]) must be employed. Furthermore, the design of the detecting telescope and the limiting of the detection spectral bandwidth are important factors to realize sufficient detection sensitivity and selectivity.

Figure 9.25 shows an ozone profile up to 6.2 [km]. The strong variation of the ozone concentration as a function of altitude is clearly visible. The enhanced ozone concentration in the layer below about 3 [km] above sea level ($r = 2.3$ [km] in the figure) can be ascribed to the UV-induced ozone production in the presence of anthropogenic precursors (NO_x).

EXERCISE

9.1 Check that a wavelength of 680 [nm] corresponds with an energy of 1.8 [eV]. What would be the energy in [J]? Notice how small this is.

10 Geophysical Methods

In this chapter we discuss how the physical properties of the earth can be used to monitor the quality of the environment. Therefore they are known as geophysical methods [116]. Of course, there is no real distinction between spectroscopy and geophysics. It is only a matter of emphasis.

In Sections 10.1 and 10.2 we discuss properties of isotopes and their use in the environmental sciences. In Section 10.1 the emphasis is on the use of radioactive elements as a clock and in Section 10.2 on the environmental information contained in the ratio of isotopes in a certain sample. In the remainder of this chapter we discuss a few simple and quick methods to find out whether soil is polluted or where an old underground waste dump is located. In Section 10.3 we focus on magnetic methods and in Section 10.4 on seismic waves.

10.1 *RADIOACTIVE CLOCKS*

In Section 5.2 on nuclear power, we have already referred to the fact that an atom consists of a cloud of Z (negatively charged) electrons around a compact nucleus with Z (positively charged) protons and N (neutral) neutrons where $A = N + Z$ is called the atomic number. The charge Z determines the chemical properties of the atom and it defines its name (H for $Z = 1$, He for $Z = 2$, X for arbitrary unspecified Z). Nuclides with the same chemical properties are still distinguished by their number of neutrons N and are indicated by ${}^{A}_{Z}\mathrm{X}$. When the symbol X is replaced by the element symbol, H, He etc., the subscript Z is superfluous and is often omitted. For the same Z, but different A, the number of neutrons $N = A - Z$ is different. These nuclides are called *isotopes*: they have the same chemical properties, but different mass.

10.1.1 *Radioactivity*

In Equation (5.18) the rule for radioactive decay has already been given as

$$N(t) = N_0\,\mathrm{e}^{-t/\tau} \tag{10.1}$$

There $\lambda = 1/\tau$ was called the decay constant and the half-life $t_{1/2}$ of a radioactive sample was, according to Equation (5.22), given by $t_{1/2} = \tau \ln 2$. In Table 5.2 we summarized the properties of β^- decay and β^+ decay, together called β decay. In terms of nuclei ${}^A_Z\mathrm{X}$ these decays can be written as

$$ {}^A_Z X \rightarrow {}^A_{Z+1} Y + \mathrm{e}^- \quad \beta^- \text{decay} \tag{10.2} $$

and similarly

$$ {}^A_Z X \rightarrow {}^A_{Z-1} Y + \mathrm{e}^+ \quad \beta^+ \text{decay} \tag{10.3} $$

Instead of emitting a positive electron the nucleus ${}^A_Z\mathrm{X}$ often also can capture a negative electron. This is called *electron-capture* and written as

$$ {}^A_Z X + \mathrm{e}^- \rightarrow {}^A_{Z-1} Y \tag{10.4} $$

Madame Sklodowska-Curie found two active emitters of radiation, which she called polonium and radium. From the emissions of the latter the word radioactivity was derived. Such emissions ionize the surrounding matter, a fact which led to Rutherford's detection of the three different components involved: α-,β- and γ-rays.

From [117], p. 26

Note that in the reactions (10.2)–(10.4), the atomic number A does not change.

When reactions like those described here occur spontaneously one has the phenomenon called *radioactivity*. The reaction equations (10.2)–(10.4) will become more transparent, if one uses a so-called *chart of the nuclides*. That is a diagram in which the number of neutrons N runs horizontally and the number Z of protons runs vertically. As real nuclei have discrete N and Z, they can be indicated as little squares in the N, Z diagram. Figure 10.1 shows the lower part of the chart of nuclides ($A < 30$). It displays the nuclei which occur in nature. Four of them are radioactive, indicated by a grey background.

The lighter three of them follow β^- decay, which is indicated by an arrow pointing to the upper left. The fourth, ^{26}Al is decaying to ^{26}Mg, sometimes by electron capture and sometimes by β^+ emission, which is represented by an arrow pointing down to the right. Alpha decay (also indicated in Table 5.2) only occurs for heavier nuclides. It is easy to deduce how the arrows would run in this case: two steps down and two steps to the left.

In nature one encounters a wide variety of half-lives. The four radioactive nuclei shown in Figure 10.1 have half-lives of a few years up to a million years (Exercise 10.2). This means that they must be produced continuously, since otherwise they would have disappeared long ago. For one of them (^{14}C) we

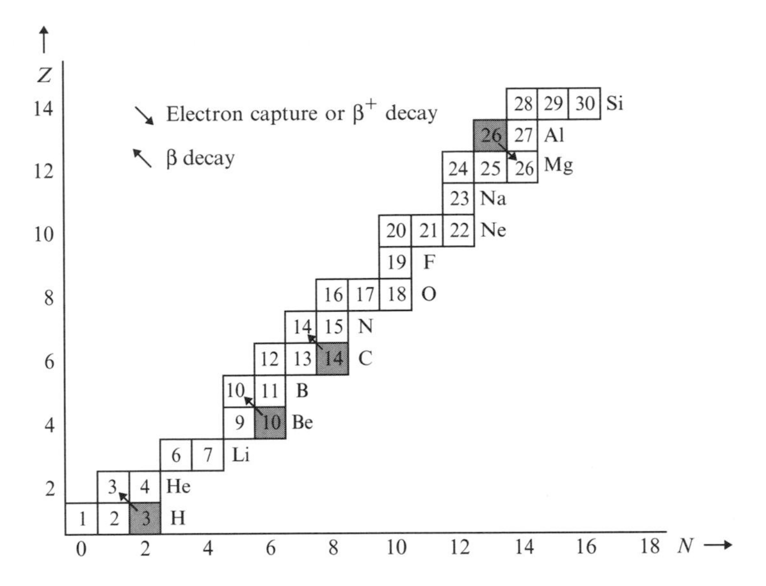

Figure 10.1 Chart of the lighter nuclides occurring in nature. A grey background indicates radioactive nuclides. The arrow for β^- decay points left upwards and for electron capture or β^+ decay points right downwards. For the four radioactive nuclides these decays are indicated

will discuss below the mechanism of continuous production. Some other radioactive nuclides have very long half-lives and may originate from the formation of the solar system, billions of years ago.

The region of stability

In Figure 10.1 we showed the nuclides with small atomic number A, that occur in nature. The nuclides not shown may be produced artificially by nuclear reactions, but they decay very quickly to the nuclides shown in the chart. The white squares in Figure 10.1 do not decay any further and they define the *region of stability*.

The stability region ends for high values of the atomic number A. In fact, ^{209}Bi is the heaviest stable, i.e. non-decaying nucleus. For the heavier nuclides, three nuclides occur in nature with very long half-lives. They are ^{232}Th, ^{235}U and ^{238}U. Their half-lives are close to or larger than the age of the solar system, which is estimated as 4.5×10^9 years.

> A mineral called xenotime grows shortly after sedimentation. It is very amenable to dating with the uranium-lead radioactive decay system. For it preferentially incorporates uranium in high concentrations during crystallization under exclusion of lead. Therefore all of the measured lead is formed from in situ decay of uranium, greatly facilitating the dating of the xenotime formation.
>
> *Science*, **285**, 2 July 1999, 58–59

It must be mentioned that nuclides heavier than ^{238}U can be produced artificially. In fact, as illustrated in Figure 5.16, in a nuclear reactor there are so many neutrons around, that ^{238}U nuclei absorb them to form ^{239}U which by two β-decays is transformed to ^{239}Pu, which has a half-life of 24 000 years.

The radioactive decay with its decay constant $\lambda = 1/\tau$ originates in the compact nucleus in the centre of the atom. Environmental conditions or chemical reactions therefore do not influence it. In that way the radioactive decay (10.1) acts as a very accurate clock, once one can determine $N(t)$ and N_0.

^{14}C as environmental scout

An example of a 'clock' with many environmental applications is the nuclide ^{14}C ([116], p. 255). Its half-life is measured as 5730 years and it decays by β-emission to ^{14}N (see Figure 10.1). Despite its relatively short half-life, it is displayed in Figure 10.1 as 'natural' occurring nuclide. This can only be true when it is formed with the same rate as it is decaying. This, in fact, is happening by a mechanism which we will describe briefly ([117], Ch. 6).

> Measuring B isotopes in the skeletons of plankton helped in determining the past CO_2 concentration in the atmosphere. The plankton usually take up carbonate to build their skeletons. But sometimes they incorporate B instead of C. The isotope composition of B depends on the pH of seawater. This among other things depends on the amount of CO_2 dissolved in the seawater as carbonic acid.
>
> *Science* **284** (11 June 1999) 1743–1746

The earth is continuously exposed to cosmic rays, most of which originate from the sun. The very high-energy cosmic-ray particles cause many reactions, which liberate neutrons. With the amply available ^{14}N in the atmosphere they cause the reaction

$$^{14}N(n, p)^{14}C \tag{10.5}$$

The notation (10.5) is widely used in nuclear physics. It means that a neutron is absorbed and a proton emitted. The net effect is that a positive charge is emitted and, just as in Equation (10.3) the nucleus changes from $Z + 1$ to Z while keeping its atomic number A. So Equation (10.5) describes the formation of ^{14}C and Equation (10.2) its decay. When the input of cosmic rays is constant and the nitrogen content of the atmosphere is also constant, an equilibrium concentration of ^{14}C will form. With the available oxygen it will react to $^{14}CO_2$ which will mix rapidly with the $^{12}CO_2$ in the atmosphere.

Plants will occasionally pick up a $^{14}CO_2$ molecule for their photosynthesis and incorporate the radioactive ^{14}C in their tissues. The ^{14}C nuclide emits β^-

radiation with a characteristic energy, which is easily measured. For a living plant the activity is measured as (13.6 ± 0.1) disintegrations per minute per gram of carbon (indicated as [dpm/g]). This can be converted to the decay constant $\lambda = 1/\tau$ of Equation (10.1) (Exercise 10.3)

> Theorists have calculated that when massive stars explode they should spew huge amounts of ^{60}Fe into interstellar space. Significant quantities have been found in metallic sediments in the South Pacific. As ^{60}Fe has a half-life of 1.5 million years, they must have been produced quite recently in a relatively nearby supernova.
>
> *New Scientist*, 10 July 1999, 5

After the death of the plant, photosynthesis stops and the plant no longer exchanges carbon with the atmosphere. Therefore the radioactivity will decrease according to Equation (10.1). Measuring the present activity of remains of an old plant will give $N(t)$. As the fraction of ^{14}C is only small, the total carbon content will not have changed much since $t = 0$. Therefore, at $t = 0$ the activity may be taken as that of a now-living plant, which has the value of (13.6 ± 0.1) [dpm/g] given above. We now easily deduce the time t since the plant died. When that is a very long time ago, as in fossil fuels, the ^{14}C will have disappeared and the clock has 'stopped'.

Example 10.1

A fossilized plant is measured to have a radioactivity of 5 [dpm/g]. What will be its age?

Answer

The basic decay reaction (10.1), which reads as $N(t) = N_0 \, \mathrm{e}^{-t/\tau}$ may be differentiated to time t. This gives

$$\frac{\mathrm{d}N}{\mathrm{d}t} = -\frac{1}{\tau} N_0 \, \mathrm{e}^{-t/\tau} \qquad (10.6)$$

The left-hand side equals the number of disintegrations per second. Relation (10.6) holds both for time t and for time $t = 0$. One may take the ratio of the two and find

$$\frac{(\mathrm{d}N/\mathrm{d}t)_{\mathrm{time}\,t}}{(\mathrm{d}N/\mathrm{d}t)_{t=0}} = \frac{(-1/\tau)N_0\mathrm{e}^{-t/\tau}}{(-1/\tau)N_0} = \mathrm{e}^{-t/\tau} \qquad (10.7)$$

The left-hand side of Equation (10.7) gives 5/(13.6). Its natural logarithm gives $\ln(5/(13.6)) = -t/\tau$. Equation (5.22) gives $\tau = 5730$ [yr] / ln 2. Finally one obtains $t = 8272$ [yr] as the age of the fossilized plant

^{14}C dating is used extensively to find the age of artefacts dug up in ancient sites. Of course it is only as reliable as the underlying assumptions. They are correct to first order only. The intensity of the cosmic radiation, for example, depends on the activity of the sun, which changes in time. Also, since the beginning of the scientific revolution, burned coal dilutes the available CO_2 with almost pure $^{12}CO_2$. Finally, the atmospheric tests of nuclear weapons in the 50's and 60's resulted in a strong extra flux of neutrons and by Equation (10.5) gave rise to a 'peak' in the ^{14}C content of the atmosphere.

This by itself has an interesting application, noted in Section 2.4. We have seen that knowledge of the ocean circulation of CO_2 is essential for a good estimate of the carbon cycle. When one measures the radioactivity of water in the oceans one should be able to identify the ^{14}C peak at a certain position and in that way determine the time that has passed since that water was in direct contact with the atmosphere.

10.2 *ISOTOPE RATIOS AS ENVIRONMENTAL SENSORS*

> Lord Kelvin in 1897 estimated the age of the earth as between 20 and 40 million years. He used the luminosity of the sun and the cooling history of the earth. In 1904 Rutherford presented a lecture with Kelvin in the audience where he showed that inclusion of radioactivity as an internal heat source slowed down the cooling of the earth and he estimated the age as more than 500 million years.
>
> From [118], pp. 2,6

Isotopes have the same number, Z, of protons and electrons but a different number, N, of neutrons. So their electronic configurations are essentially the same, but their mass is somewhat different. Roughly speaking one could say that isotopes exhibit the same chemistry as their chemical properties are determined by their electronic configurations. In the same manner those physical properties in which the mass of the atom plays a role, are somewhat different.

The ratios in which isotopes are found in nature are fairly constant. In fact, in tables of elements one will find their relative abundance given to fractions of a percent. This accuracy reflects their constancy through nature. This must be due to the fact that they were all formed in the interior of stars, dispersed by supernovae explosions and well mixed before the cloud of matter which formed the earth started to cool down. The mass difference of the isotopes may influence physical and chemical processes a little bit. In certain molecules or materials they then give rise to tiny deviations from the average, called *fractionation*, for which we discuss the mechanism and give a few examples below.

10.2.1 *Fractionation*

One usually distinguishes two effects: kinetic and thermodynamic ([118], p. 323). The first is easiest to understand as follows. The nuclides are practically always embedded as atoms in a molecule, which may exhibit vibrations and rotations. In a classical approximation the force between the atoms is proportional to the deviation $x = x(t)$ from their equilibrium distance:

$$-kx = \mu \frac{d^2x}{dt^2} \tag{10.8}$$

where k is the force strength and μ an effective mass. At the right-hand side of Equation (10.8) one finds the familiar expression mass times acceleration. The force in Equation (10.8) has already been met in Equation (8.39) and it will result in the same frequency (8.40). As shown in Exercise 10.4, the frequency f becomes $f = (2\pi)^{-1}\sqrt{(k/\mu)}$. A molecule consisting of atoms with a lower mass will have a higher frequency than one, which contains higher masses. It has more opportunity to react with the surroundings and is more reactive. In quantum mechanics the argument can be made more precise as follows. The vibrational energy of a molecule at low temperatures equals $hf/2$. With a higher frequency f (and therefore a lower mass μ) it is easier to cross reaction barriers and the molecule becomes more reactive. At higher temperatures this effect becomes less pronounced as the heat energy kT also helps in crossing barriers.

The thermodynamic effect of fractionation has to do with the fact that systems strive for a state with the lowest possible free energy.* Because of the mass difference between isotopes, this results in a slightly different distribution of isotopes over the possible states and consequently a reaction may occur in which isotopes are exchanged between chemical compounds. The corresponding reaction constants K can be measured.

In practice, the lighter isotopes are not only a little bit more reactive in a chemical sense, but, their physical

> It stands to reason that the isotopic effects are largest when the masses are small, for then the relative difference between both masses is largest. So for heavy nuclides the isotope effect is almost negligible. For the gas UF_6 for example the difference in diffusion for ^{235}U and ^{238}U is very small. It therefore was a technological 'tour de force' to use diffusion to separate both uranium isotopes at the end of the Second World War as part of the atomic bomb program.
>
> From: [118], pp 7,8

*More precisely stated: for systems with a constant temperature and pressure, the Gibbs free energy G tends to a minimum. See the end of Section 4.2.

properties are also different. The water molecule H_2O with all isotopes the lightest possible ($^1H^1H^{16}O$) will evaporate more easily than water molecules where one of the isotopes is replaced by a heavier brother (e.g. $^1H^1H^{18}O$ or $^2H^1H^{16}O$). Also, the lightest CO_2 molecule ($^{12}C^{16}O^{16}O$) has a faster diffusion in a medium than a heavier one such as $^{13}C^{16}O^{16}O$. In both cases this reflects the fact that a lighter particle has a bigger mean free path between two collisions with its surroundings than a heavier one. We will encounter both examples below.

10.2.2 *The ^{13}C effect*

Carbon has two stable isotopes: ^{12}C with an abundance of 98.89 percent and ^{13}C with an abundance of 1.11 percent. These numbers should be regarded as an average, as the isotopic effects just described give variations in the terrestrial ^{13}C content by some 10 percent. These variations are expressed by a parameter $\delta^{13}C$, which is defined as follows

$$\delta^{13}C = \frac{(^{13}C/^{12}C)_{\text{sample}} - (^{13}C/^{12}C)_{\text{standard}}}{(^{13}C/^{12}C)_{\text{standard}}} \times 1000 \qquad (10.9)$$

This result is given per millage (i.e. per thousand). The standard was agreed internationally, based on work at the University of Chicago and is called PDB[†] It was unfortunately chosen in such a way that most samples have a negative $\delta^{13}C$ value. Atmospheric CO_2, for example, has a value of $\delta^{13}C \approx -7‰$ (i.e. 7/1000).

Photosynthesis causes a fractionation of carbon in plants ([118], p. 380). In a first step the atmospheric CO_2 passes the walls of the cells. Here ^{12}C goes quicker, which results in a lower value of $\delta^{13}C$. In the next steps, enzymes catalyse diverse chemical reactions which differentiate between the $\delta^{13}C$ values for the different parts of the plants, also dependent on the type of plant. Terrestrial plants have $\delta^{13}C$ values in the range $-24‰$ to $-34‰$, while desert plants and tropical grass range between $-6‰$ and $-19‰$. Fossil fuels originate from plants and have values of $-25‰$ for coal and $-28‰$ for petroleum while for natural gas it varies widely in the order of $-40‰$ to $-70‰$. From these data it is clear that the burning of fossil fuels during the last century has changed the $\delta^{13}C$ value of the atmosphere to more negative values.

One must be careful in generalizing the above data to other carbon compounds. In recent years, for example, a thorough study has been made of the

[†]See [119], Section 1.5. Recently the standard was replaced by a Vienna-PDB, or VPDB, which is defined in such a way that the results are essentially the same.

methane emissions into the atmosphere because, as we saw in Table 3.1, methane is an important greenhouse gas. For methane in the atmosphere, the measured $\delta^{13}C$ value is around −47‰ ([120], p. 6). This value is quite different from the value of −7‰ we mentioned for carbon from CO_2 in the atmosphere. The reason is that CO_2 in the atmosphere is enriched in the heavier molecules that only penetrate the walls of the cells relatively slowly. Methane, on the contrary, originates from the cells themselves and therefore from the lighter molecules that could penetrate the cell walls more easily.

For methane also, the precise negative $\delta^{13}C$ values depend on the conditions. Cows, eating plants release methane with $\delta^{13}C$ between −55‰ and −65‰. Wetlands have rotting plants and generate values between −50‰ and −80‰. Fossil fuels and biomass have $\delta^{13}C$ values of −15‰ to −50‰, perhaps contrary to what is expected as they originate from photosynthesis.

> The existing estimates of European methane emissions, from inventory calculations, may be incorrect both in total and in specific source estimates. Atmospheric studies of local areas and regions are needed on a continuing basis to validate emission inventory calculations. In particular, emission reduction policies must be validated.
>
> From [120], p. 24

Measurements of $\delta^{13}C$ values for methane in the atmosphere are helpful in determining its origin. Figure 10.2 shows the international stations where the methane concentrations are monitored. Although the differences in $\delta^{13}C$ values of the emissions are small, nevertheless it is possible to deduce that, for example, in the Netherlands 15 percent of the methane emissions originate from fossil fuel sources and 85 percent from biotic sources (landfills, cows and

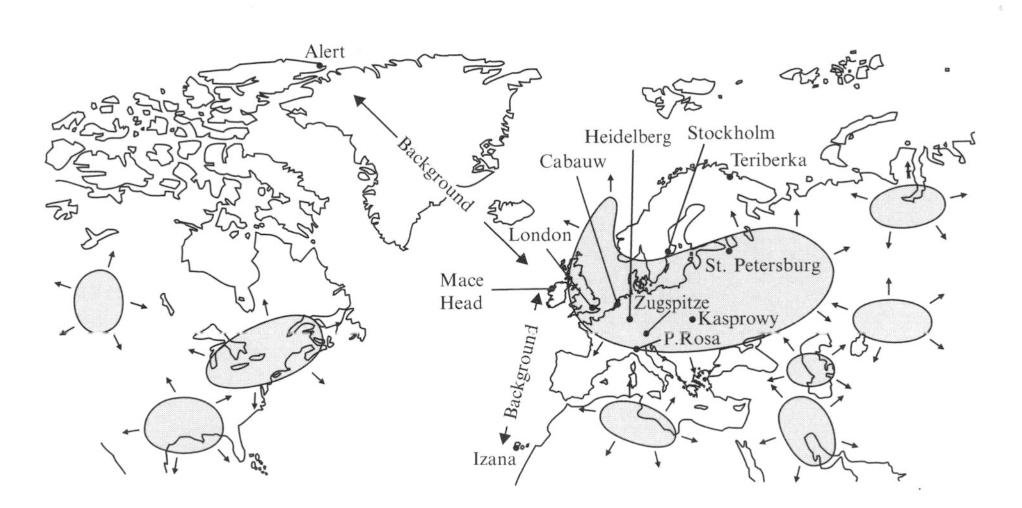

Figure 10.2 Methane source regions in the Northern Hemisphere. Shown are the major monitoring stations, and in grey colours, the regions from where most of the methane originates. Reproduced with permission of the European Commission from [120], p. 4

wetlands). Similarly, from measurements around London it was concluded that leakage of gas is happening during storage and transmission – something that can be prevented.

It is clear that measurements of the small differences shown here require very accurate equipment and careful analysis. It is also clear that measurements cannot cover a complete territory. One always has to select certain sites and has to rely on transport models to generalize the results of the measurements.

10.2.3 *Water with heavy isotopes:* ^{18}O *and* ^{2}H *(or D)*

> Geochemists have standard samples, which they use for comparison. For Carbon it is the proportions of ^{12}C and ^{13}C that were originally found in a Mesozoic fossil, a belemnite from a Cretaceous formation at Peedee in South Carolina. Belemnites crystallize their hard parts directly from seawater, so the values for this fossil represent the oceans of the Mesozoic.
>
> Inside Science, *New Scientist*, 18 September 1999, p. 2

Water with its chemical formula H_2O has two constituting elements: hydrogen and oxygen. Hydrogen has two stable isotopes: ^{1}H with an abundance of 99.985 percent and ^{2}H with an abundance of 0.015 percent. The latter is called deuterium and frequently written as ^{2}D, or simply as D. Oxygen has three stable isotopes: ^{16}O with an abundance of 99.76 percent, ^{17}O with an abundance of 0.038 percent and ^{18}O with an abundance of 0.20 percent. From these numbers it is clear that almost all water atoms will consist of the lightest atoms ^{1}H and ^{16}O. In a much smaller concentration one will find water with one of the three atoms replaced by a heavier nuclide. In practice, one considers two cases: water with ^{16}O replaced by ^{18}O and water with ^{1}H replaced by ^{2}D.

Just as with ^{13}C, the fractionation of the molecules is expressed by comparing a sample with a standard, yielding similar equations

$$\delta^{18}O = \frac{(^{18}O/^{16}O)_{sample} - (^{18}O/^{16}O)_{standard}}{(^{18}O/^{16}O)_{standard}} \times 1000 \qquad (10.10)$$

and

$$\delta D = \frac{(D/H)_{sample} - (D/H)_{standard}}{(D/H)_{standard}} \times 1000 \qquad (10.11)$$

The best known application of isotope methods for water is in establishing the surface air temperature in Greenland and the Antarctic a long time ago, by studying the snow and ice which was precipitated at that time. The bulk of the precipitation originates from evaporation in the tropical oceans. So one has to start the discussions there, following the schematic presentation of Figure 10.3.

The main phenomena are evaporation and condensation. The differences between the various H_2O molecules originate from their masses and are expressed physically in different vapour pressures: the lighter molecule has the higher vapour pressure and prefers the vapour phase above the liquid one. In this respect, water enriched in ^{18}O or ^{2}D behaves in a similar way. There is in fact a linear relationship between $\delta^{18}O$ and δD ([117], p. 379 or [118], p. 328), so for simplicity we will discuss $\delta^{18}O$ only.

> Tritium (^{3}H) is produced in the upper atmosphere by interaction of fast cosmic ray neutrons with stable ^{14}N. It decays with a half-life of 12.26 years. The concentration of tritium in various parts of the hydrosphere has increased as a result of the explosion of hydrogen bombs between 1954 and 1963. This artificially produced tritium has been used to study mixing rates in the atmosphere and the oceans and trace the movement of groundwater.
>
> From [118], p. 330

In Figure 10.3 water evaporates from the warm oceans and becomes vapour enriched in ^{16}O. The moist air mass moves toward higher latitudes, it cools and loses water in the form of rain or snow. That precipitation is enriched in ^{18}O, as it prefers the liquid phase above the vapour phase. The remaining humid air consequently is enriched even more in ^{16}O.

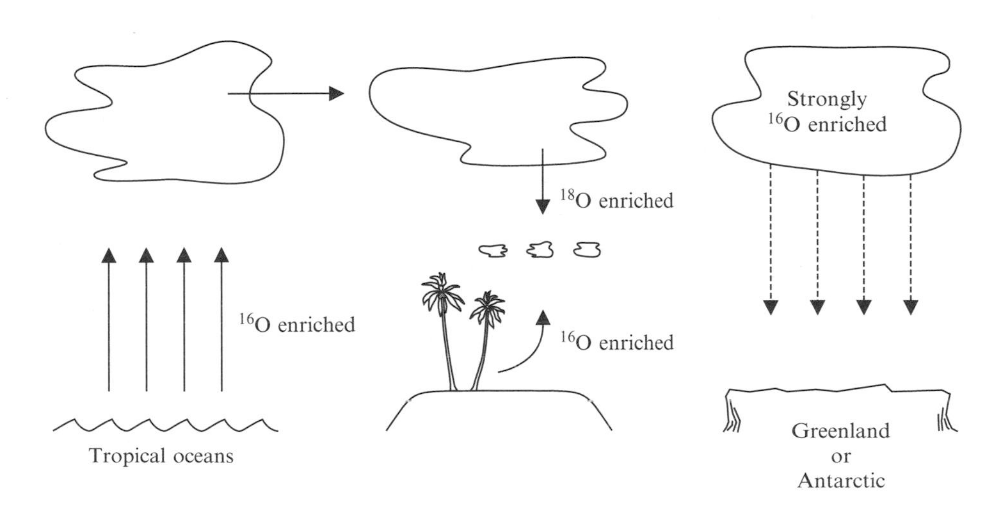

Figure 10.3 Precipitation on Greenland or the Antarctic. The oceans evaporate light water enriched in ^{16}O. Clouds on their way to the cold, rain out heavier water enriched in ^{18}O. The raindrops partly evaporate water, again enriched in ^{16}O. Also plants give off water enriched in ^{16}O. At the end the clouds are strongly enriched in ^{16}O. The values of $\delta^{18}O$ of the snow and ice correlate with the local surface air temperature

Not all of the precipitation reaches the surface of the earth. Part of it evaporates again in the lower, warmer air. That part is enriched in ^{16}O and joins the already enriched clouds. In the meantime the plants at higher latitudes pick up water from the atmosphere and evaporate it again, even further enriching the evaporated water in ^{16}O. Finally, the water in the air when it reaches the Arctic and Antarctic regions, is strongly enriched in ^{16}O. The water precipitates as snow and ice and steadily accumulates at the ground level where in the course of ages, it forms a layer of thousands of metres of depth.

At high temperatures the fractionation between lighter and heavier isotopes is virtually absent, while at low temperatures it may be appreciable. So, it is understandable that there will be a relationship between the $\delta^{18}O$ value of the precipitation and the temperature at ground level in the Arctic or Antarctic. However, because of the rather complicated processes indicated in Figure 10.3, one might expect that many other parameters would enter as well, such as temperatures in the tropics or in the medium latitudes. Surprisingly, many effects appear to cancel and $\delta^{18}O$ for the mean annual precipitation exhibits a linear relationship with the mean annual surface air temperature T in the high latitude places of precipitation. The empirical relationship for the present is shown in Figure 10.4. It is linear and is written as ([117], p. 380)

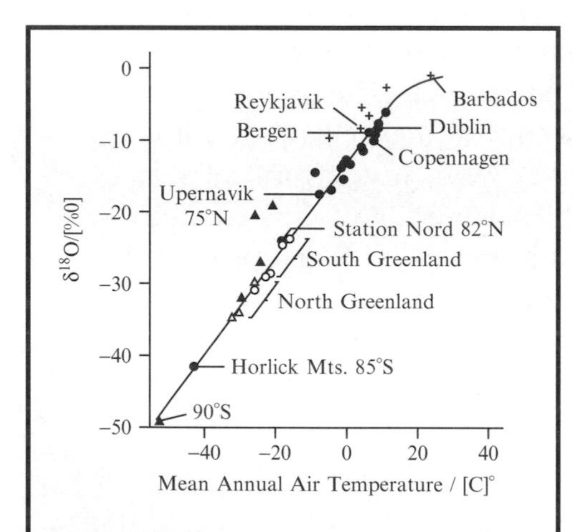

Figure 10.4 The annual mean value of $\delta^{18}O$ in precipitation as a function of the mean surface air temperature. Data points are close to (10.12). With permission of Munksgaard Publ. Copenhagen, Denmark taken from [121], Figure 3

$$\delta^{18}O = \alpha T - \beta \qquad (10.12)$$

where T is expressed in [°C] and $\delta^{18}O$ in [‰]. The pioneering work of Dansgaard [121] represented in Figure 10.4 gave parameters $\alpha = 0.695$ [‰ $^{\circ}C^{-1}$] and $\beta = 13.6$ [‰].

By means of relation (10.12) $\delta^{18}O$ acts as a *paleo-thermometer* and it is used to determine the surface air temperature at which samples, at present deep down under the snow caps, were formed. For that situation the parameters α and β have to be found independently, but from Figure 10.4 it is plausible that the relationship was linear in the past as well. The procedure is given in [122]. The values of $\delta^{18}O$ and the present temperature of the sample were measured; from the layering of the snow the time of formation was deduced, which dated back to 100 000 years ago. Then some values of α and β were assumed, giving with Equation (10.12), some temperature T at that time.

Next, the interchange of heat with the surroundings was calculated to obtain the present temperature. The correct values of α and β were then obtained through a search process as those values for which the calculated present temperature coincided with the measured present temperature. Reference [122] finds $\alpha = 0.327$ [‰°C^{-1}] and $\beta = 24.5$ [‰]. If one assumes that these values also hold for other locations and longer periods of time, the paleo-thermometer helps to deduce the climate variations in the past (see Chapter 3, Figure 3.1, which was obtained with the older parameters of [121]).

A final warning is appropriate. A relation such as Equation (10.12) is applicable only for the cases for which it was derived. It may certainly not be applied to determine a temperature at which water attached to minerals in rock formations, was formed.

10.3 MAGNETIC SURVEYS

A magnetic field is described by the *magnetic flux density* **B** and measured in units of Tesla [T]. In adition to **B** there is another physical quantity, called the *magnetic field intensity* **H**. In vacuum, or, for all practical purposes in air as well, they are simply proportional:

$$\mathbf{B} = \mu_0 \mathbf{H} \tag{10.13}$$

The factor μ_0 is called the permeability of free space and has the value

$$\mu_0 = 4\pi \times 10^{-7} [\text{T m A}^{-1}] \tag{10.14}$$

Its numerical value and its dimension are due to the way in which the international system of units SI has been constructed. The physical difference between the **B** and **H** fields will become clear below.

It is well known that a small circular current, sketched in Figure 10.5, causes a field **B**, for which the field lines run in the way shown. On the axis of the *loop* the magnetic field points along that axis and diminishes in magnitude B when the distance d to the plane of the loop increases:

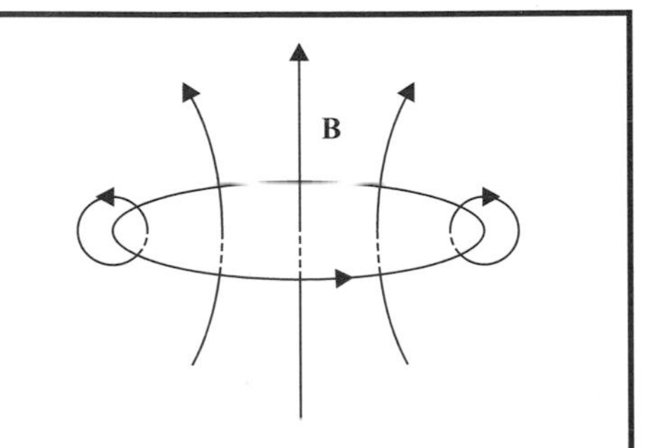

Figure 10.5 The magnetic flux density **B** of a circular electric current (a loop)

$$B \propto \frac{1}{d^3} \tag{10.15}$$

The direction of the **B**-field is such that a screwdriver, which rotates in the same direction as the current, proceeds in the direction of the field. Close to the loop the field lines are closed, resembling circles, again following the screwdriver rule. In between the loop and the circle they deviate widely, but ultimately will be closed again. In this section we describe how measurements of the **B** field may indicate the position of old dumps and polluted soil.

10.3.1 Earth's magnetic field

The magnetic field of the earth is shown schematically in Figure 10.6. It has the same behaviour as the field of a simple loop, which was given in Figure 10.6. From this similarity it is inferred that the earth's magnetic field is caused by circular electric currents in the iron interior of the earth. From Figure 10.6 it is immediately clear that at almost all places on earth the magnetic field will have a horizontal and a vertical component. In the Northern Hemisphere the vertical component points down and in the Southern Hemisphere it points up.

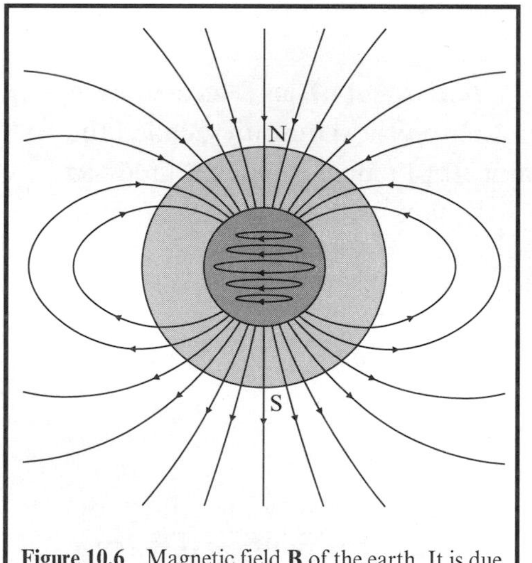

Figure 10.6 Magnetic field **B** of the earth. It is due to electric currents in the interior. Symbols N and S indicate the magnetic North and South Poles

The earth's magnetic field slowly changes in time with respect to the earth itself. Therefore, data on the earth is magnetic field are given with a time attached. In Coimbra, Portugal, for example, the horizontal component was 25 [μT] and the downward vertical component was 36 [μT], in the year 1990 ([15], Section 2.7).

On an atomic scale all materials may be described as positive atomic nuclei with a cloud of electrons. Usually all electrons move randomly and do not give rise to an external magnetic field. Some materials, however, like iron (Fe) exhibit so-called *ferromagnetism*. In these materials some of the electronic currents or loops are aligned in an external magnetic field. And, even when the external field is taken away, the orientation of the loops remains. This causes a *magnetization* **M** of the material, which results in a magnetic flux density

$$\mathbf{B} = \mu_0 \mathbf{M} \tag{10.16}$$

This equation holds inside the material. With distance d from the material the field B decreases as in Equation (10.15).

A compass needle consists of a little piece of magnetized iron, which orients itself to the earth's magnetic field. Its horizontal component points to magnetic North, which is in the neighbourhood of geographic north, but wanders around in time.

When, besides the magnetization (10.16) originating from ferromagnetic materials, an external field (10.13) is present, both fields add and the total field becomes

$$\mathbf{B} = \mu_0(\mathbf{H} + \mathbf{M}) \tag{10.17}$$

Finding dumps

Figure 10.7 shows an abandoned waste dump, in which some cars and metal containers are buried. In the course of time the earth's magnetic field will orient their microscopic **B**-fields, resulting in a magnetization **M** [123]. Above ground level the earth's magnetic field and the field originating from the magnetization **M** have to be added. The resulting **B**-field may be measured at different heights above ground level, using a magnetometer as described in [116], Box 11.1. The deviation from the earth's magnetic field is shown in Figure 10.7 for several heights above ground level. Because of the attenuation of the dump field with distance, according to Equation (10.15), the deviation from the earth magnetic field weakens with height. Note that, near ground level, the measured dump field is about 1 percent of the earth's magnetic field and variations of 1 [nT] can be measured. It is also interesting

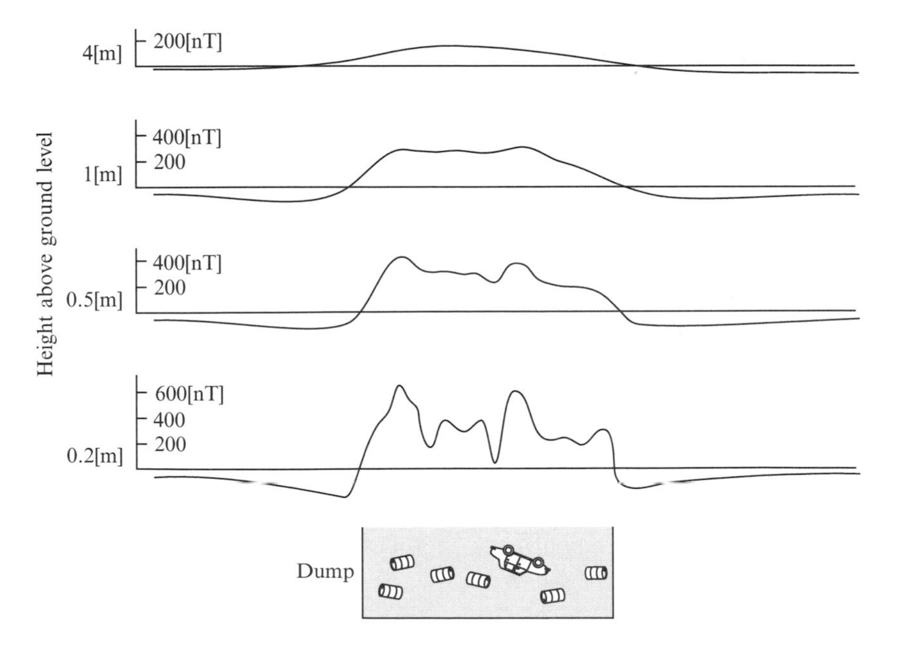

Figure 10.7 A waste dump in which iron objects are buried changes the magnetic field above ground level. The figure shows the deviations from the earth's magnetic field. At 0.2 [m] height the details of the underground are clearly visible. At 4 [m] only the outline of the dump is visible. Reproduced by permission of Springer Verlag from [123], p. 10

that already at 0.5 [m] signals from the different objects mix, thereby blurring the fine structure. These features are clearly observed in Figure 10.7. The top curve shows little detail, but the presence of the dump is still manifest.

Even more details than in Figure 10.7 may be observed, if one subtracts two of the lower curves and divides by the distance. In that way variations in the **B** field due to the composition and structure of the soil disappear. This method was applied to a real case in Germany and the resulting signal in [nT/m] is shown in Figure 10.8. The circumference of the dump (indicated in black) is very clearly observed. The peaks at the right indicate an underground iron gas mains. The other peaks should indicate cars or iron containers.

10.3.2 *Magnetic susceptibility*

Materials, which are not ferromagnetic still may respond to the application of an external magnetic field. In the case of *paramagnetic* materials, the microscopic electric current loops orient themselves to the field as long as the external field is present. In this way the external field is reinforced. In the case of *diamagnetic* materials, current are created which just counteract the applied field. Both effects may, in a first approximation, be described by

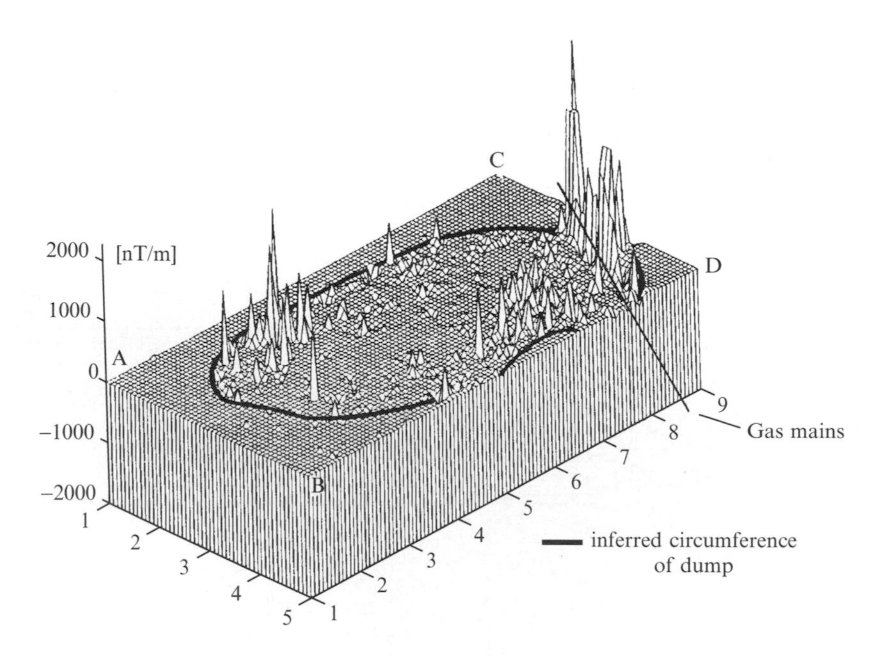

Figure 10.8 Magnetic investigation of a dump in Germany. The vertical coordinate is the change in magnetic field intensity [nT/m]. On the right, an iron gas mains is visible. The other peaks should represent iron containing materials in the dump. Reproduced with permission of Springer-Verlag from [123], p. 52

$$\mathbf{M} = \chi \mathbf{H} \tag{10.18}$$

where χ is called the *magnetic susceptibility* of the material. For paramagnetic substances $\chi > 0$ and for diamagnetic materials $\chi < 0$. When the external field **H** disappears, the resulting magnetization **M** also vanishes.

Finding pollution by metals

Measuring the **B** field of top soil in an externally applied field may give a quick and easy way to check whether the soil is polluted with heavy metals. Figure 10.9 shows the geography of a little stream in the North of Turkey in which a plant is discharging its effluent [124]. The plant is situated about 1 [km] from the Black Sea. Most heavy metals have $\chi < 0$, while clean sand has $\chi > 0$.

The results of the measurements along the stream are displayed in Figure 10.10 as a function of the distance d to the first measurement point. That point is located about 150 [m] upstream from the factory outlet. The dashed curve in Figure 10.10 is measured at the soil surface, while the full curve is measured from soil samples, analysed in the laboratory.

One may notice that, from $d = 0$ the susceptibility already starts to decrease, which means that pollution with heavy metals catches up. Apparently

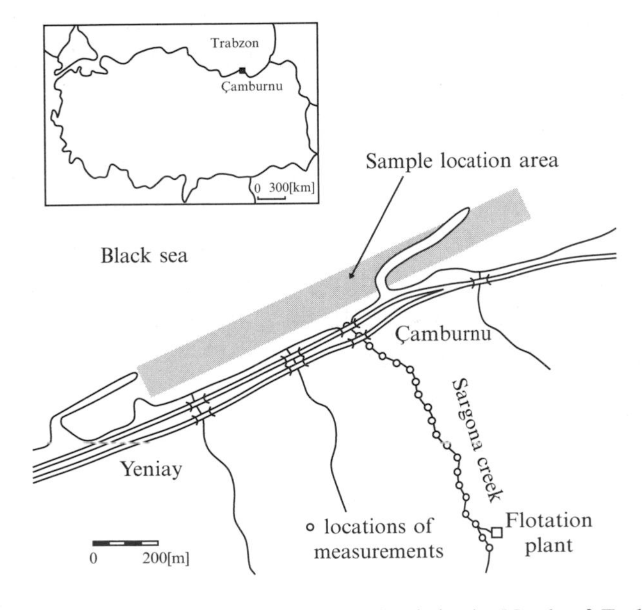

Figure 10.9 A plant is discharging into the Sargona Creek in the North of Turkey. Susceptibility measurements were made from a little upstream of the plant down to the Black Sea (the open circles). The grey area in that sea indicates where sand samples were taken and investigated in the laboratory. Reproduced by permission of Karadeniz Technical University from [124], p. 1199

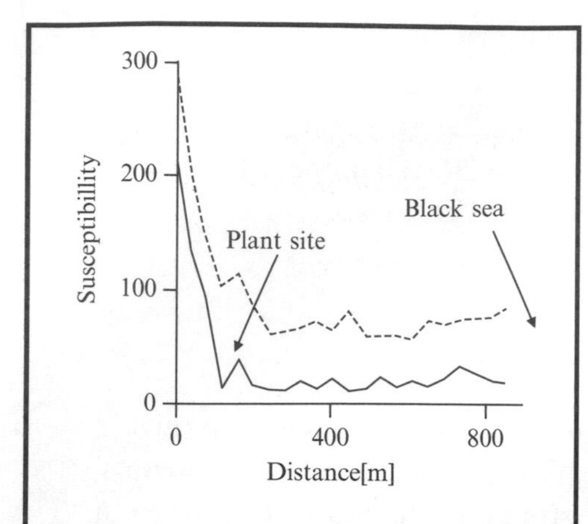

Figure 10.10 The susceptibility (arbitrary units) at the locations of Figure 10.9. The dashed curve is the surface susceptibility and the full curve is measured in the laboratory from samples taken. Reproduced by permission of Karadeniz Technical University from [124], p. 1200

the pollution is able to move somewhat upstream by turbulent diffusion (Chapter 7). The full curve, representing the laboratory measurements clearly is somewhat more accurate than the surface measurements on location. The fact that the susceptibility is lower may mean that metals have penetrated the top soil.

The area indicated in grey in Figure 10.9 indicates a region in the Black Sea which is close to the coast. There the analysis of samples taken gives an indication of the locations where the heavy metals are deposited on the sea floor. After a quick and easy measurement, demonstrated in Figure 10.10 one clearly has to perform a chemical analysis of the samples taken in order to know precisely which metals are present and in what concentrations.

10.4 SEISMIC PROBING

Put a metal plate on the ground and hit it with a heavy sledgehammer. In the open air this causes a sound wave, which propagates in all directions. In a first approximation the path from the location of the bang to an observer may be described by a straight line. One might speak of a 'sound ray' in analogy with a light ray.

In Section 8.1 we described sound as a pressure fluctuation in the air. When the fluctuation passes, the air particles oscillate a little bit along the direction of propagation. One may compare this with the compression of a spring, shown on the left of Figure 10.11, which may move along the spring from left to right. Also, here the spring particles oscillate a little way along the direction of propagation. This is called a *longitudinal wave*. Sound in air propagates as such a longitudinal wave.

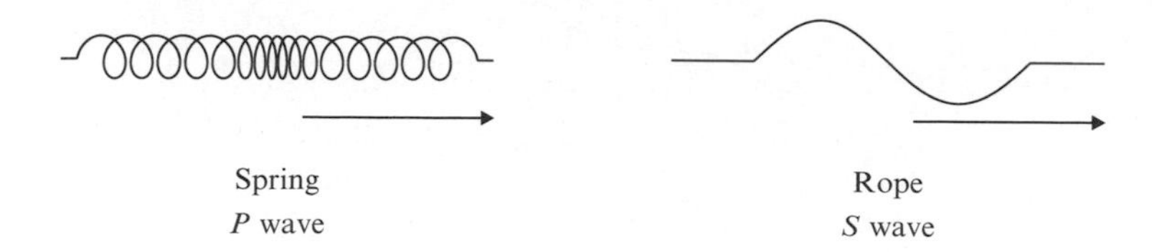

Figure 10.11 A propagating compression in a spring is an example of a longitudinal wave: the particles move in the direction of propagation. The undulating rope is an example of a transverse wave: the particles move perpendicular to the direction of propagation. In seismology the longitudinal waves are called *P* or primary waves, the transverse are called *S* or secondary waves

10.4.1 *Seismic waves*

The bang from the hammer will also cause propagating waves in the underground ([116], Section 4.5.2). These seismic waves may be *longitudinal*, called *P waves*. They also may be *transverse*, called *S waves*. In the latter case, the motion of the particles is perpendicular to the direction of propagation of the wave. The standard example of a transverse wave is the undulation of a rope; shown on the right of Figure 10.11. Transverse waves easily pass a solid piece of matter like rock. In loose soil, humid sand or gravel they will quickly be damped. As we are interested in the first 10 [m] or 20 [m] of top soil we will ignore the *S* waves in the remainder of this chapter. Also, one may use an experimental set-up, where only *P* waves are relevant. Consider a vertical peg driven deep into the ground. Hit the peg horizontally with a hammer. In the direction of the hit, only longitudinal waves will be excited ([116], p. 45, Figure 4.5)

> Earthquakes, deep in the earth, create both longitudinal P waves and transverse S waves. The velocity of propagation of the P waves is highest, so they arrive first. That is why P stands for Primary and S for Secondary. The S waves are most destructive, so in earthquake regions people may save themselves by leaving their houses as soon as they 'feel' the arrival of the first P wave.

Refraction

Seismic waves propagate along a straight line, called a seismic ray. This is completely analogous to light rays and consequently one would expect the same laws of physics (the laws of geometrical optics) to apply. It is well known that a light ray, passing the interface between a medium with a low velocity of light like water to a medium with a higher velocity of light like air, exhibits a phenomenon called *refraction*. This is illustrated on the top of Figure 10.12, where the ray first has an angle θ_1 with respect to the normal to the interface and, down in the second medium, has a larger angle θ_2. This phenomenon is observed when one looks at a straight stick put into a pond. It looks broken.

Snell's law There exists a relation between the angles θ_1 and θ_2 and the velocities in the two media. This is given by *Snell's law*

$$\frac{\sin\theta_1}{\sin\theta_2} = \frac{u_1}{u_2} \tag{10.19}$$

This implies that, for increasing θ_1, θ_2 will also increase. Further, as $u_2 > u_1$ also $\theta_2 > \theta_1$. Taken all together this means that for increasing θ_1 a point will

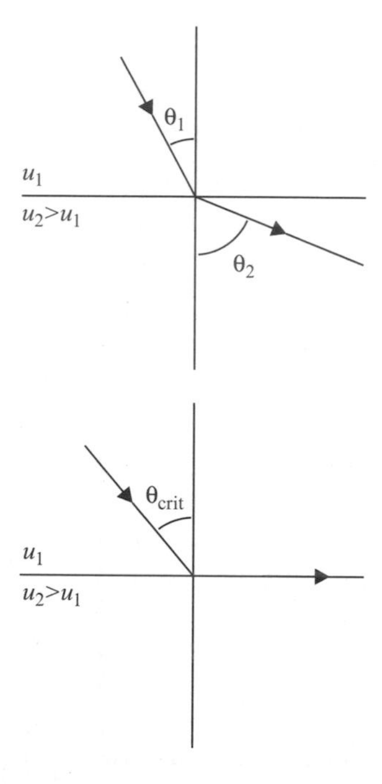

Figure 10.12 Refraction from a medium with a smaller velocity to one with a higher velocity. Snell's law gives the critical angle, where the refracted wave runs along the interface of the media

be reached where $\theta_2 = 90°$. In this case the refracted ray runs precisely along the interface of the two media. This is shown on the bottom of Figure 10.12 and the corresponding angle θ_1 is called the critical angle θ_{crit}. As $\sin\theta_2 = \sin 90° = 1$ it follows from Equation (10.19) that the critical angle is given by

$$\frac{\sin\theta_{crit}}{1} = \frac{u_1}{u_2} \tag{10.20}$$

or

$$\sin\theta_{crit} = \frac{u_1}{u_2} \tag{10.21}$$

10.4.2 *Analysis of the underground by refraction*

Figure 10.13 shows a schematic structure of the top ground. The top layer is homogeneous and coloured light grey, while the lower soil is also homogeneous and coloured in a darker grey. The velocity of seismic waves in the top layer would be u_1 and in the lower layer it would have a higher value, u_2.

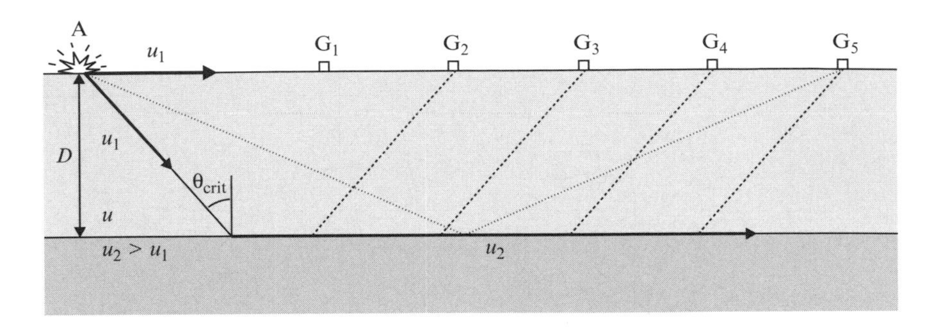

Figure 10.13 On the left (point A) seismic waves are generated. At the critical angle θ_{crit} the refracted ray follows the interface. Geophones G_1, G_2 etc. pick up and distinguish both signals and may deduce the thickness D of the top layer. The lightly dotted line indicates a wave reflected at the interface

Figure 10.13 also shows the experimental set-up ([123], p. 33). On the left, at A, a bang or small explosion generates seismic waves through the ground. Some run along the surface with velocity u_1 and are detected by so-called geophones G_1, G_2, G_3 etc. at distances of about 1 to 5 [m] (The principle of geophones is described in [116], Section 4.2). A fraction of the seismic waves penetrate into the soil and will reach the interface with the next layer at different angles. For angles smaller than the critical, many of them will continue down, as on the left of Figure 10.12. But waves arriving at the critical angle will follow the interface with the much higher velocity u_2, which is emphasized by the lengths of the velocity vectors in Figure 10.13.

The refracted seismic ray along the interface is accompanied by a propagating wave, which means that the signal of the passing wave may be detected by the geophones at the surface. This is indicated by the dotted lines in Figure 10.13*. The first two geophones will pick up the surface signal, first G_1, next G_2. From these data one may deduce the velocity u_1. Geophones further away, say G_{10} and G_{11} will pick up the refracted signal first. From these data one may deduce u_2. With Equation (10.21) one will find the critical angle θ_{crit}. Then, by looking at the time difference by which the direct and the refracted signal arrive at geophone G_{11}, say, one may deduce the thickness D of the layer (a more sophisticated procedure is given in Exercise 10.6).

By installing a series of geophones it becomes possible to perform the measurement accurately. It also allows to distinguish the two signals just described from the waves which are approaching the interface with a large angle and are reflected. One of these reflected waves is indicated with a lightly dotted line in Figure 10.13.

In real life one does not have smooth horizontal layers, as depicted in Figure 10.13. Especially if there is a waste dump hidden somewhere in the

*One may show that their angle with the vertical again is the critical angle ([116], Sect. 6.1.2)

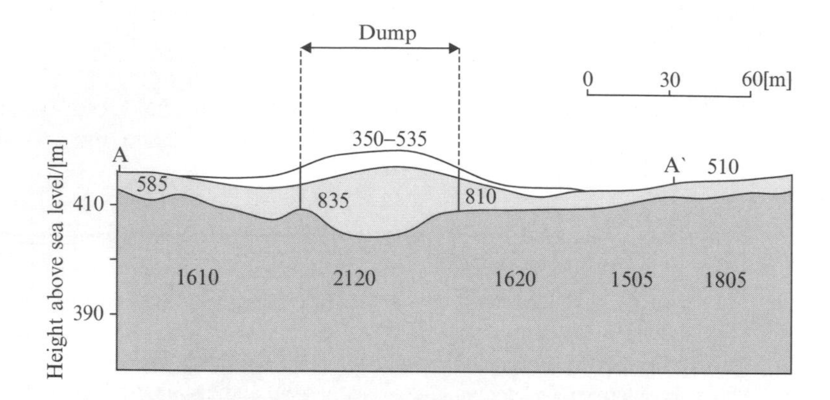

Figure 10.14 Seismic measurement of a realistic dump. One notices the dump, indicated by a depression in the middle layer. The numbers indicate the measured velocities of the seismic waves. They are lowest in the loose top soil and highest in the rigid, perhaps rocky underground. Reproduced by permission of Springer Verlag from [123], p. 73, Bild 3.18

underground that will rather resemble a pond filled with rubbish and covered with top soil. It therefore looks like Figure 10.14. This figure was constructed after several measurements with lines of geophones, which were carefully analysed. The numbers in the figure indicate the velocity of the seismic waves, which were deduced. In the deeper underground they range between 1500 [m s^{-1}] and 2100 [m s^{-1}], which implies that the geophones should be able to measure times shorter than a millisecond. At a higher level in the soil, the velocities are considerably lower, consistent with an interface between the two main layers, indicated by a curved line. The bottom of the dump is clearly seen by a depression of that curve.

EXERCISES

10.1 In radioactive elements, a positive electron is often emitted rather than a negative one. Look at equations (10.3) and (10.4). Verify by a quick glance at a table of radioactive elements (for example in [15]) whether β^+ emission and electron capture occur hand in hand.

10.2 Write down the four radioactive nuclides of Figure 10.1, which occur in nature. Consult a table of radioactive elements (for example [15]) and find their half-lives. Can they have survived from the beginning of earth's history? Also look at the decay of ^{36}Cl, not shown in the graph. What is remarkable here?

10.3 The ^{14}C activity in plant tissues is estimated as $A = (13.56 \pm 0.07)$ disintegrations per minute per gram of carbon. What is the ^{14}C fraction in the sample, knowing that its half-life is 5730 years.

10.4 For a simple classical harmonic oscillator, described in Equation (10.8) deduce the classical frequency $f = (2\pi)^{-1}\sqrt{(k/\mu)}$

10.5 Find the components of the earth's magnetic field for a place near the location where you are living. Use [15], Section 2.7 (or a similar table) and the time variation given there.

10.6 In the discussion of Figure 10.13 we outlined a way of finding the thickness D of the top layer. The usual procedure is more sophisticated and goes as follows. (i) Consider a series of geophones at equal distances from each other and from the initial bang at location A ($t = 0$). Draw a graph with, horizontally, the distance d from A and vertically, the time t at which the direct surface seismic wave arrives. This should be linear with $t = d/u_1$. (ii) The refracted wave will exhibit a linear relationship on the same graph with $t = t_L + d/u_2$. Show this gives an interpretation of t_L without writing down the equation. (iii) Suggest how the intersection of both lines will determine depth D.

11 Science and Society

In the preface we wrote that environmental physics deals with problems arising from the interaction between man and his natural environment. We went even one step further in saying that we not only want to understand the problems but also to analyse and mitigate them. It will take public money to perform the necessary research and a political will to implement possible solutions.

> Perhaps science comes to be seen as Strangelove bent on complete control. Maybe this explains the European outcry against genetically modified foods. Perhaps it is not so much in the food as in the way it was forced onto our plates.
>
> *New Scientist* 5 June 1999, 3

Students from the natural sciences have to realize that scientific analysis is only one of the factors in political decision making. Moreover, scientists also have different points of view, both on scientific questions and on their political implications. That may be apparent from the appetisers that are scattered throughout the text. A close look at these appetisers and the sources from which they are taken, will reveal that they reflect different views: some are more optimistic about the possibilities of technology than others; some take a more political stance than others.

Our own belief is that the clever use of technology is able to provide an acceptable standard of living for all of the world's population. Several examples of technological solutions or technological 'fixes' were given in the previous chapters. But, we would add, a global solution requires a modest use of natural resources in the high-income countries, as there are limits to the possibilities of science and technology.

For physicists, such as the present authors, writing about society is tricky, since physicists are trained in solving well-defined problems to which there is a unique solution. With social issues, even the definition of the problem is a matter of value judgement. Also the selection of information and arguments in a chapter like this one is open to discussion. We are aware of all this and are open to suggestions from our readers who may have differing or complementary views.

We start in Section 11.1 with energy resources and energy use. In Section 11.2 we briefly discuss how one deals with the health effects of pollution. We proceed in Section 11.3 with environmental policy and the difficulties of implementing them, which leads us in Section 11.4 to question our way of thinking and our way of life. The last section, 11.5, is devoted to the responsibility of the scientist in dealing with the environment.

11.1 *FINITE RESOURCES OF ENERGY*

Resources of land and fossil fuels are finite. So, eventually they will be exhausted. The question is when. In 1798, more than 200 years ago, the British economist Malthus published his 'Essay on the principles of population'. He argued that most people will remain poor as the population doubles every 25 years, while the output of agriculture and industry will increase more slowly with time. More recently, in 1972, the Club of Rome made a similar point in analysing the finiteness and exhaustion of many materials [125]. Still, in the Western world, people are not daily confronted with limitation of material resources. Therefore, we take the example of energy, which is an essential resource anyway, to look into this point.

> According to the UN's population division in New York, fertility rates are now below the long-term replacement level of 2.1 in 61 countries, including most of Europe. It looks increasingly as if the 21st century may see the beginnings of a population implosion. The most immediate impact will be an ageing population and a grey, conformist society.
>
> *New Scientist*, 2 October 1999, 20–21

11.1.1 *Scarce resources*

In Figure 11.1 the energy consumption per person is shown for India, the USA and for the world as a whole, from the beginning of the industrial revolution around 1800 up to the year 1998 ([54], [57], [126]). On the left axis both the units [10^9 J/year] and Watt [W] are indicated. This latter unit is more illuminating, as we have seen in the beginning of Chapter 4, that manpower, maintained for some length of time, does not produce much more than about 20 [W] of mechanical work, when averaged over a day. This has to be compared with, for example, the 1000 [W] of power, which a small electric heater will use.

From Figure 11.1 one observes that, on a world scale, the energy consumption per capita is growing steadily, with a dip around 1970 due to the 'energy crisis' of that time. Note that after such a calamity the steady growth catches up again. In Figure 11.1 the growth of the world population is also indicated

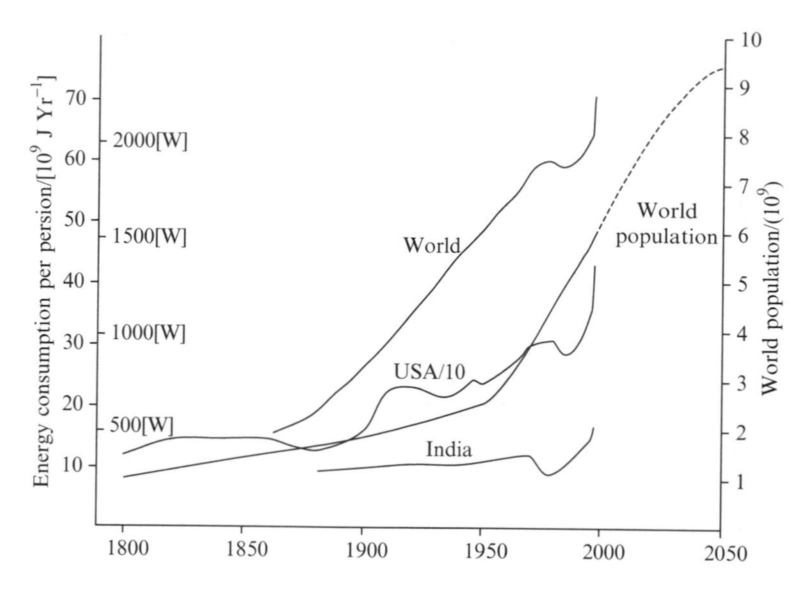

Figure 11.1 Energy consumption per person from 1800 to 1970 [126] and up to 1998 ([54], [57]) with the scale indicated on the left axis. Population data are given for the world as a whole on the right axis ([54], [127]). Note that the curve for the USA has to be multiplied by 10

with a projection up to 2050 [127]. The total energy use is the product of the world population and the energy consumption per person and this number is increasing rapidly. Let us look at the consequences for resources, emission of greenhouse gases and pollution in general.

Table 11.1 shows aggregated data for the energy consumption in 1998 and the 'easy' resources at the end of that year. It refers to all fossil fuels together, to nuclear and renewables. Nuclear and most of the renewables produce electricity directly, so the data of Table 4.4 were used as input to Table 11.1.

The 'easy resources' referred to in Table 11.1 are resources which are exploitable at current market prices with the technology of today. It is remarkable that between 1990 (data not given here) and 1998 the 'easy resources' have increased, despite a total consumption of about 2400 [10^{18}J]

Table 11.1 Energy consumption (1998) compared with the (equivalent) easy resources in 1998. Data are based on [37], [54] and [57]*

	Consumption/[10^{18}J yr^{-1}]	Easy resources
Fossil	344	43500 [10^{18} J]
Nuclear	26	2250 [10^{18} J]
Renewables	55	176 [10^{18} J/yr]
Total	425	

*Fossil resources are based on [54], nuclear resources on [57], which refers to 1995. Renewable resources were estimated in [1], Table 8.4.

Different oil fields have different crudes. Some are heavy, viscous and packed with toxic contaminants. These are usually left underground. Oil companies are now turning to bacteria to devour the thickest crudes locally. When they are broken down into lighter fractions and the contaminants are removed as well, they are easy to pump up and deal with. The wastewater containing sulphur and organic salts has to go somewhere, however.

New Scientist, 18 December 1999, 33–35

in this period. As the market prices hardly changed, the increase must be due to exploration and improved technology. Indeed, a lot of research is done to increase the percentage of oil that one can extract from a field in an economically viable way.

In view of the earlier successes of technology, one might hope that the world does not quickly run out of fossil fuel resources. Yet, burning fossil fuel produces CO_2, which is, as we have seen, a major greenhouse gas. Reducing the speed of climate change will require a drastic reduction of CO_2 emissions (Figure 3.14), which is expensive. Moreover, keeping the emissions of other greenhouse gases or pollutants like SO_2 within limits, has its own costs as well. Therefore, it is wise to look at renewable energies for at least part of the world's energy supply.

In Table 11.1 the easy resources of renewable energies were estimated as $176 \times [10^{18}\text{J yr}^{-1}]$. This is based on estimates† for 2025, which may only be reached with strong government support. Much of the increase from $55 \times [10^{18}\text{J yr}^{-1}]$ in 1998 to $176 \times [10^{18}\text{J yr}^{-1}]$ in 2025 can be attributed to biomass combustion, about $75 \times [10^{18}\text{J yr}^{-1}]$. Of course the biomass has to be grown as rapidly as it is consumed. To keep greenhouse gases or other pollutants out of the atmosphere with biomass combustion will be as costly as for fossil fuel combustion unless direct means of conversion of photosynthesis into electric power are found and proven to be cost-effective. Although that combustion is renewable and will ease the stress on resources, it does not solve the pollution problem.

Shell International Renewables: We are involved in renewable energy, because we think they will comprise a steadily increasing portion of the energy mix in the future. Our energy scenarios project a share of 30% to 50% of global energy consumption by 2050.

New Energy 1/2000, 8–9

Therefore, attention should be focussed on clean solar energy and wind energy, which in 1998 only contributed $0.01 \times [10^{18}\text{J yr}^{-1}]$ and $0.24 \times [10^{18}\text{J yr}^{-1}]$ respectively (Table 4.4). Their potentials in 2025 are estimated as $19 \times [10^{18}\text{J yr}^{-1}]$ and $8 \times [10^{18}\text{J yr}^{-1}]$ respectively. Certainly this is worthwhile, but is no quick and easy solution.

†Ref. [37], p. 88 gives a range of $130 \times [10^{18}\text{J yr}^{-1}]$ to $230 \times [10^{18}\text{J yr}^{-1}]$.

The question therefore arises whether it is possible to maintain a reasonable standard of living with lower energy consumption. To judge this, in Figure 11.2 the energy consumption per person and the Gross National Product (GNP)[‡] per person are presented for a number of countries. The GNP is defined as the total production of a country during a year, expressed in the market prices of that year ([128], Chapter 10). It is generally regarded as a first approximation to the wealth of a country.

From Figure 11.2 one notices that the data points are widely scattered. Even comparable countries like Britain and Italy or the Netherlands and Belgium show notable differences in their energy consumption per capita. An industrial country like Japan is as wealthy as the United States but with a much lower energy consumption. Switzerland seems to perform even better than Japan, but that may be erroneous as Switzerland is small, earning a lot of its GNP by international (financial) services – and one may not extrapolate that to a world scale. Even so, it is clear that a more efficient use of energy is one of the options many countries should explore. From an

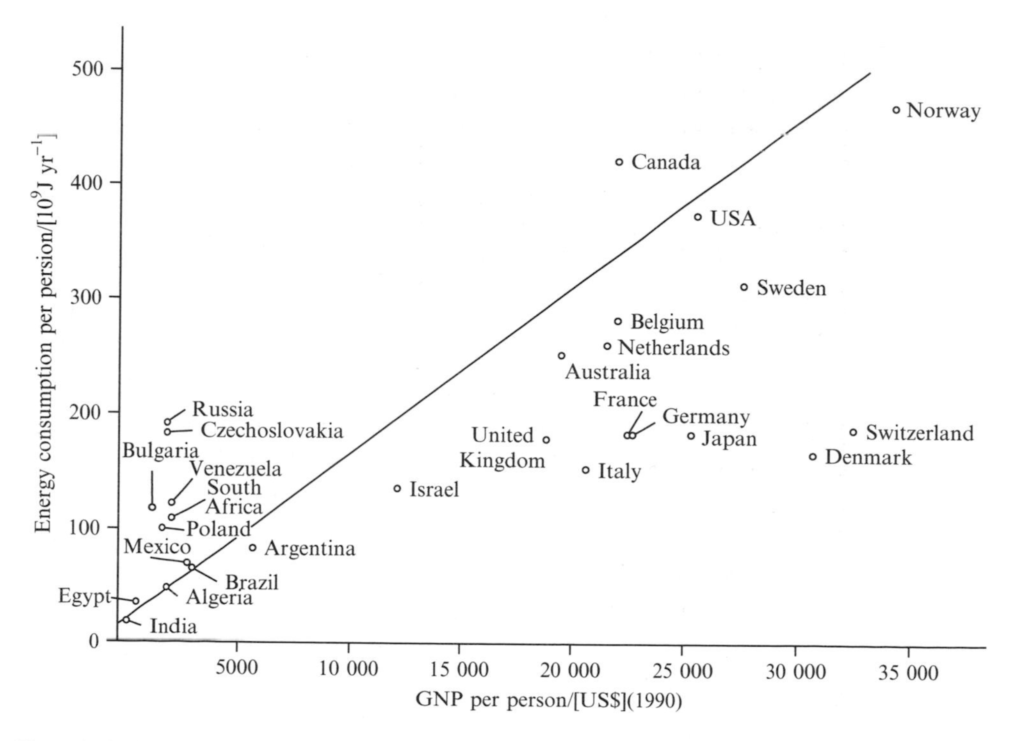

Figure 11.2 Energy consumption and GNP, both per person for a variety of countries in 1998. Data are taken from [53] with use of [57]. The US $ represent the purchasing power of 1990

[‡]The US $ in Figure 11.2 represent the purchasing power of 1990. Tables often use $ of a particular year in order to compare the trends of, for example, GNP over the years.

environmental point of view, it is better to take Japan or Germany as a model than the USA*.

11.2 POLLUTION AND HEALTH EFFECTS

Emissions from energy conversion may have effects on health, which can be direct or indirect. *Indirect effects* are caused, for example, by emission of greenhouse gases or sulfur oxides, which damage the ecosystem by climate change or acid rain, respectively. An example of an emission, which causes *direct health effects* is the discharge of hydrocarbons by cars, which produce ozone at ground level (Section 9.4). The petrochemical industry itself is emitting PAHs and PACs, some of which are carcinogenic and for which detecting techniques were discussed in Section 9.3.

A government will try to protect the health of its population by regulating the maximum concentrations of certain compounds in food, air and water. It will also regulate the emissions from industry and polluting vehicles. The emissions from a chimney, for example, may be related to the air quality in the neighbourhood (Sect. 7.2); this gives an indication as to what emissions are acceptable. For air quality and health hazards, see [10].

> Air pollution due to dust and smoke from wood and coal fires had been recognized as bad for the health in London, 1285. A royal proclamation prohibiting use of soft coal in kilns was issued, but was not being observed.
>
> In the first century BC it was recognized that transportation of drinking water in lead pipes was bad for the health.
>
> [130], pp 33,34

In order to regulate, government agencies need to know what concentrations are harmful. The answer to this question is far from trivial and has tremendous economic implications. To start with, if industry is forced to reduce its emissions by a factor of 10, it will need extra filters at the outlets and methods to make the caught chemical compounds harmless (if possible). The cost ultimately will have to be paid by the consumer. Similarly, drinking water utilities have to push down the concentrations of all kinds of chemicals below the allowed level. Also here, the cost ultimately has to be met by the consumer. It will be clear that, at the end of the day, a compromise has to be found in which judgements have to be made about which health risks are acceptable [130].

The discussion is made even more complicated as it is difficult to find out the difference between harmless and harmful intakes in the human body. One

*Also in the USA there are many scientists arguing for reducing energy use. Ref [129] for example gives many technologies showing how that can be done. Many of them are discussed in this book.

cup of coffee, for example, contains 100 [mg] of caffeine, which provides a stimulus for the drinker and generally is considered harmless. An intake of 10 [g] of caffeine, corresponding with 100 cups may well be lethal to an average person. But if the 100 cups are distributed over 100 days (one cup a day) it again is considered harmless ([131], p. 43). In this example it is not only the dose itself which counts, but also the time factor: does the body have the time to get rid of the potentially hostile materials.

Another telling example are pharmaceuticals, which are healing in small, prescribed concentrations, but often are damaging in larger concentrations.

The scientific problem is how to find out what concentrations of certain chemicals in food, air or water are harmful, harmless or perhaps even healing. The traditional method is the use of *dose-effect* or dose-response relations. It is important that the effect is measurable in a quantitative way, such as harm to a specified organ, a cancer or death. For this purpose the dose d is supplied to experimental animals under controlled conditions. It is expressed in [μg] of chemical per [kg] of bodyweight: [$\mu g\ kg^{-1}$] or in [μg] per [kg] of bodyweight per day: [$\mu g\ kg^{-1} day^{-1}$].

One of the most studied effects of a chemical is the mortality rate during the lifetime of an experimental animal, which may be in the order of two years. So, in principle, one has to study a group of animals that receive a certain dose with a comparable group that does not receive the dose, put them in the same circumstances (food, light, air etc.) and compare both groups as to their mortality.

Such a procedure yields a curve like the one shown on the left-hand side of Figure 11.3. The horizontal axis is the logarithm of the dose d and the vertical

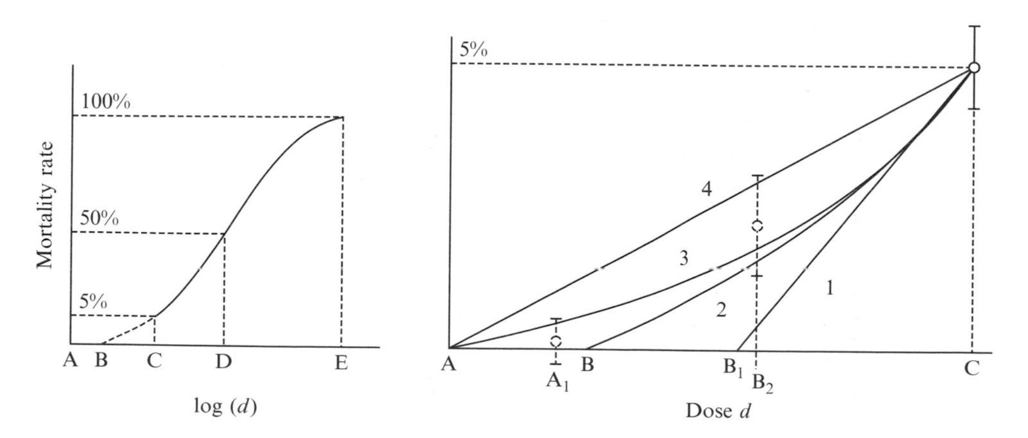

Figure 11.3 The left figure shows a standard graph of induced mortality rate against the logarithm of dose d, which means that the dose rate may span several factors of 10. The right figure shows the part AC of the left graph on a linear scale. In order to decide between the four extrapolations one needs to supplement measurements with theories and models

axis is the induced mortality rate. Point E has a mortality rate of 100 percent and point D a rate of 50 percent. The corresponding dose is called LD_{50}.

Although practically all curves show the behaviour of Figure 11.3, the number of factors of 10 along the horizontal axis may be quite different. For example, inhaling the gas CO for a number of hours in a concentration in the air of 100 [mg m^{-3}] causes a headache, but not much more, while a concentration of 250 [mg m^{-3}] is lethal when inhaled for a few hours. These two numbers are rather close, which suggests that, for this case, the curve in Figure 11.3 will be rather steep.

In Figure 11.3 the curve is solid from high doses down to point C and dashed below down to point B. This means that the curve is measured down to C and extrapolated below that dose. The reason is that it is very difficult, if not impossible, to measure effects accurately at low concentrations. To illustrate this the region from A to C is enlarged on a linear scale on the right of Figure 11.3. At dose B_2 one might find the hypothetical dashed data point and at point A_1, after a measurement with perhaps many experimental animals, another dashed data point. The error bars in this example will cross the axis, which means that the measurement is not conclusive.

> Science investigates and attempts to explain natural phenomena; it is cautious, incremental and truth seeking. Government, in its capacity as regulator, seeks to affect human behaviour and settle human disputes; it is episodic and peremptory and pursues resolution rather than truth.
>
> [132], p. 1011

If the chemical compound under investigation is *not carcinogenic*, point B or perhaps A_1, where no significant effect is found in test animals is taken as a starting point for regulations. A safety factor of 10 is often applied to account for possible differences between man and animal and another factor of 10 to account for the fact that part of the human population is more vulnerable than others: e.g. the old, sick and very young. After applying this factor of 100 an acceptable daily intake is found [132].

The safety factors of 10 are taken rather arbitrarily. Modern toxicology investigates the mechanisms by which a certain chemical may do harm to the human body and then looks into which of the most common experimental

> Chloroform, the most common chemical by-product of water disinfecting, causes cancer in laboratory animals when administered at very high doses. The scientific issue concerns whether there is a dose below which chloroform is safe.
>
> EPA guidelines for assessing cancer risk extrapolate linearly the harmful effects at high doses, so that any dose is considered harmful. But a non-zero safe 'threshold' could theoretically be set if enough were known about a chemical's mode of action.
>
> *Nature* **399**, 24 June 1999, 718

animals, rats, mice or other rodents, exhibit a similar mechanism. Depending on the chemical compound, some animals are more like humans than others– which gives a choice of criterion and may lead to better estimates of safety factors.

If the chemical compound is *carcinogenic*, the damage presumably is caused by an attack of the chemical (or a reactive metabolic product) on the DNA of the organism. This, in fact, is similar to the mechanism by which ionizing radiation may induce cancers. It is assumed, or at least regarded as possible, that any amount of chemical or radiation has a certain probability of starting the development of cancer. In other words there is, in principle, no safe threshold. In terms of the right side of Figure 11.3 the measured dose-effect curve would go from D to C. As conventional animal experiments cannot distinguish between the convex curve 3 or the straight line 4, the straight line is assumed in order to be 'on the safe side'.* This is also the way in which radiation risks†, quoted in Table 5.5 are found by the ICRP [65]. Also, in this case, toxicological research may lay better foundations for a choice between the different extrapolations and delineate in which cases there may be a threshold and in which cases there is none.

11.2.1 *Acceptable dose*

> It's gone completely unnoticed, but it is undoubtedly a victory for animal rights. The LD_{50} test will soon be abandoned by the leading industrial countries. In each test at least 20 animals receive escalating doses until half are dead. Less drastic tests were agreed on and the aim is to find tests that betray a chemical's toxicity even before it causes pain.
>
> *New Scientist*, 9 December 2000, 3, 6

In the way just described it is possible to calculate a dose-related probability for an individual to catch a tumour. Finally the question has to be answered as to what probability is acceptable. To begin with, it is assumed that the tumour is indeed lethal. Then the argument goes as follows.

The average lifetime of a human being is less than 100 years, therefore the lifetime death probability is in the order 10^{-2}. The extra death probability induced by society should be considerably lower. The annual chance of death for a healthy young male is in the order of 10^{-4}. It is generally accepted that

*One may show that for low doses the curve should be linear [133]. But curve 3 is also linear for very small concentrations, so this argument does not distinguish between curves 3 and 4. Note that the precise interpretation of Figure 11.3 (right) should be the extra risk of cancer, induced by the chemical compound or the radiation, in addition to the 'natural' incidence.

†A resource letter on effects of ionizing radiation at low doses can be found at http://phys4.harvard.edu/~wilson/resource_letter.html.

the individual death risk from a certain technological, chemical[‡] or radiation exposure should be definitely lower than this number, so 10^{-6} is taken [132].

11.2.2 *Accidents*

> Besides animal experiments, accidents in human society may be used to deduce the damage certain compounds or levels of radioactivity will do to humans. For the survivors of the A-bomb explosions on Hiroshima and Nagasaki in 1945 one has followed their health until present. For the category which received dose levels of 100 mGy or lower no excess cancer incidence has been found. The advised upper dose limit is put as 1 [mSv yr^{-1}], (our Table 5.5)
>
> *Pugwash Newsletter*, June 2000, 43

Up to this point we have discussed health effects for 'normal use'. But accidents may happen and do happen. It is often human nature which is at the heart of accidents such as people who are not obeying safety instructions or managers who, for financial reasons, do not want to comply with them. For the workers where the accident occurs, or people nearby, the health consequences may be disastrous. For people further away dose-response relations such as those discussed above may help to calculate the consequences.

11.3 *ENVIRONMENTAL POLICY*

The market by itself will not force industries or individuals to follow a behaviour that is environmentally benign. Industries will be afraid to lose competition with their competitors who do not spend money to reduce emissions, and individuals tend to leave unpleasant measures to their neighbours*. Therefore, governments are developing environmental policies, which include regulations on emissions and sometimes (eco-) taxes on environmentally unfriendly activities. Increasingly, environmental policies are a subject of international agreements.

The need for regulations and their enforcing may be illustrated by a variant of the tragedy of the commons ([134], p. 91). If an individual has some waste to get rid of, he may throw it in a river or burn it in the atmosphere. The value of the environment may deteriorate a little bit by this individual action, but for the individual it would be more costly to purify the waste before release. This is rational behaviour for an individual, but when everybody is doing it,

‡For medical applications and therapies, the situation is different. There the risk of applying a therapy has to be weighed against its possible benefit

*In popular literature this is called NIMBY (= Not In My Back Yard)

> The best outcome of negotiations on the World Trade Organization would be to start to build other international structures, needed to ensure that freer trade and faster growth are not achieved at the expense of the environment nor by allowing the strong to dictate terms to the weak. That sounds like the beginning of global government but that surely is where we are heading.
>
> *New Scientist*, 4 December 1999, 3

the gross result may be disastrous. The individual will not change his behaviour voluntarily, unless he knows that others are doing likewise. This is only achievable by agreements or regulations, which ultimately should be enforceable.

Environmental policy tries to maintain or improve the quality of the natural environment. It looks at the consequences of activities to human health, it also aims at keeping the environment in good condition for future agriculture or industry. This, by its own right, poses an upper limit to the changes and the rate of change with which humans interfere in natural processes.

> Environmental Protection as practised today is unsustainable. The balance should shift from the monitoring and control of pollutants and wastes to a new focus based on resource use. A ten-fold increase in energy and resource productivities should be the strategic goal. Already industrialised societies should achieve this while halving total resource flows.
>
> [135]

The environmental impact E_{imp} of human activities already may be so large that the GNP, defined in Section 11.1 no longer is a good measure for the well-being of people. Because, in the GNP, the costs of cleaning up polluted land are also counted as positive. Although it is not easy to define, a term like Net Economic Welfare, NEW, may be a better indicator of the 'wealth of nations' than the GNP. The NEW would subtract hidden ecological costs and perhaps social costs as well ([128], p. 195).

In a qualitative way, the ecological costs may be described by the environmental impact E_{imp} of human activities, which may be written as

$$E_{imp} = \frac{\text{impact}}{\text{unit of resource}} \times \frac{\text{units of resource}}{\text{person}} \times \text{number of persons} \qquad (11.1)$$

Let us look at this equation from the right to the left ([134], Chapter 20). The *number of persons* increases with the population growth, shown in Figure 11.1, and is difficult to control. Certain churches and social movements are opposed to any family planning, so in practice one has to wait until the industrialized society 'by itself' leads to smaller families.

The resource use per person, the middle term in Equation (11.1), is seen as a sign of affluence. Putting in limitations on average affluence by high taxes is difficult to achieve in a democracy where the bulk of the population will always strive after the lifestyle of the 'happy few' with their big spending. On a global scale, the question is being asked whether the European Union is able to develop a specific, European lifestyle or whether the 'American way of life' will be copied as a model ([136], p.5).

The implication is seen from Figure 11.2. A European lifestyle only makes sense when the resource use per person remains much smaller than in the USA. And would developing countries develop their own, perhaps environmentally friendly, lifestyle? If so, these countries have to decide for themselves. (Reference [135] proposes drastic measures, especially in the industrialized world).

> In the United States, per capita GNP rose by 49% during 1976–1998, whereas per capita 'genuine progress' (the economy's output with environmental and social costs subtracted and added weight is given to education, health, etc.) declined by 30%.
>
> *Science* **287**, 31 March 2000, 2419

As the reduction of resource use per person is hard to achieve by taxation or similar means, governments are turning to technology. We have seen that it is possible to restrict the resource use without sacrificing lifestyle too much. Putting in renewable energies instead of fossil fuels, for example, does reduce the use of exhaustible energy per person. For the time being government subsidies have to support this, which ultimately is taxpayer's money. Therefore, indirectly, money is reallocated from direct consumption to investments in solar energy installations, for example.

Another point is that new technologies like the Personal Computer and internet may change the working life of people. Doing work from home reduces the need for commuting between home and the working place. If the liberated time is used for driving to a recreation centre, no resources will be saved. So it remains to be seen how this will work out.

As a last example of technological improvements we mention the increasing energy efficiency of car engines and car designs. The growing use of air conditioning and other electric devices in cars, however, cancels this effect. We therefore believe that the market alone will not act in an environmentally friendly way and that government measures, including direct or indirect taxation are necessary if one wants to reduce the resource use per person.

> We are suffering from the actions of industrial nations who are employing delaying tactics rather than trying to help us. In the West you are spending millions of dollars a year to protect endangered species. Soon, in the South Pacific Kiribati Islands, we will be an endangered species too.
>
> *New Scientist*, 12 February 2000, 44–47

Most governments avoid difficult controversial questions and focus on the first term in Equation (11.1): the *environmental impact per unit use of resources*. Here, technology should help by reducing emissions. The second term in Equation (11.1) may be reduced by improving the efficiencies with which resources are used. However, as noted before, people may spend this gain on more consumption, i.e. more resource use. In the end, the middle term of Equation (11.1) will not decrease.

The official policy objective of many governments is a *sustainable development*. We will discuss this concept first. Then we turn to environmental policy on a global scale, where we take as an example the UN Framework Convention on Climate Change with the Kyoto Protocols.

11.3.1 Sustainable development

The basic definition of this concept dates from the Brundtland report ([137], p. 87; [134], pp 3, 145) which says:

'. . . *sustainable development* = development that meets the needs of the present without compromising the ability of future generations to meet their own needs.'

The difficulty with this definition is the concept of 'need' that will reflect something like the essential resource use per person, the middle term of Equation (11.1). The question is what is an essential or basic human need? Most people believe that the essential minimum is not much lower than the average of the society they happen to live in. Given the enormous differences in energy use on a global scale shown in Figure 11.2, this will be difficult to agree on. If one would take the US data as the standard for all people and all generations, technology will have difficulty in complying.

> The transition to sustainability requires active co-operation between business and environmental science. Needed is a scientifically based vision of a 'sustainable' business enterprise. Many environmental changes are foreseeable. Environmental science can help to quantify the relative merits of alternative courses of action. Virtually every major fishery in the world has been drastically degraded, some irreparably. If complete scientific knowledge had been available to the industry at the outset, would it have regulated itself to prevent the overfishing that took place? Not without strongly enforced legislation. Major corporations, committed to long term involvement support such legislation.
>
> *Nature* **403**, 20 January 2000, 245

The Brundtland definition obviously has to be taken as the starting point of a discussion and indeed it has worked out that way. Australian environmental groups, for example, specified a few of the points, saying that ([134], p. 5):

- conservation of bio-diversity should be a fundamental constraint on all economic activity. The non-evolutionary loss of species and of genetic diversity needs to be halted.
- natural capital (a healthy and clean environment) must be maintained from one generation to the next. Only that income which can be sustained indefinitely, taking account of the bio-diversity principle, should be taken.
- policy decisions should err on the side of caution, placing the burden of proof on technological and industrial developments to demonstrate that they are ecologically sustainable.

> No country in the world has been able to de-link its economy from the generation of carbon dioxide because all countries continue to depend on fossil fuels. Therefore, as long as the world remains bound to fossil fuels, poor countries need the maximum 'environmental space' for their future economic growth. Keeping this in mind, the government of India has proposed '*equitable per capita entitlements*' to the atmosphere as a principle for trading credits.
>
> *CSE*, 15 April 2000

Major international environmental groups put similar aims in their policy statements, but add a paragraph on personal attitudes and practices, saying ([134], p. 8):

- people must re-examine their values and alter their behaviour in order to adopt the ethic for living in a sustainable way. Society must promote values that support the new ethic and discourage those that are incompatible with a sustainable way of life.

> George Bush, USA president (1992): The American lifestyle is not negotiable.
>
> [134], 217

This statement addresses the resource use per person in Equation (11.1). So, this point is on the political agenda. As always, governments are acting cautiously and slowly, as they are under pressure from business interests and the preferences of the population. Environmental groups act as a countervailing power and all struggle in the political arena where, in democracies, the decisions are taken. A national government, supported by parliament, will put decisions into practice. In Europe, the European Union has a strong regulatory power in environmental matters, which national governments have to accept, so the checks and balances operate on the level of the EU. On a global scale, regulations work differently. Let us look at the example of climate change.

11.3.2 *Climate change*

We have two stories: one optimistic and one pessimistic.

Abating the ozone hole

We start with the good news: the protection of the ozone layer. In Section 9.4 we mentioned the discovery of the ozone hole in 1979 and the analysis that it originated from the CFCs, released as a consequence of human activities. In 1985 an international agreement was reached in Vienna, which was extended by the so-called Montreal Protocol in 1987*.

The Montreal Protocol requires parties to gradually phase out the production and use of CFCs. The developing countries were allowed a time delay until the beginning of the 21st century. In practicc, CFCs might bc rcplaccd by less harmful HCFCs. These in turn should be phased out between 2020 for developed countries and 2040 for developing countries. The latter countries are supported by a fund from which, between 1991 and 1999, around 10^9 dollars was spent.

> Trade sanctions are a very inegalitarian tool. They can be used only by the rich countries against the poor countries. If Bangla Desh were to impose trade sanctions against the world's biggest polluter of greenhouse gases (the USA) which are causing global warming and which are predicted to drown nearly a third of the country in the next century, everybody would laugh at the attempt.
>
> *CSE*, 30 Dec 1999

Countries seem to obey the obligations from the Montreal Protocol, perhaps because the hole is undoubtedly real, is measured continuously and may be watched in its development on the internet†. Another reason may be the public fear of skin cancer, which makes international agreements easier to accept. From the fact that the Montreal protocol is put into practice one would expect the ozone hole to gradually disappear with the CFCs in the environment. Unfortunately, there are indications that the global warming of the lower atmosphere introduces a slight cooling of the higher ozone region at

*For the most recent information the international ozone secretariat can be reached by http://www.unep.org/ozone. Older information we took from [112].
†http://earthobservatory.nasa.gov

25 [km] of altitude [138]. If that is right it may take longer before the ozone concentration returns to pre-industrial values. Nevertheless, the Montreal Protocol must be seen as one of the successes of international environmental policy.

Reducing climate change

The ozone hole can be measured accurately and the resulting increase in harmful UV radiation can be calculated accurately. Climate change is a more complicated concept. For one thing, from history it is known that climate is changing all the time due to natural causes. One cannot calculate the natural fluctuations, as one does not understand them in detail. Therefore, it is inherently difficult to separate human-induced change from natural causes. Also, reducing climate change requires measures which will interfere deeply in human society, including all three terms in Equation (11.1). It is no surprise that slowing down the rate of climate change is meeting much more opposition than the fight against the ozone hole. Let us review the situation.

> BP Amoco, the world's second largest oil company has bought a large solar energy company. It must be remembered that in the late 70's when the oil crisis was hitting the world, oil companies were also buying solar energy companies. Their interest was to control these emerging technologies and to kill their growth.
>
> Centre for Science and the Environment, India, press release 13 May 1999

Following alerts by scientists, the World Meteorological Organization and the United Nations Environmental programme jointly established the Intergovernmental Panel on Climate Change in 1988. This panel published lengthy and detailed assessments on climate change in 1990, 1995 and 2001 and is expected to continue to do so‡. The panel is using sophisticated models of which the concepts were discussed in Chapter 3 of this book. The results of the first assessment confirmed that there was reason to worry, which in 1992 led to the adoption of the United Nations Framework Convention on Climate Change. Subsequent negotiations resulted in the Kyoto Protocol of December 1997 [139].

Kyoto Protocol By the Kyoto Protocol the industrialised world accepted reductions in greenhouse gases leading to a level of 5.2 percent below 1990 levels in the year 2012. The percentage differs from country to country, but is on average 8 percent for the EU countries, 7 percent for the USA and 6 percent

‡The history and the latest publications can be found at http://www.ipcc.ch/

for Japan and Canada*. The Kyoto Protocol defines greenhouse gases as the gases which are indicated with an asterisk (*) in Table 3.1. In discussing Table 3.1 we noted that the different greenhouse gases have different Global Warming Potentials (the effect relative to carbon). These potentials depend on the time horizon that one is considering. This means that the future effect of a greenhouse gas depends on the particular gas and on the time since the gas was emitted. The Protocol leaves the determination of this time to future discussions.

> Because the United States will probably be unable to achieve all of its target reductions at home, it can fully comply with its Kyoto obligations only by relying on the protocol's flexibility mechanisms. The most attractive trading partners for the United States are Russia and the Ukraine, to which the protocol bequeathed a bonanza of greenhouse-gas emission allowances that they won't need to use at home. This bonanza is called 'hot air'.
>
> *Nature* **402**, 18 November 1999, 233–34

The Protocol accepts that developing countries have to catch up in their development. So, for the time being, no quantitative ceilings for these countries have been determined. Countries that are undergoing the process of transition to a market economy, such as Russia and the Ukraine form a third category. Some of them, including the two mentioned, are allowed higher ceilings for the year 2012.

The most straightforward way of reducing emission, would be to do it at home. That is not only painful, but as some believe, also sub-optimal. For with a certain amount of money available, the reduction of emissions in an industrialized country will be much less than in a developing country. The reason is that the installations in the developed world usually have higher energy efficiency than in the developing world. In virtue of this argument some mechanisms are being developed that distribute the pain and may be more economic. This roughly will work as follows.

- *Joint Implementation* (JI). Developed countries may try to achieve their objectives jointly. That must lead to additional savings above what would be achieved otherwise.

- *Clean Development Mechanism* (CDM). Developed countries may invest in energy savings in developing countries. The emissions saved there are seen as 'credits' to the investing country.

- *Emissions Trading*. Countries not able or not willing to achieve reductions, may 'buy' the reductions from other countries in the form of tradeable emission credits.

*The student may look up the number for his or her own country in [139].

> The US State department is selling the Kyoto Protocol to the Senate as 'economically effective'. The Clean Development Mechanism (CDM) would imply that emissions can be bought for 3 $ to 20 $ for each tonne of carbon saved. Taking measures 'at home' would cost $ 125. So, using CDM would mean investing further in fossil fuel, locking out the cleaner energy sources like solar and wind. This will actually push the world towards an ecological catastrophe.
>
> CSE, 17 March 2000

> In Israel all new buildings are required by law to have a solar installation for hot water. Using solar heating for household hot water alone saves over 5 percent on the country's electricity consumption. The government's target is to obtain the cheapest solar [J] throughout the lifetime (the guarantee period) of the solar water heater.
>
> *Renewable Energy World*, March/April 2000, 92–99

Note that these mechanisms will require accurate bookkeeping of all greenhouse emissions which should take into account their global warming potentials. Also, an international body of inspectors should check the accounts carefully and systematically, otherwise everybody will suspect the neighbouring country of cheating. These technical problems may be solved. There are, however, more fundamental criticisms.

The Kyoto Protocol has come under criticism from two sides. The government of India, for example, does not accept the small reduction of the 1990 'status quo' as a final objective in 2012 [140]. They want 'equal breathing space for Indians and others' or more precisely 'equitable per capita' rights of emissions into the atmosphere as starting point.

Selling emission rights from developing countries to high-income countries, who then would hardly reduce emissions, would mean that after 10 or 20 years, in the next round of negotiations, the developing countries would have a hard time to reduce further. When the calculations leading to Figure 3.14 are correct, the later reductions have to go much further than has been written down in the Kyoto Protocol. By that time the still more powerful developed world could easily spend money on their further reductions, while the developed world will have to pay heavily for extra reductions from a lower income. This point of view is supported by scientists from the United States [141].

On the other hand, the US senate does not accept any limitations in US emissions, when countries like India and China go free without any commitments themselves. They do not want to ratify the Protocol, which means that it will not become legally binding to any state.† Added to that, many Americans do not want to accept expensive measures, when it has not been shown beyond doubt that global change is real and has disastrous consequences.

†This is the way in which the protocol is formulated

These doubts are reinforced by well-written publications, which are obtainable free of charge [142].

> The Kyoto agenda shuffled along at the recent climate change meeting in Bonn (1999), but climate change mitigation was still not in sight. Governments at the Bonn meeting were carefully negotiating ways to wriggle out of commitments to control climate change.
>
> CSE, 15 Dec 1999

Our own position is that the evidence that human induced global change is real, accumulates and becomes increasingly convincing. For policy decisions it is relevant to know that when global change is shown 'beyond any reasonable doubt', it is too late to take adequate measures. For the heat capacity of the oceans will continue to heat up the atmosphere for a long time even when zero emissions would be achieved today (Section 3.2). So, we are in favour of the Kyoto Protocol and its successors.

We also accept the position taken by the Indian government, among others, that its population has the same rights as other earth inhabitants. It is unrealistic to believe that differences in wealth and power will disappear soon, if at all. Therefore we chose for a society in which the person who directly or indirectly spoils the atmosphere with greenhouse gases should pay for it. We therefore end up with a 'greenhouse tax' on all products and services. The polluter should pay, we believe. Not very realistic at present perhaps, but yet, it is our position.

In the mean time the discussions on the implementation of the Kyoto Protocol continue and, up to the time of writing, have not led to practical agreements. By July 2001 the main stumbling blocks were [143]:

- should countries gain carbon credits for 'watching their trees grow'?
- should countries have to reduce their CO_2 emissions primarily at home?
- is nuclear power an acceptable 'green' technology?
- who pays for developing countries to adopt green energy policies?
- what penalties should there be for breaching the protocol?

11.4 THINKING ABOUT THE ENVIRONMENT AND HUMAN LIFE

In the definition of sustainable development given earlier, a few ideas were taken for granted: the position of man, the feasibility of management and the political power relations. Let us look at them and quote the definition once more:

Sustainable development = development that meets the needs of the present without compromising the ability of future generations to meet their own needs.

11.4.1 The central position of man

The 'needs of the present' and the 'future generations' almost without a thought have been taken as referring to man, the collective of human beings. This is not trivial. There are people who take the position that animals and plants have an intrinsic value. It would be acceptable for them for natural evolution to change the species, but man should not substitute deer for elephants or a stretch of grass for an old tree ([134], Chapter 14). Indeed, there are cities* where the local government has to give permission for the felling of every tree and where the local population is active in trying to 'save' any tree in their neighbourhood.

The present authors take the position that man is more important than plants or animals. We also believe that the environment should be treated with respect, as it has a value by itself. This position would require careful environmental management, by which decisions like the cutting of a tree indeed have to be weighed against the benefits.

> It is in the interests of Western societies to help protect indigenous cultures. By 1994, a pharmacist had identified over 119 plant-derived substances that are used globally as drugs. Many come from plants used in traditional medicine.
>
> *Science*, **287**, 7 January 2000, 44–45

This position is probably close to that of the advocates of sustainable development. It presupposes that environmental management is feasible, which brings us to our next point.

11.4.2 The feasibility of environmental management

Management presupposes a good understanding of the relation between cause and effect. In ordinary life much is based on experience. Mankind is now interfering with nature on a scale and with a rate of change that is unprecedented. One therefore needs computer models by which to check the consequences of industrial or agricultural developments and the results of government measures.

*An example is the city of Amsterdam, where the authors are working. Environmental protection up to this micro scale is widespread.

The sand pile metaphor

Is it possible to model the natural environment? Nature is, we believe, to be seen as a complex system, a system whose properties cannot be fully explained by its component parts. It therefore will be difficult, if not impossible, to predict accurately what will happen if one disturbs the system in a certain way. Accurate predictions and fully-fledged management certainly will be impossible if the natural environment may be regarded as a *self-organized critical state*. This can be explained by the famous metaphor of the sand pile, which we show in Figure 11.4 [144].

> Romans used to build huge aquaducts that ran for tens of miles to bring water to their settlements.
>
> These aquaducts represent the utter stupidity of the Romans. Rome was built on the river Tiber. The city did not need an aquaduct. But the Romans polluted the river and then had to bring water from far-off places. As a result, water outlets were few, and the elite swiped off most of it using a system of slaves. Traditional Japanese never threw their muck into a river. They would use a dry bin for human excreta and then use it on the fields.
>
> CSE, 15 February 2000

On the left of Figure 11.4 one observes the beginning of a sand pile. After some time one gets the situation on the right. Any addition to the pile may start an avalanche in which the complete pile participates. Adding grains will on average keep the pile in, what is called a critical self-organized state. It is *self-organized* as all interactions within the pile were gradually building up while the pile was formed – and *critical* as any minute change may produce the maximum effect. Experiments with grains of rice, which may be marked more easily than grains of sand, have shown

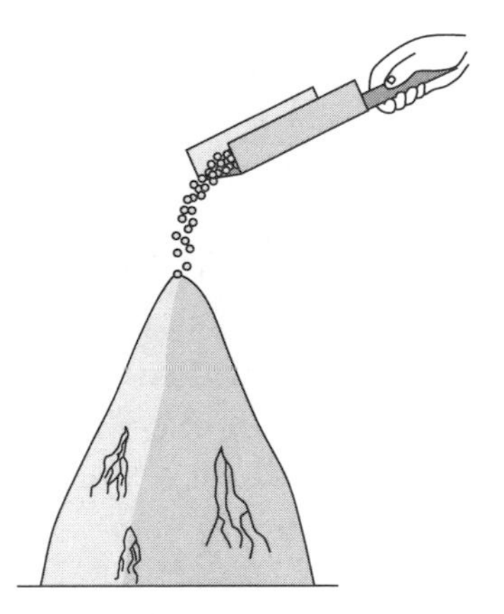

Figure 11.4 The metaphor of the sand pile. On the left the pile is being formed. The interactions between the grains are organizing it to the critical self-organized state on the right. There, any minute addition of a grain may result in an avalanche in which all grains participate [144]

that a grain far from the start of the avalanche may be affected and follow an unpredictable course before leaving the pile.

The message should not be that modelling is nonsense. In that case we should not have written this textbook. Many aspects, such as the transport of pollutants in a river after a calamity can be modelled with quite accurate results. The reason why this modelling is reliable is that, in this case, cause and effect are clear and may be separated from other variables in nature.

Modelling climate change is already close to the sand pile metaphor. The message we get is that we should be careful and cautious in performing deliberate large-scale experiments with nature. Certainly we should refrain from interfering in complex cases where many variables are involved. Again the student should realize that we are giving judgements to which one may object.

> Setting a thief to catch a thief always was a dodgy proposition. And so, we suggest, is unleashing one form of pollution to neutralise another. That's the problem with an ingenious plan hatched by Australian scientists to fertilise the oceans as a way of curbing global warming. The idea is to pipe nitrogen-rich fertiliser into the sea to grow plankton that will soak up CO_2 from the water and ultimately from the atmosphere.
>
> It would happen near the coast of Chile, which country then could claim credits under the Kyoto protocol and sell them to industrial countries.
>
> *New Scientist*, **8** April 2000, 3, 18

Discounting

With due caution and on a modest scale perhaps, human society has to manage the change of the environment. That will require a weighing of costs and benefits. As a rule, the benefits during the first few years will be much larger than the costs involved. Otherwise one would not have started the process in the first place.

Difficulties in judgement arise because many of the costs will be in the future. For example, for a nuclear power station the costs of its dismantling will occur 30 to 100 years into the future. Environmental costs are more difficult to quantify. But it is clear that in cutting down a forest for timber, the foresters will reap the benefits immediately. The costs in terms of erosion of the land or even in terms of climate change have to be paid in the future.

The way in which economists deal with future costs (or benefits) is the method of *discounting*, which we explain in the simplest possible way. Assume that a cost of F has to be paid after t years. In order to have the money available by then it is sufficient to put an amount of money P in the bank now, add the interest to it at the end of the first year and leave it in the bank, drawing a little more interest in the second year. This compounding of

> Waste is a by-product of our way of life in the industrial countries. Its volume is growing all the time, to say nothing of its complexity or toxicity. In today's Europe, waste is being generated at a rate of over 1 [kg] per person per day! The way we manage this is often outdated and represents a serious health and environmental hazard that can only be reduced by drastic changes in production methods and consumption habits.
>
> EU, Environmental Waste fact sheet, p. 1

> The polluters would have us believe that we are all just common travellers on 'Spaceship Earth', when in fact a few of them are at the controls and the rest of us are choking on their exhaust.
>
> [134], p. 286

interest goes on for t years, at which time the capital should have grown to the required F. The relation between the *present value* P and the *future value* F will be

$$F = (1 + i)^t P \qquad (11.2)$$

The factor in front of the present value P strongly depends on the values taken for the interest rate i and the length of time t. With $i = 0.08$ and $t = 25$ years one finds a factor of 6.85. In reality one has complications like inflation, but Equation (11.2) demonstrates that the method of discounting reduces the weight of future costs. Some people say that it discriminates against future generations ([134], p. 55).

One could ask other questions as well. Is the money P in fact put in a bank under government control. Is the interest rate i realistic or taken too low. It is even more complicated when the time t is very long. Such is the case with dismantling nuclear power stations (100 years after the end-of-life) or with nuclear waste (thousands of years). Will the bank in which the money was put still exist or will there be a reliable government guaranteeing the loans? The answers one will give here are decisive for the kind of environmental management that results.

Politics

In any form of environmental management the questions we pose in this chapter have to be answered, if not explicitly, then implicitly. The decisions to be taken, the choices to be made, are part of the political process.

We do not believe that in this world a few people are in control, who are spoiling our environment with disregard for other people and for their own descendants. We represent the view that the dominant culture of our society is to blame. That culture does not want to question the resource use per person of Equation (11.1). It avoids difficult decisions and is geared toward individual needs much less than social needs. It is true that interest groups, which profit

> The US city of Seattle will adopt renewable energy and energy efficiency measures to reduce its greenhouse gas (GHG) emissions to zero. The city Council has unanimously adopted a proposal to meet future electricity needs with no net emissions of GHG, by relying on existing hydropower and new wind, geothermal, solar and landfill gas facilities, together with energy conservation measures. Seattle politicians hope the commitment will challenge other communities to tackle global warming.
>
> *Renewable Energy World*, May/June 2000, 17

from this dominant culture, are trying to keep the culture going. They are using all means, from commercials to scientific publications*.

There are countervailing powers, fortunately. Environmental groups are participating in the discussions, presenting well-thought alternatives. Social groups, not only intellectuals in high-income countries, but also groups in low-income countries are questioning our western way of life. We admit that power relations have to be changed. That can only be a solution if we are prepared to question our dominant culture and our way-of-life at the same time. This discussion should be an intrinsic part of the political process as well.

11.5 THE RESPONSIBILITY OF THE SCIENTIST

> Sound science is about the best possible way to answer a given question; to present with rigour the certainties and uncertainties of knowledge, and the assumptions underlying certain conclusions. But, crucially, it is not a method for deciding which questions should be posed, or for determining the acceptable risks and desirable benefits of technologies. The public should be involved in the formulation of strategies, rather than merely being consulted on drafted proposals.
>
> *Nature* **400**, 5 August 1999, 499

Students who have studied this book will be interested in environmental problems and they may choose a job in which the knowledge within this book – and of more advanced texts – can be used. They will find that a solution, which may seem sound from a scientific perspective, is not accepted by the organization they are working for, or even not by the public at large. Still, the natural scientist with his or her expert knowledge has an essential role to play in the public discussions [145], [146]. Interaction with the public means that the scientist must be open to a possible

*The sympathy of the authors is with the environmental groups. We do not give references to interest groups that want to maintain the status quo. It might be unfair to discredit people with a few words.

change in the problem definition or to seemingly sub-optimal solutions. To act in a responsible way means that we have to know a little about the nature of science and its control. We will sketch our views in a few broad strokes, starting from a historical perspective.

11.5.1 *The nature of science*

Science, in the strict sense of natural science, has an undisputed body of knowledge. Its data are published in ever expanding tables of physical and chemical constants, which any scientist uses without question. The conductivity of a piece of copper wire at a certain temperature and pressure is the same everywhere.

Also, experimental methods are rigorous. Students are trained in students' labs on careful experimental set-ups and a precise control of the experimental circumstances. The experiments aim at measuring a certain property (conductivity of copper for example) or at testing a certain theory (the occurrence of a certain gas in the solar atmosphere, for example). During an experiment, they will carefully vary a single parameter while keeping the others constant. They know that by applying the methods in this way, they will get objective and reproducible results, even when the experiments are performed at the other side of the earth.

The method of science becomes less straightforward when the phenomenon to be studied cannot be isolated from the rest of nature and cannot be put on a laboratory table. Climate is an example in case. One can study the composite gases under controlled circumstances in the laboratory, but one cannot reproduce an atmosphere of 100 [km] of thickness with all kinds of motions horizontally and vertically. Here, one has to rely on computer models with many variables, which are checked against their accuracy in predicting past behaviour of the atmosphere and the oceans. Scientific judgements and considered opinions form part of the final assessment of the outcome of the modelling, which is the forecasted climate change.

> One of the main authors of the 1995 IPCC report was wrongly accused of doctoring the final report and removing references to uncertainties about whether humans were warming the planet. This author was accused of 'scientific cleansing' by the Global Climate Coalition, a body which represents the interests of the American oil and car companies.
>
> *New Scientist*, **4** March 2000, 43

We met something similar in discussing Figure 11.3 on dose-effect relations. As the effect of very small doses may be smaller than the natural occurrence of cancers, one cannot measure them directly. Scientists would say that they disappear in the noise. So, one has to rely on considered scientific opinion as to the health safety of certain chemicals in the environment. It is the responsibility of the scientist to share his or her arguments with the public, and, if necessary, accept a sub-optimal outcome of the debate.

11.5.2 *The control of science*

> One in three UK scientists working in government or recently privatized laboratories has been asked to alter research findings. Reasons were: to suit the customer's preferred outcome, to obtain further contracts or to prevent publication.
>
> *Nature* **403**, 17 February 2000, 689

The previous two examples (climate and low doses) make it clear that the results of considered opinions have great social consequences. If climate is changing rapidly with disastrous consequences because of human interference, expensive measures have to be taken; and if a low dose of certain chemicals or radioactive materials is harmless, it will save the industry a lot of money on emission controls. So a question arises about the independence and objectivity of scientific judgements. Namely, who decides on the scientific questions to ask and puts up the money to find the answers.

From the 17th to 19th century, scientists of independent means carried out most scientific research. The 19th century saw the rise of university laboratories where scientific research was funded by a government that accepted the academic freedom of the university professors in doing their jobs. In the 20th century the military implications of science became manifest and, at the same time, scientific research became increasingly expensive.

> Pledge initiated by Student Pugwash, USA. I promise to work for a better world, where science and technology are used in socially responsible ways. I will not use my education for any purpose intended to harm human beings or the environment. Throughout my career, I will consider the ethical implications of my work before I take action. While demands placed upon me may be great, I sign this declaration because I recognize that individual responsibility is the first step on the path to peace.
>
> *Science* **286**, 19 November 1999, 1475

Many academic scientists deplored the military use of science and took a pledge not to participate in it. Later on this was extended to all harmful environmental applications of science [147], [148]. Scientists who take one of these pledges, accept individual responsibility for their actions. We see this as an admirable first step, but not more. For it is not always clear what application is harmful. Also, there may be disagreements on the harm of applications, arising from differing value judgements. Finally, when an individual scientist is under pressure from superiors or financiers, it is difficult to 'keep his or her back straight'. So, in our opinion, pledges have to be backed by professional organizations, preferably an organization with a very broad membership.

The high cost of science resulted in high government expenditure on science and, towards the end of the 20th century, in an increasing demand for social relevance. In other words, the taxpayers, represented by parliament and government require value for money. Added to this, university science is supplementing its income by contract work. This happens for industry, for government and for public interest groups who want contra-expertise to combat government decisions.

Academic science and science in large government funded institutions therefore resembles industrial research. In both cases the influence is top-down. The top, consisting of a small selection of academic or industrial scientists, decides where the money flows and individual scientists do not have much choice but to follow.

> The role of science – with all its attendant uncertainties – must be to illuminate political choices, not enforce them. Like Winston Churchill said: Scientists should be on tap, not on top.
>
> *Nature*, 6 January 2000, 6

11.5.3 A new social contract between science and society

This leaves us with the following groups:

- a public that wants science and technology to reduce the first two factors of Equation (11.1), the environmental impact of the use of resources and the resource use per person, without sacrificing (much) of its standard of living. The public knows that wishful thinking is not enough, so it accepts the need for reliable knowledge
- interest groups that want scientific evidence to support their views (industry, government, public interest groups)
- individual scientists, who just want to match their intelligence and skills against nature or perhaps only want to make a decent living
- leaders of research groups who spend a lot of time obtaining money from all kinds of sources
- professional societies who want to maintain the integrity of their profession

> Are the public any good in judging the economic and health aspects of science? Yes, say the British Lords. Given the full story, people use their common sense to make sophisticated judgements about science.
>
> Openness in these matters is the only way forward. As the US has found out about gene therapy and Britain about the whole of science, anything less than absolute frankness will lead to disaster.
>
> *New Scientist*, 18 March 2000, 3
>
> It is vital that parliamentarians obtain and use unbiased information on technical matters and on people's concerns. The provision of unbiased scientific information, including an honest admission of the uncertainties involved, will be recognized as an essential tool in the process of scrutinizing governments. Although it is possible to prevent bad decisions being made on the basis of bad science, good science does not necessarily present an obvious political pathway. The role of science is essential but not necessarily definitive.
>
> *Nature* **403**, 27 January 2000, 357–359

We therefore support the view that a new contract between science and society should be signed, metaphorically speaking [145]. Such a contract should accommodate the interests of the five groups indicated above. A contract implies an agreement, which in turn implies that all five should perhaps compromise a little bit. Let us finish this book by suggesting, like an old-fashioned schoolmaster, what the five groups should do:

- the public has to accept that there are limits to the personal use of resources and inherent limitations in the predictability of complex systems

- interest groups who pay for science have to accept that scientists will give 'fair' judgements that they will not always like

- individual scientists have to recognize the social context of their work; they would accept ethical and social considerations as natural to science. This does not mean that every scientist should deal with social questions for all of his working life

- research leaders should fight to keep part of their funding without 'strings attached'. Basic research without straightforward applications or social relevance keeps one sharp, maintains scientific standards and may lead in unexpected new directions

- professional societies should not only ban and punish scientific misconduct; they should operate on the brink between science and society. They should accept that scientific assessments might reflect values or preferences of the scientists. This means that political views may enter scientific assessments, which may result in ambiguous judgements.

> Quality control checks on the size of fuel packets in Sellafield, UK, have been faked. Workers had 'cut and pasted' data from previous checks.
>
> *New Scientist*, 26 February 2000, 18

EXERCISES AND ESSAY QUESTIONS

Some of the questions below require some argument. They may resemble more of an essay than a physics exercise. In such cases use some text to explain your views, as your teacher cannot read your thoughts. However, be concise and to the point.

> How far should environmental scientists go in interpreting their findings for policy makers? Many ecologists express deep reservations about crossing the blurry line that separates scientific meaning from social values. They worry that to the public, environmental scientists are becoming indistinguishable from environmental activists.
>
> Others however, feel that they have a moral responsibility as a citizen to make people aware of what science means.
>
> *Science* **287**, 18 February 2000, 1188–1192

11.1 Look at the appetisers of Section 11.1, or any other section. Analyse them: some may contain well thought scientific opinions, others contain value judgements. If the latter is the case, do you agree with the judgement?

11.2 Figure 11.2 contains information about a lot of countries. Look at two countries you know, perhaps your own country. Can you explain where the differences in the data originate? Can you explain the extraordinary point for Russia?

11.3 It is sometimes argued that the criteria of the Australian environmental groups [134], mentioned in Section

11.3 lead to conservatism and an under use of technology. What is your point of view?

11.4 By now, the student will know more about climate change than many politicians, albeit less than a few of them. What would be your attitude towards the Kyoto Protocol, given the uncertainties involved?

11.5 The sub-section on discounting (Equation (11.2)) ended with several questions. What would be your answers – and what would be the consequences of your position?

11.6 Consider the appetiser from Science **287**, 18 February 2000, 1188–1192 on the previous page. What would be your position in this dilemma?

Appendix A: Numerical and Physical Data

Solar 'constant'	$S = 1367\,[\mathrm{W\ m^{-2} = J\ s^{-1}\ m^{-2}}]$
Avogadro's number	$N_A = 6.022 \times 10^{23}\,[\mathrm{mol^{-1}}]$
Stefan-Boltzmann constant	$\sigma = 5.671 \times 10^{-8}\,[\mathrm{W\ m^{-2}K^{-4}}]$
Velocity of light	$c = 3.0 \times 10^{8}\,[\mathrm{m\ s^{-1}}]$
Planck's constant	$h = 6.626 \times 10^{-34}\,[\mathrm{J\ s}]$
Planck's constant/2π	$\hbar = 1.0546 \times 10^{-34}\,[\mathrm{J\ s}]$
Universal gas constant	$R = 8.3145\,[\mathrm{J\ K^{-1}\ mol^{-1}}]$
Boltzmann constant	$k = 1.38 \times 10^{-23}\,[\mathrm{J\ K^{-1}}]$
Energy content	
1 tonne coal equivalent	$0.0293 \times 10^{12}\,[\mathrm{J}]$
1 tonne oil equivalent	$0.04187 \times 10^{12}\,[\mathrm{J}]$
1 [m^3] natural gas	$0.039021 \times 10^{9}\,[\mathrm{J}]$
1 electronvolt [eV]	$1.602 \times 10^{-19}\,[\mathrm{J}]$
1 [kWh]	$3.6 \times 10^{6}\,[\mathrm{J}]$
Earth	
Radius	$R = 6.37 \times 10^{6}\,[\mathrm{m}]$
Angular velocity	$\omega = 7.292 \times 10^{-5}\,[\mathrm{s^{-1}}]$
Gravitation	$g = 9.8\,[\mathrm{m\ s^{-2}}]$
Air	
Specific gas constant dry air	$R_{air} = 287\,[\mathrm{J\ K^{-1}\ kg^{-1}}]$
Specific heat	$c_p = 1004\,[\mathrm{J\ kg^{-1}K^{-1}}]$
Density (20 [°C], 10^5 [Pa])	$\rho = 1.184\,[\mathrm{kg\ m^{-3}}]$
Density (25 [°C], 10^5 [Pa])	$\rho = 1.161\,[\mathrm{kg\ m^{-3}}]$
Kinematic viscosity	$v = 14 \times 10^{-6}\,[\mathrm{m^2\ s^{-1}}]$
1 [atm] pressure	$p = 1.01325 \times 10^{5}\,[\mathrm{Pa = N\ m^{-2}}]$
Water	
Specific heat (26 [°C])	$c_p = 4179\,[\mathrm{J\ kg^{-1}\ K^{-1}}]$
Heat of evaporation (25 [°C])	$\Delta H = 2.4 \times 10^{6}\,[\mathrm{J\ kg^{-1}}]$
Viscosity (25 [°C])	$\mu = 0.89 \times 10^{3}\,[\mathrm{Pa\ s}]$
Chemistry	
1 mole substance (mol weight *M*)	$M \times 10^{-3}\,[\mathrm{kg}]$
molar concentration	$1\ \mathrm{M} = M \times 10^{-3}\,[\mathrm{kg\ litre^{-1}}]$
Elementary particles	
mass electron	$m_e = 9.096 \times 10^{-31}\,[\mathrm{kg}]$
charge electron	$e = 1.602 \times 10^{-19}\,[\mathrm{C}]$
mass proton	$m_p = 1.6710 \times 10^{-27}\,[\mathrm{kg}]$
mass neutron	$m_n = 1.6724 \times 10^{-27}\,[\mathrm{kg}]$

Appendix B: Simple Vector Algebra

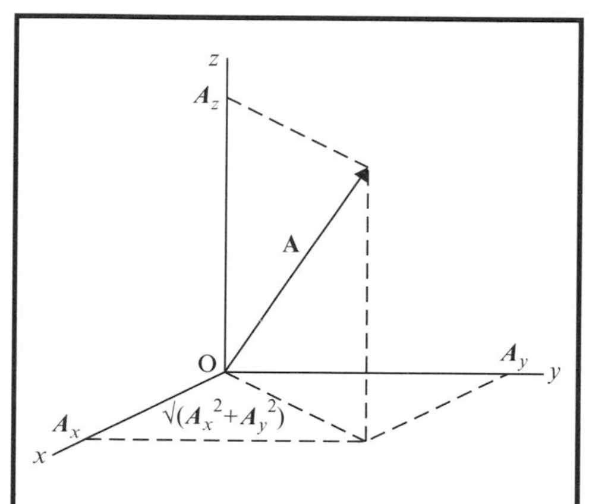

Fig. B.1 A vector **A** and its three components in an orthogonal x, y, z frame

In this Appendix we summarize a few simple vector relations, which are proven and elaborated in standard basic courses in mathematics or mathematical physics.

Vectors summarize physical entities that have a magnitude and a direction, such as a force or a velocity. These are indicated by bold printed letters **A**, **B** etc. Also the orientation relative to a fixed point in space may be interpreted as a vector **r**, such as was done in Figure 2.10. We usually consider vectors in a three-dimensional space with three orthogonal axis. They then have components (A_x, A_y, A_z) or (A_1, A_2, A_3). This is sketched in Figure B.1. In a figure or a hand written text, vectors are usually denoted by an arrow on top of their symbol, but in our figures we will use bold print.

The length of a vector **A** is indicated by the symbol itself, without an arrow or bold print, A. From Figure B.1 one will see that the length may be found by twice applying Pythagoras' rule, which gives

$$A^2 = A_x^2 + A_y^2 + A_z^2 \tag{B.1}$$

Adding or subtracting of vectors

Two vectors may be added in the form $\mathbf{C} = \mathbf{A} + \mathbf{B}$ or subtracted as $\mathbf{D} = \mathbf{A} - \mathbf{B}$. This is indicated in Figure B.2. Adding means putting the tail of **B** at the head of **A** and drawing the sum vector from the tail of **A** to the head of **B**. Subtracting can most early be remembered by first drawing $-\mathbf{B}$ as the opposite of vector **B**, as $\mathbf{B} - \mathbf{B} = 0$. Next, one adds **A** and $-\mathbf{B}$ to $\mathbf{D} = \mathbf{A} - \mathbf{B}$. The components just behave like algebraic numbers:

$$\begin{aligned} C_x &= A_x + B_x \\ C_y &= A_y + B_y \\ C_z &= A_z + B_z \end{aligned} \tag{B.2}$$

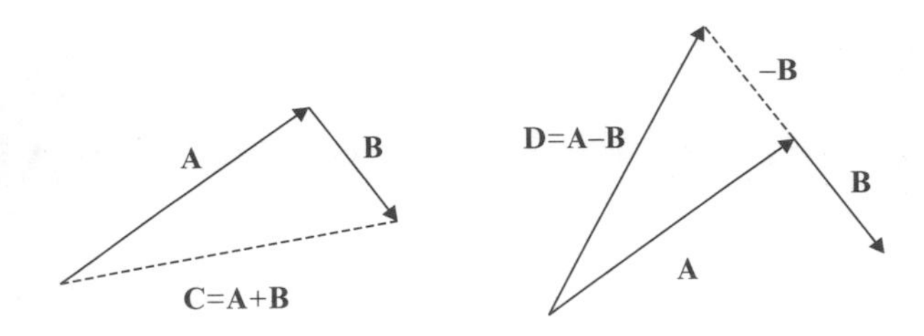

Fig B.2 Adding two vectors (left) and subtracting two vectors (right)

$$
\begin{aligned}
D_x &= A_x - B_x \\
D_y &= A_y - B_y \\
D_z &= A_z - B_z
\end{aligned} \tag{B.3}
$$

Multiplication of vectors

The scalar product

The scalar product of two vectors **A** and **B** is defined as a number $\mathbf{A.B} = A\ B \cos\alpha$, where α is the angle between the two vectors. The word 'scalar' in scalar product indicates that the result is not a vector. It may, however, be expressed in the components of **A** and **B** by

$$\mathbf{A.B} = A\ B \cos\alpha = A_x B_x + A_y B_y + A_z B_z \tag{B.4}$$

Work

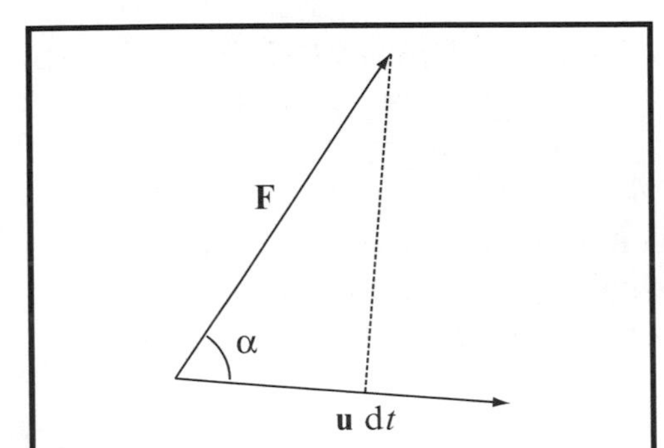

Fig B.3 The work performed by force **F** is the product of the path times the projection of **F** on the path

An example of a scalar product of two vectors is the work performed in a unit of time. Consider a particle with mass m and velocity $\mathbf{u}$. In a small time $\mathrm{d}t$ it moves over a distance $u\,\mathrm{d}t$, where u is the length of the vector $\mathbf{u}$. Let a force **F** operate on the particle. The work $\mathrm{d}W$ it performs is given by the projection $F\cos\alpha$ of the force on the particle path times the length of the path. So it becomes

$$\mathrm{d}W = F\cos\alpha\, u\,\mathrm{d}t = \mathbf{F.u}\,\mathrm{d}t \tag{B.5}$$

The work per unit of time becomes the scalar product **F.u.**

The vector product

The vector product of two vectors **A** and **B** is defined as a vector $\mathbf{C} = \mathbf{A} \times \mathbf{B}$ with the following properties

- it is perpendicular to the plane defined by **A** and **B**
- its direction is such that a screwdriver which is turned from **A** to **B** will go in the direction of C*
- the length is $C = A\ B\ \sin\alpha$, where α again is the angle between the two vectors. This, incidentally, equals the surface area of the parallelogram defined by **A** and **B**

(B.6)

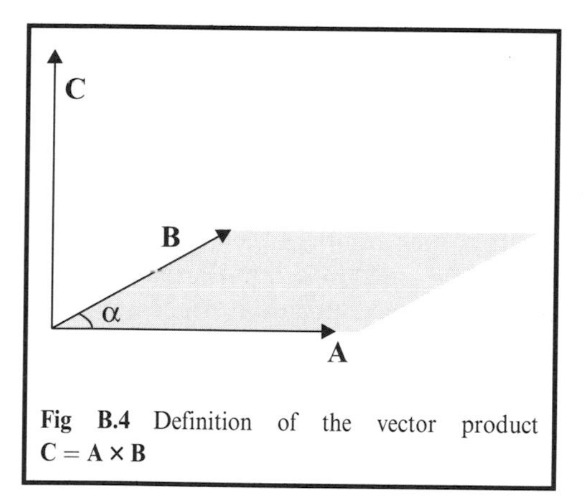

Fig B.4 Definition of the vector product $\mathbf{C} = \mathbf{A} \times \mathbf{B}$

Vector multiplication is sketched in Figure B.4. The vector **C**, being a vector (hence the name vector multiplication) has three components

$$\begin{aligned} C_x &= A_y B_z - A_z B_y \\ C_y &= A_z B_x - A_x B_z \\ C_z &= A_x B_y - A_y B_x \end{aligned} \tag{B.7}$$

This is not difficult to remember if one notes the structure of equation (B.7). In the first line one just has the order x, y, z of the alphabet. The same holds for the second and third lines when after z, one starts all over again. One calls this a cyclic permutation. Incidentally, reading along the columns in equation (B.7) one notes the cyclic structure.

From definitions (B.6) or (B.7) we note an unexpected behaviour of the vector product:

$$\mathbf{A} \times \mathbf{B} = -\mathbf{B} \times \mathbf{A} \tag{B.8}$$

So definition (B.6) implies that turning the screwdriver in the other direction drives the screw in the opposite direction, giving a minus sign. Interchange of A and B in equation (B.7) gives a minus sign as well.

*Usually this is called the corkscrew rule.

Angular Momentum

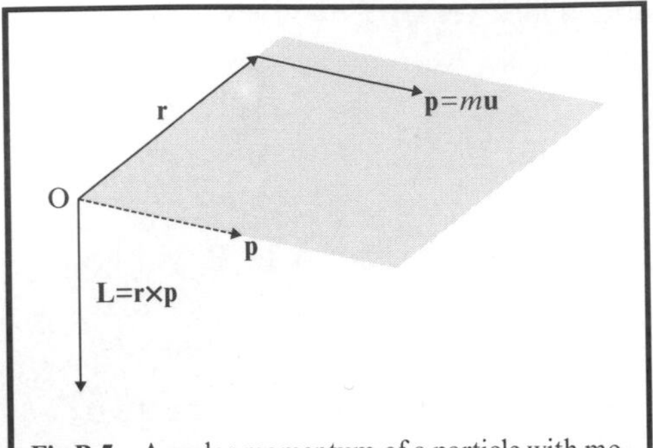

Fig B.5 Angular momentum of a particle with momentum **p** with respect to a fixed point O. To apply Figure B.3 the momentum is shifted to O

An example of the vector product, which is often encountered, is the angular momentum of a particle (or a parcel of fluid) with respect to a certain point O. This is illustrated in Figure B.5. At a certain moment the particle has a position, defined by the position vector **r** from O. Its velocity is **u** and its momentum with mass m becomes $\mathbf{p} = m\mathbf{u}$. In the figure one may move **p** to O, just for convenience. Then the vector multiplication, using the screwdriver shows that the angular momentum

$$\mathbf{L} = \mathbf{r} \times \mathbf{p} \tag{B.9}$$

points downwards.

Time differentiation

Finally, we note that in physics many quantities are dependent of time. In such a case the vectors are written as a function of time $\mathbf{A} = \mathbf{A}(\mathrm{t})$ etc. In fact, this notation also comprises three functions of time $A_x = A_x(t)$ and the same for the other two components. A vector may also be a function of position, indicated as $\mathbf{A} = \mathbf{A}(\mathbf{r}) = \mathbf{A}(x, y, z)$ or it may be a function of both position and time $\mathbf{A} = \mathbf{A}(\mathbf{r}, t) = \mathbf{A}(x, y, z, t)$.

Vector differentiation

Differentiation of a vector with respect to time is trivial. It is done precisely as a mathematical function. For example, as in equation (2.10)

$$\mathbf{u}(t) = \frac{d\mathbf{r}}{dt} = \mathrm{Lim}_{\Delta t \to 0} \frac{\Delta \mathbf{r}}{\Delta t} = \mathrm{Lim}_{\Delta t \to 0} \frac{\mathbf{r}(t + \Delta t) - \mathbf{r}(t)}{\Delta t} \tag{2.10}$$

It will be clear that one may just look at the x-component of $\mathbf{r}(t) = x(t)$, differentiate it with respect to the time and find the x-component of $\mathbf{u}(t) = u_x(t)$.

Differentiation with respect to position is somewhat more complicated. First look at the differentiation of a scalar function such as $p(x, y, z)$.

The gradient

We define the gradient of the scalar function $p(x, y, z)$ as a vector, denoted by

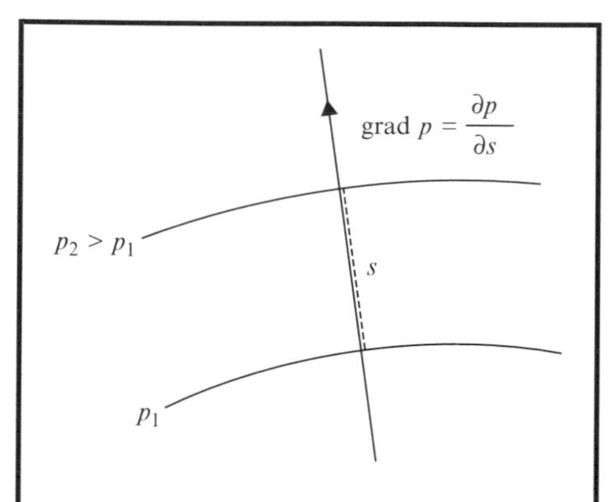

Fig B.6 Definition of grad p. It points perpendicular to the surfaces of equal p toward increasing p and measures the rate of change of p along that line

$$\text{grad } p = \nabla p = \frac{dp}{d\mathbf{r}} \tag{B.10}$$

It has the x-component

$$(\text{grad } p)_x = \frac{dp}{dx} \tag{B.11}$$

and similarly for the y- and z-components. So the three components of grad p are just the three partial derivatives, which one obtains by differentiating to one variable, keeping the others constant. For physical applications it is important to know the following properties:

- grad p points perpendicular to the surfaces with constant value of p, the iso-p surfaces
- its direction is toward increasing value of p
- its length, or magnitude, may be found by taking a length parameter s along the perpendicular line and then differentiating along the line with respect to s, so

$$|\text{grad } p| = \frac{dp}{ds} \tag{B.12}$$

This last property may be understood by choosing the x-axis along the line perpendicular to the iso-p surfaces. In that case (B.12) reduces to (B.11). So the length of grad p is the change of p per unit length along the perpendicular line.

Appendix C: Miscellaneous Sources and Websites

Below we first give the websites, which proved useful for us in preparing appetisers. We add some sites, useful for finding the most recent information on graphs and tables. Then we give information on the journals quoted or summarized.

Websites

For European policy statements we consulted the European Commission, Directorate General 11, which deals with environmental issues:

http://europa.eu.int/comm/environment/

For the European parliament we consulted:

http://www.europarl.eu.int/dg7/ftcom

For the Intergovernmental Panel on Climate Change, technical reports may be downloaded as well as summaries for policy makers:

http://www.ipcc.ch/
http://www.ipcc.ch/pub/techrep.htm

For the Centre for Science and Environment, New Delhi, India, as typical of an industrializing country we used:

http://www.oneworld.org/cse/html/dte/

The United States Environmental Protection Agency has a comprehensive website with official and citizen information included:

http://www.epa.gov/

Recent energy data may be found in the International Energy Annual:

http://www.eia.doe.gov/iea

For health effects and key papers in connection with health, consult the World Health Organization:

http://www.who.int/

The authors' department has opened a website where the radiation balance of Figure 3.7 is modelled and where the experiments of Appendix D are being described in more detail:

http://www.nat.vu.nl/EnvPhysExp

Journals

Bulletin of the Atomic Scientists, Educational Foundation for Nuclear Science, 6042 S. Kimbark Avenue, Chicago, III. 60637 USA

New Energy, Bundesverband Wind Energie e. V., Herrenteichsstr. 1, D-49074 Osnabrück, Germany.

Nature, MacMillan Magazines, 4 Crinan Street, London N1 9XW, England

New Scientist, Reed Business Information, 151 Wardour Street, London W IV 4 BN, England

Physics Today, American Institute of Physics, One Physics Ellipse, College Park, MD 20740–3843, USA

Renewable Energy World, James & James, 35–37 William Road, London NW1 3ER, England

Science, American Association for the Advancement of Science, 1200 New York Avenue, NW, Washington DC 20005, USA

Scientific American, 415 Madison Avenue, New York, NY 10017–1111, USA

Appendix D: Experiments in a Students' Lab

Below, a short description is given of a few environmental experiments that can be done at an average students' lab. More details, including a cost estimate may be found at the web http://www.nat.vu.nl/EnvPhysExp

Determine the hydraulic conductivity

Groundwater flow can be described by Darcy's law (Section 7.4). The hydraulic conductivity (k) of a sample can be experimentally determined by measuring the discharge of the fluid $Q\,[m^3 s^{-1}]$ which is needed to maintain a water level difference Δh across a vertically placed sample with length L and cross-section A.

Determine the thermal conductivity of sand

Heat transfer by means of conductivity (Section 4.1) is caused by local temperature differences in a material. Kinetic energy is transferred due to collisions between atoms. The heat flow (q'') from a higher to a lower temperature is, according to Fourier's law, a relation between the temperature gradient (∇T) and the thermal conductivity $\lambda\,[W\ m^{-1} K^{-1}]$. For a cylindrical system this becomes a one-dimensional relationship from which the thermal conductivity of the material can be determined by measuring the radial temperature profile in the steady state.

Heat transfer by radiation and convection

Heat transfer is always caused by local temperature differences (Section 4.1). Three mechanisms can be distinguished: conduction, convection and radiation. In this experiment, heat transfer (q'') by means of radiation and additional parallel convection (pressure dependent) can be investigated.

Thermal emissivity (ε) can be determined from time-dependent temperature measurements of a constantly heated metal rod (T_1) in a vacuum environment surrounded by a surface with temperature T_2. With Newton's law of cooling (4.10) the convective heat-transfer coefficient (h) can be determined from the additional heat loss of the constantly heated metal rod.

Laser Doppler Anemometry

Laser Doppler Anemometry (LDA) is a technology to measure local velocities of very tiny particles in a flow. This technique is based on measuring scattered laser light from particles that pass through a series of interference fringes (a pattern of light and dark stripes).

The great advantages of this technique are that there is no physical contact with the flow being measured so no disturbances are introduced; also a high spatial resolution can be created by focusing the two laser beams. This makes LDA a valuable measuring technique with various applications. For example, airflow measurements within combustion engines and aeroplane engines to improve fuel efficiency reduce pollution and aeroplane noise.

Radon in the environment

Radioactive Uranium can be found in soil and soil products (Section 5.2). One of the decay products is the inert gas radon. This can emanate into the surrounding environment. For health reasons it is important to determine the amount of radioactive radon gas in the environment. This can be done by measuring the time-dependent radon daughter concentrations $C(t)$. These growth curves are dependent on temperature and humidity, which therefore are monitored.

Laser remote sensing

Laser spectroscopy (Chapter 9) is a powerful method to investigate fluorescence of the photosynthetic system of green plants. Wavelength- and time-dependent fluorescence can be monitored at a distance. This results in basic information on fluorescence in general and on the process of photosynthesis in particular, which could give an indication about the health of green plants. Because of very weak signals, improvement of the signal-to-noise ratio by lock-in technique plays a very important role.

References

We number the references throughout the book in the order where they first appear. To make it easy to find where a reference is used for the first time, we indicate the chapter.

CHAPTER 1

[1] (*) Egbert Boeker and Rienk van Grondelle, *Environmental Physics*, Second Edition, John Wiley, Chichester, UK, 1999. This book contains more mathematics and more derivations than the present one. As it is meant for students in physics it is somewhat more limited in scope as it goes into more depth.

[2] (*) Thomas E. Graedel and Paul J. Crutzen, *Atmosphere, Climate and Change*, Scientific American Library, W. H. Freeman, New York, USA, 1995. This book is very readable. It has many illustrations on climate, pollution and instrumentation and contains outspoken views on required policy measures. It is useful for Chapter 2, 3, 9, 10 and 11.

[3] (*)J. L. Monteith and M. H. Unsworth, *Principles of Environmental Physics*, Second Edition, Edward Arnold, London, UK, 1990. This book focuses on agricultural applications. It contains a lot of material and has a broad scope, useful for Chapters 2 to 6. It does not question the societal context of the field.

[4] J. L. Chapman and M. J. Reiss, *Ecology: Principles and Applications*, Cambridge University Press, 1995

[5] K. V. Ellis, *Surface Water Pollution and its Control*, MacMillan, 1989

[6] R. S. Scorer, *Meteorology of Air Pollution, Implications for the Environment and its Future*, Ellis Horwood, Chichester, UK, 1990. This book contains some very beautiful photographs of smoke stacks, with a clear explanation.

[7] Gwyneth Howells, *Acid Rain and Acid Waters*, Ellis Horwood, New York, 1990

[8] Michael Treshow and Franklin K. Anderson, *Plant Stress from Air Pollution*, John Wiley & Sons, Chichester, 1991

[9] C. W. Fetter, *Contaminant Hydrogeology*, MacMillan, New York, 1993

[10] Thad Godish, *Air Quality*, Lewis, Chelsea, MI, USA, 1991

[11] John C. Fisher, *Energy Crisis in Perspective*, John Wiley, New York, 1974

[12] Martin Bennett and Peter James, eds., *Sustainable Measures, Evaluation and Reporting of Environmental and Social Performance*, Greenleaf, Sheffield, UK, 1999

[13] S. L. Valley, ed., *Handbook of Geophysics and Space Environments*, McGraw-Hill, New York, 1965

[14] Brian Easlee, *Liberation and the Aims of Science. An Essay on Obstacles to the Building of a Beautiful World*, Chatto & Windus, London, 1973

[15] G. W. C. Kaye and T. H. Laby, *Tables of Physical and Chemical Constants*, 16th edition, Longman, Burnt Mill, UK 1995

[16] Gerhard Zuba, Air Pollution Modeling in complex terrain, in: *Computer Science for Environmental Protection*, eds. M. Hälker and A. Jaeschke, Springer, Berlin, 1991, p. 375–84
[17] ESCOBA, *A European Multidisciplinary Study of the Global Carbon Cycle in Ocean, Atmosphere and Biosphere*, European Commission 1999, EUR 16989 EN, 24 pages.
[18] J. T. Houghton, L. G. Meira Filho, B. A. Callander, N. Harris, A. Kattenberg and K. Maskell, eds., *Climate Change 1995, The Science of Climate Change*, International Panel on Climate Change, Cambridge University Press, 1996

CHAPTER 2

[19] J. Buisman, *Duizend Jaar Weer, Wind en Water in de Lage Landen*, A. F. V. van Engelen, ed., part 3, Van Wijnen, Franeker, Netherlands (in Dutch). This book is based on letters exchanged by monks in the Middle Ages and on the qualitative weather reports of the time.
[20] (*) Dennis L. Hartmann, *Global Physical Climatology*, Academic Press, San Diego, 1994. This book discusses climate physics in a comprehensive way. Its level is somewhat more advanced than the present book, but physical oriented students should be able to read it.
[21] Robin McIlveen, *Fundamentals of Weather and Climate*, Stanley Thornes, Cheltenham 1998. A very readable introduction to meteorology, which may be used as a quick reference text.
[22] R. R. Rogers and M. K. Yau, *A Short Course in Cloud Physics*, 3rd ed., Pergamon Press, Oxford, 1991
[23] Henk A. Dijkstra, Oceaancirculatie en klimaatverandering, *Ned T. Natuurkunde* **63/4** (1997) 79–83 (in Dutch)
[24] J. David Neelin and Mojib Latif, El Niño Dynamics, *Physics Today*, December 1998, 32–36
[25] M. J. Rodwell, D. P. Rowell and C. K. Folland, Oceanic Forcing of the Wintertime North Atlantic Oscillation and European Climate, *Nature* **398**, 25 March 1999, 320–23

CHAPTER 3

[26] Christian-Dietrich Schönwiese and Bernd Diekmann, *Der Treibhauseffekt*, Deutsche Verlags-Anstalt, Stuttgart, 1987 (in German)
[27] A. L. Berger, Milankovitch effects on long-term climatic changes, in: R. Fantechi, G. Maracchi and M. E. Almeida-Teixeira, Eds, *Environment and the Quality of Life*, European Commission, Brussels, 1991, EUR 11943 EN, pp 29–45
[28] Daniel B. Karner and Richard A. Muller, A Causality Problem for Milankovitch, *Science* **288** (23 June, 2000), 2143–44
[29] George C. Reid, Solar Total Irradiance Variations and the Global Sea Surface Temperature Record, *Journal of Geophysical Research* **96**, D2, February 20, 1991, 2835–44
[30] J. Lean, Contribution of Ultraviolet Irradiance Variations to Changes in the Sun's Total Irradiance, *Science* **244**, 4 April 1989, 197–200

[31] L. W. Alvarez, W. Alvarez, F. Asaro and H. V. Michel, Extraterrestrial cause for the Cretaceous-Tertiary extinction, *Science* **208**, 1980, 1095–98

[32] J. Smit, The Global Stratigraphy of the Cretaceous–Tertiary Boundary Impact Ejecta, *Ann. Rev. Earth Planet Sci.* **27**, 1999, 75–113

[33] K. Hutter, H. Blatter and A. Ohmura, *Climatic Changes, Ice Sheet Dynamics and Sea Level Variations*, ETH-Zürich, 1990

[34] M. E. Schlesinger, Equilibrium and transient climatic warming induced by increased atmospheric CO_2, *Climate Dynamics* **1** (1986) 35–51

[35] Kyoto Protocol to the *United Nations Framework Convention on Climate Change*, December 1997, FCCC/CP/1997/L.7/Add.1

[36] J. T. Houghton, B. A. Callander and S. K. Varney, eds., *Climate Change 1992, The Supplementary Report to the IPCC Scientific Assessment*, IPCC, Cambridge University Press, Cambridge, UK, 1992

[37] Robert T. Watson, Marufu C. Zinyowera and Richard H. Moss, eds., *Climate Change 1995, Impacts, Adaptations and Mitigation of Climate Change: Scientific-Technical Analyses*, IPCC, Cambridge University Press, 1996, Cambridge, UK

[38] Paul R. Epstein, Climate and Health, *Science* **285**, 16 July 1999, 347–48

[39] Münich Reinsurance Company, quoted in *New Energy*, 4/2000, p. 14,

[40] Vladimir Petoukhov, Andrey Ganopolski, Victor Brovkin, Martin Claussen, Alexej Eliseev, Claudia Kubatzki and Stefan Rahmstorf, Climber-2: A Climate System Model of Intermediate Complexity, part I: Model Description and Performance for Present Climate, *Climate Dynamics* **16** (2000) 1–17

[41] Kathryn S. Brown, Taking Global Warming to The People, *Science* **283**, 5 March 1999, 1440–41

[42] Nicola Jones, Sunblock, *New Scientist*, 23 September 2000, No. 2257, 28–31

CHAPTER 4

[43] Eugene S. Ferguson, The Measurement of the Man-Day, *Scientific American* **224** (1971) 96–103

[44] Robert E. Schofield, *The Lunar Society of Birmingham, A Social History of Provincial Science and Industry in Eighteenth-Century England*, Clarendon Press, Oxford, 1963

[45] Frank P. Incropera and David P. DeWitt, *Introduction to Heat Transfer*, John Wiley & Sons, New York, 1990

[46] D. S. L. Cardwell, *Technology, Science and History*, Heinemann, London, 1972

[47] (*) Mark W. Zemanski and Richard H. Dittman, *Heat and Thermodynamics*, McGraw-Hill, Singapore, 1981. This book is a standard text for the physics of thermodynamics. It is accurate and very detailed and contains many examples which are far outside the scope of the present book.

[48] *Efficient Use of Energy*, American Institute of Physics, AIP Conference Proceedings no. 25, New York, 1975

[49] P. W. Atkins, *Physical Chemistry*, Fifth Edition, Oxford University Press, 1995

[50] European Union, *Air Quality: Facts and Trends*, EU, DG XI, from internet (see Appendix. C)

[51] (*) Attilio Bisio and Sharon Boots, eds., *The Wiley Encyclopedia of Energy and Environment*, 2 vols., John Wiley, New York, 1997. This encyclopaedia contains much information with an emphasis on the American standards and regulations. The

information is scattered over several articles and has to be accessed by using the alphabetic index.

[52] (*) John H. Seinfeld and Spyros N. Pandis, *Atmospheric Chemistry and Physics: from Air Pollution to Climate Change*, John Wiley, New York, 1998. This comprehensive, standard text is written on a more advanced level than the present book. It is very informative and authoritative and discusses air transport of pollution at length.

[53] Archie W. Culp Jr., *Principles of Energy Conversion*, McGraw-Hill, New York, 1991

[54] Department of Energy, USA, *International Energy Annual* 1998, see www.eia.doe.gov/emeu/iea/

[55] Danish Wind Turbine Manufacturers Association, www.windpower.dk/stat/tab19.htm

[56] Wim Sinke, using various compilations, private communication, ECN, Netherlands

[57] United Nations, *Energy Statistics Yearbook* 1995, United Nations, New York, 1997 and *Energy Statistics Yearbook 1996*, United Nations, New York, 1998. Because of the long collection time of the data and the careful printing, the yearbook is published at least two years after the last year it refers to. It also contains the data of the previous four years.

[58] Esso Oil Company, PR Dept, *Essoscope*, 1998, 12–15

[59] The future of fuel cells, special issue, *Scientific American*, July 1999, 56–75

[60] Alan Jenkins, End of the acid reign?, *Nature*, **401**, 7 October 1999, 537–38

[61] Howard Herzog, Baldur Eliasson and Olav Kaarstad, Capturing Greenhouse Gases, *Scientific American*, February 2000, 54–61

CHAPTER 5

[62] Tomas Markvart, ed., *Solar Electricity*, John Wiley, Chichester, England, 1994

[63] John F. Walker and Nicholas Jenkins, *Wind Energy Technology*, John Wiley, Chichester, England, 1997

[64] Johannes Schmidl, Penetrating the markets, Biomass and commercial distribution, *Renewable Energy World*, May 99, 106–115

[65] International Commission on Radiological Protection, ICRP publication 60, 1990 *Recommendations of the International Commission on Radiological Protection*, Pergamon, Oxford, 1991

[66] S. Glasstone and P. J. Dolan, *The Effects of Nuclear Weapons*, US Department of Defense and Energy, Washington, DC, USA, 1977

[67] Wolf Häfele, Energy from nuclear power, *Perspectives in Energy*, 1991, **1**, 13–40

[68] Carlo Rubbia, The Energy Amplifier, CERN-OPEN-94-001, in: Proceedings LNS, Gif-sur-Yvette, 1994 LNS Ph 94-12 (115–123). This short summary may be downloaded from CERN.

CHAPTER 6

[69] A. Baumann, R. Ferguson, I. Fells and R. Hill, A methodological framework for calculating the external costs of energy technologies, *Proceedings 10th EC*

Photovoltaic Solar energy Conference, A. Luque, Ed., Kluwer, Dordrecht, 1991, 834–837; also reproduced in [20], p. 177
[70] Adam Serchuk, The environmental imperative for renewable energy, *Renewable Energy World*, July–August 2000, 32–41
[71] Karl Mallon, Iain MacGill and Nick Milton, PV-Breaking the Solar Impasse, *Renewable Energy World*, November 1999, 95–96
[72] C. Jäger, The Challenge of Global Water Management, in: E. Ehlers and T. Krafft, eds. *Understanding the Earth System: Compartments, processes and Interactions*, Springer Verlag, Heidelberg, 2001, p. 125–135. See also p. 118
[73] Ernst von Weizsäcker, Amory B. Lovins, and L. Hunter Lovins, *Factor Four, Doubling Wealth – Halving Resource Use*, Earthscan, London, 1998. This book is written for the general public and has a rather optimistic character. It contains many examples of reduction of energy and materials without reducing wealth.
[74] C. Gallo, M. Sala and A. A. M. Sayigh, eds., *Architecture: Comfort and Energy*, Elsevier, Oxford, 1998. Also published as *Renewable and Sustainable Energy Reviews*, **2** (1998). This book contains many examples of traditional buildings and homes adapting to climate, and of some modern buildings where one has learned the lesson.
[75] Federico M. Butera, *Principles of Thermal Comfort*, in: [74], pp. 39–66
[76] Ali Sayigh and A. Hamid Marafia, *Vernacular and Contemporary Buildings in Qatar*, in: [74], pp 25–37
[77] Cettina Gallo, *The Utilization of Microclimate Elements*, in: [74], pp. 89–114

CHAPTER 7

[78] Charles Kittel, *Introduction to Solid State Physics*, 7th Ed., John Wiley, New York, 1996
[79] Stephen Pond and George L. Pickard, *Introductory Dynamical Oceanography*, 2nd Ed., Pergamon Press, Oxford, 1991
[80] L. P. B. M. Jansen and M. M. C. G. Warmoeskerken, *Transport Phenomena Data Companion*, Edward Arnold, London, UK, 1987
[81] A. J. Mee, *Physical Chemistry*, 3rd Ed., Heinemann, London, 1947. This is one of the classic texts with many examples from old practice.
[82] Harold F. Hemond and Elisabeth J. Fechner, *Chemical Fate and Transport in the Environment*, Academic Press, San Diego, 1994
[83] Roger N. Reeve, *Environmental Analysis*, John Wiley, Chichester, 1994
[84] Roy Lewis, *Dispersion in Estuaries and Coastal Waters*, John Wiley, Chichester, 1997
[85] Gilbert M. Masters, *Introduction to Environmental Engineering and Science*, Prentice Hall, Englewood Cliffs, USA, This book is written on the level of the present one, with many more details and examples – chemistry oriented.
[86] K. V. Ellis, *Surface Water Pollution and its Control*, MacMillan, London, 1989
[87] A. van Mazijk, *One-dimensional Approach of Transport Phenomena of Dissolved Matter in Rivers*, thesis Delft, 1996
[88] Joseph S. Devinny, Lorne G. Everett, James C. S. Lu, Robert L. Stollar, *Subsurface Migration of Hazardous Wastes*, Van Nostrand Reinhold, New York, 1990
[89] A. Verruyt, *Theory of Groundwater Flow*, MacMillan, London, 1982

CHAPTER 8

[90] Lawrence E. Kinsler, Austin R. Frey, Alan B. Coppens and James V. Saunders, *Fundamentals of Acoustics*, John Wiley, New York, 1982. This book discusses acoustic waves in many diverse media, also underwater.

[91] P. A. Nelson and S. J. Elliott, *Active Control of Sound*, Academic Press, London, 1992. This book gives a short introduction to acoustics and quickly starts discussing modern and advanced techniques to reduce sound in places where it is not desired.

[92] Nederlandse Stichting Geluidhinder, Geluidhinder, Isolatie, Buitenlawaai, (in Dutch), Delft, 1988.

[93] International Organisation for Standardisation, *Acoustics-Normal Equal-Loudness Level Contours*, ISO 226: 1987 (E). The complete standard may be purchased from the ISO member bodies or directly from the ISO Central Secretariat, Case Postal 56, 1211 Geneva 20, Switzerland. Copyright remains with ISO

[94] Commission of the European Communities, *Proposal for a Directive of the European Parliament and of the Council, Relating to the Assessment and Management of Environmental Noise*, Brussels, 26 July 2000, COM(2000) 468 final (see the web).

[95] Brigitta Berglund and Thomas Lindvall, Eds., Community Noise, *Archives of the Center for Sensory Research*, 1995, 2(1), 1–195. See the website of the World Health Organization

[96] A. C. Verhoeven, *Bouwfysica*, Lecture Notes, T U Delft, 1980, (in Dutch)

CHAPTER 9

[97] John Ziman, *The Force of Knowledge, The Scientific Dimension of Society*, Cambridge University Press, Cambridge 1976. This book, written by one of the pioneers on 'Science and Society' contains many beautiful pictures and interesting anecdotes.

[98] H. Naus, W. Ubachs, P. F. Levelt, O. L. Polyansky, N. F. Zobov and J. Tennyson, Cavity-ring-down spectroscopy on water vapour in the range 555–604 nm, *Journal of Molecular Spectroscopy* **205** (2001) 117–21

[99] Dr. A. Kylling, NILU, Kjeller, Norway, private communication.

[100] Fred Sigernes, private communication and (http://unis31.unis.no/tibet/NILUV.JPG)

[101] Pu Bu Ci Ren, Fred Sigernes and Yngvar Gjessing, Ground-Based Measurements of Solar Ultraviolet Radiation in Tibet: Preliminary Results, to be published (http://unis31.unis.no/tibet/lhasauv.html)

[102] Measured by the second author

[103] M. Suh, F. Ariese, G. J. Small, R. Jankowiak, A. Hewer and D. H. Phillips, Formation and persistence of benzo[a] pyrene-DNA adducts in mouse epidermis *in vivo*: importance of adduct conformation, *Carcinogenesis* **16** (1995) 2561–69

[104] *Information for Sustainability*, Proceedings of the 27th International Symposium on Remote Sensing of Environment, June 8–12, 1988, Tromso, Norway

[105] A. G. Dekker, L. van Liere, E. Seyhan and T. J. Malthus (in Dutch), Remote Sensing en waterkwaliteit: een helder beeld van troebel water, *Gea* **23** (1990) nr. 3, 86–90

[106] A. G. Dekker and H. J. Hoogenboom (in Dutch), Remote sensing met hoge resolutie van Vecht en Amsterdam-Rijnkanaal, Rijkswaterstaat, directie Noord Holland, RPV 93.03, October 1993

[107] D. J. Wuebbles and J. M. Calm, An Environmental Rationale for Retention of Endangered Chemicals, *Science* **278** (1997) 1090–91
[108] J. A. Parrish, R. R. Anderson, F. Urbach and D. Pitts, *UV-A: Biological Effects of Ultraviolet Radiation with Emphasis on Human responses to Longwave Ultraviolet*, Plenum Press, New York, 1978
[109] Tapani Koskela, FMI, Finland (Table 1 in Molina and Molina, *Journal Geophysical Research*, **91** (1986) 14501–08)
[110] J. Jagger, *NRC Report*, 1982
[111] J. Jagger, *Solar UV Actions on Living Cells*, Praeger Special Studies, Praeger Publishing Division of Greenwood Press Inc., Westport, Conn, 1985
[112] E. K. Duursma, *Ozone Hole(s) 2000–2100* A. H. Heineken Foundation for the Environment, Amsterdam, 2000
[113] R. P. Kane, Ozone Depletion, related UV-B changes and increased skin cancer incidence, Int. *J. Climatol* **18** (1998) 457–72
[114] Network detection Change Status, NASA, Washington, 1990, Figure 1, p. 26
[115] Uwe Brinkman, ed., Continuous Monitoring of the Atmosphere Using Dial, *Lambda Highlights*, No. 30/31, October 1991

CHAPTER 10

[116] Alan E. Mussett and M. Aftab Khan, *Looking into the Earth, An Introduction to Geological Geophysics*, Cambridge University Press, Cambridge, England, 2000. This book is well written and gives a rather complete overview of geophysical methods. Its level of mathematics is comparable with the present text.
[117] Robert Bowen, *Isotopes in the Earth Sciences*, Elsevier Applied Science, London, 1988
[118] Gunter Faure, *Principles of Isotope Geology*, John Wiley, New York, 1977
[119] Jochen Hoefs, *Stable Isotope Geochemistry*, Fourth Edition, Springer, Berlin, 1997
[120] Environment and Climate Program, *Quantification of the west European methane budget by atmospheric measurements*, European Commission 1997, EUR 17511 EN, 24 pages
[121] W. Dansgaard, Stable Isotopes in Precipitation, *Tellus* 16(4) (1964) 436–468
[122] Kurt M. Cuffey, Gary D. Clow, Richard B. Alley, Minze Stuiver, Edwin D. Waddington and Richard W. Saltus, Large Arctic Temperature Change at the Wisconsin-Holocene Glacial Transition, *Science* 270 (20 October 1995), 455–458
[123] Dieter Vogelsang, *Geophysik an Altlasten*, Springer, Berlin, 1991 (in German)
[124] Kenan Gelisli, Ali Aydin and Omer Alptekin, Investigations of environmental pollution using magnetic susceptibility measurements, pp 1197–1205 in: T. Ayhan, I. Dincer, I. Olgun, S. Dost and B. Cuhadaroglu, Eds., *Proceedings of the TIEES-96, First Trabzon International Energy and Environment Symposium*, Trabzon, Turkey, 1996

CHAPTER 11

[125] Dennis L. Meadows, *The Limits to Growth, A Report for the Club of Rome Project on the Predicament of Mankind*, Universe Books, New York, 1972
[126] Chancey Starr, Energy and Power, *Scientific American* **244** (September 1971) 37.

[127] Eduard Bose, MY T. VU, Ernest Massiah and Rodolfo A. Bulatao, *World Population Projections*, World Bank/John Hopkins, Washington DC, 1994, p. 60
[128] Paul A. Samuelson, *Economics*, Ninth Ed., McGraw Hill, Tokyo, 1973
[129] Arthur H. Rosenfeld, Tina M. Kaarsberg and Joseph Romm, Technologies to Reduce Carbon Dioxide Emissions in the Next Decade, *Physics Today*, November 2000, 29–34
[130] Dennis J. Paustenbach, ed., *The Risk Assessment of Environmental and Human Health Hazards: A textbook of Case Studies*, John Wiley, New York, 1989
[131] J. H. Koeman, Algemene Inleiding in de Toxicology (in Dutch), Pudoc, Wageningen, 1983
[132] Richard A. Merrill, Regulatory Toxicology, Chapter 34 in: Curtis D. Klaassen, ed., *Casarett and Doull's Toxicology: the Basic Science of Poisons*, McGraw Hill, New York, 1996, 5th (rev)ed.
[133] Kenny S. Crump, An improved procedure for low-dose carcinogenic risk assessment from animal data, *J. of Environmental Pathology, Toxicology and Oncology*, **5** (4–5), 1984, 339–48
[134] Sharon Beder, *The Nature of Sustainable Development*, Earth Foundation, Scribe, Newham, Australia, 1993. This well written book from an Australian perspective offers a welcome addition to the dominating European-American literature.
[135] Paul Weaver, ed., *Factor 10*, Greenleaf, UK, 2000
[136] European Commission, *Lifestyles, Participation and Environment*, Report on a DG XII Workshop, EUR 18494, September 1998
[137] World Commission on Environment and Development, *Our Common Future*, Chairman Gro Harlem Brundtland, Oxford University Press, Oxford, 1987
[138] A. E. Waibel *et al.*, Arctic Ozone Loss due to Denitrification, *Science* **283** (26 March 1999) 2064–69
[139] Kyoto Protocol to the United Nations Framework Convention on Climate Change, December 1997. On the web the text and comments may be found for example at http://www.cnn.com/SPECIALS/1997/global warming/stories/treaty
[140] Centre for Science and Environment, 15 April 2000, http://www.oneworld.org/cse/
[141] Paul Baer, John Harte, Barbara Haya, Antonia V. Herzog, John Holdren, Nathan E. Hultman, Daniel M. Kammen, Richard B. Norgaard and Leigh Raymond, Equity and greenhouse gas responsibility, *Science* **289**, 29 September 2000, 2287
[142] George C. Marshall Institute, *A guide to Global warming, Questions and Answers on Climate Change*, January 2000, www.marshall.org
[143] Fred Pearce, Climate Change, *New Scientist*, 2 December 2000, 4–5
[144] Per Bak, *How Nature Works, The Science of Self-organized Criticality*, Oxford University Press, Oxford, 1997
[145] Michael Gibbons, Science's new social contract with society, *Nature* **402** SUPP, 2 December 1999, C81–C84
[146] John Ziman, Why must scientists become more ethically sensitive than they used to be?, *Science* **282**, 4 December 1998, 1813–14
[147] Student Pugwash USA, *Science* **286**, 19 November 1999, 1475
[148] President European Physical Society, *Europhysics News*, Nov/Dec 1999, 137

Subject Index